Renewable and Sustainable Energy

Renewable and Sustainable Energy

Edited by **Ted Weyland**

SYRAWOOD
PUBLISHING HOUSE

New York

Published by Syrawood Publishing House,
750 Third Avenue, 9th Floor,
New York, NY 10017, USA
www.syrawoodpublishinghouse.com

Renewable and Sustainable Energy
Edited by Ted Weyland

© 2016 Syrawood Publishing House

International Standard Book Number: 978-1-68286-111-0 (Hardback)

Printed in the United States of America.

Contents

Preface

The book provides a detailed study on renewable and sustainable energy and covers full range of renewable and sustainable energy technologies. The text outlines technological principles behind deriving power from solar, wind and hydro energy sources. It aims to elucidate the environmental impacts and future prospects of renewable energy harvesting. The chapters included in this book have undergone a tremendous reviewing process and makes this book a reliable source of information. The book provides a valuable insight into the field of renewable and sustainable energy. This book will serve as a resource guide for environmentalists, engineers, students, researchers and professionals.

This book is a result of research of several months to collate the most relevant data in the field.

When I was approached with the idea of this book and the proposal to edit it, I was overwhelmed. It gave me an opportunity to reach out to all those who share a common interest with me in this field. I had 3 main parameters for editing this text:

1. Accuracy – The data and information provided in this book should be up-to-date and valuable to the readers.
2. Structure – The data must be presented in a structured format for easy understanding and better grasping of the readers.
3. Universal Approach – This book not only targets students but also experts and innovators in the field, thus my aim was to present topics which are of use to all.

Thus, it took me a couple of months to finish the editing of this book.

I would like to make a special mention of my publisher who considered me worthy of this opportunity and also supported me throughout the editing process. I would also like to thank the editing team at the back-end who extended their help whenever required.

Editor

Photoreaction of Au/TiO$_2$ for hydrogen production from renewables: a review on the synergistic effect between anatase and rutile phases of TiO$_2$

K. Connelly · A. K. Wahab · Hicham Idriss

Abstract The review focus is on hydrogen production from renewables using photocatalysis. In particular we focus on the role of synergism on the reaction rate. Among the most studied examples for this phenomenon are catalysts based on TiO$_2$. TiO$_2$ exists in two common phases: anatase and rutile, with the latter more thermodynamically stable. For hydrogen production the photocatalyst is often composed of nano-size precious metals deposited on TiO$_2$ (such as Pt, Pd, or Au). It has been observed by many researchers over a decade that M/TiO$_2$ rutile is far less active than M/TiO$_2$ anatase. Yet, the presence of the two phases together results in considerable enhancement of the reaction rate when compared to M/TiO$_2$ anatase alone. The main reason for this is the increase of the charge carriers' lifetime allowing for electron transfer to hydrogen ions and hole transfer to oxygen ions (and/or the sacrificial agent used). In this work we review the few proposed models, so far, explaining the way by which this charge transfer occurs across both phases.

Keywords TiO$_2$ anatase · TiO$_2$ rutile · Hydrogen production · Synergism · Synergistic effect · Ethanol reactions · Water splitting

K. Connelly · H. Idriss (✉)
Department of Chemistry,
University of Aberdeen,
Aberdeen AB24 3UE, UK
e-mail: idrissh@sabic.com; h.idriss@abdn.ac.uk

A. K. Wahab · H. Idriss
SABIC T&I and CRI,
Riyadh and KAUST,
Riyadh, Saudi Arabia

Introduction

TiO$_2$ has many properties that make it effective for use as a photocatalyst. It is cost effective, abundant, has good surface stability, is non-corrosive, environmentally friendly and has great versatility in its application [1]. Furthermore, due to the position of the conduction band (CB) and valance band (VB) of TiO$_2$ in relation to a large selection of redox potentials, TiO$_2$ also shows activity for a large number of surface reactions [2]. Research into the use of TiO$_2$ as a means for generating hydrogen was initially accelerated by the work of Fujishima and Honda in the 1970s. This work involved photocatalysis to break down H$_2$O using an electrochemical cell with a TiO$_2$ electrode connected to a platinum electrode via an external circuit. Fujishima and Honda found that when the TiO$_2$ electrode was exposed to UV light, a current was established. This was due to the formation of electron-hole pairs, with the current direction indicating oxidation occurred at the TiO$_2$ electrode and reduction at the platinum electrode [3]. Since this discovery there has been a proliferation of work conducted in the field of photocatalysis, using TiO$_2$ and other semiconductors, for the intention of hydrogen generation.

Currently the most economically and energetically viable hydrogen production methods are from unsustainable hydrocarbon reforming. However, even methane reforming is currently around four times more expensive than current methods for producing gasoline with the same energy value [4]. The more integral problem with steam methane reforming is that it is clearly unsustainable in the long term. More sustainable methods, such as those utilising renewable feedstock's such as biomass or water, are presently unfeasible because of economics, incomplete technologies and/or low production yields. However, due to the inexhaustible global supply of water, as well as the greater accessibility of biomass to the majority of the population in comparison to hydrocarbons, the benefits of using a purely renewable feedstock for H_2 production have far reaching implications.

To date there has been a far greater amount of work conducted using powdered TiO_2-based photocatalytic materials than on single crystals. This is because working with single crystals under ultra-high vacuum conditions is currently rather limited because of the technical difficulties associated with conducting these types of experiments. Another limiting factor is the availability of TiO_2 anatase single crystals compared to the rutile ones. As will be discussed in the case of the hydrogen production from organic compounds (as well as water), the rutile phase is largely inactive. This is not necessarily the case for photo-oxidation of organic compounds using molecular oxygen. Because of the very high affinity of O_2 to electrons from the CB of TiO_2, both anatase and rutile have shown high activity, albeit the anatase phase was generally seen to be more active. Photoreaction/oxidation of alcohols, carboxylic acids, acetaldehyde and acetone as well as for smaller molecules such as H_2O, O_2 and CO over the single crystal rutile TiO_2 (110) surface has been studied. Most of these results have been discussed in recent reviews [5–9]. Other recent reviews on photoreaction of powder systems are available and these include those of references [10, 11].

In this review, after a brief presentation of the photocatalytic process with renewables, we will focus on the synergistic effect between TiO_2 anatase and TiO_2 rutile for photoreaction as it is an intriguing phenomenon that may yield considerable activity once well understood.

Au/TiO$_2$ nanoparticles

TiO_2 alone is not efficient for the photo production of hydrogen from water with or without sacrificial agents (such as methanol, ethanol or glycols) [12]. The effect of the addition of noble metals to TiO_2 has been studied by many groups as it considerably increased the hydrogen production rate. Among the most interesting of them is gold nanoparticle interaction with TiO_2. This review will focus on Au/TiO_2 because the available data allow for a systematic extraction of information. The introduction of Au nanoparticles (NPs) dispersed onto TiO_2 has been found to increase reaction efficiency by facilitating electron transfer and therefore inhibiting electron-hole recombination, as shown in Fig. 1, and by reducing the overpotentials for H_2 generation [13, 14].

The Fermi level (E_f) of Au NPs lies slightly lower than the E_f of TiO_2 and therefore photo-excited electrons can migrate from TiO_2 to surface Au NPs until the two E_fs are aligned with each other. This then allows for photo-

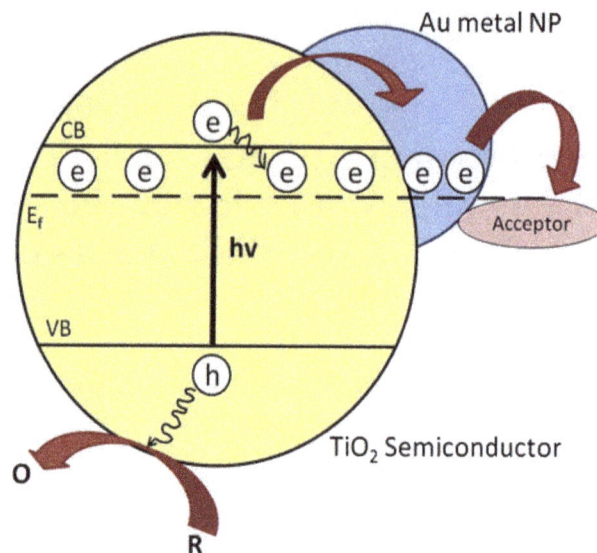

Fig. 1 Enhancement of photo-excitation due to Au nano particles acting as an electron acceptor. Adapted from [14]

generated holes in TiO_2 to migrate to the surface rather than recombining with electrons. The resultant Schottky barrier formed at the Au/TiO_2 interface serves as an efficient electron trap enabling charge separation, and inhibition of recombination, as well as the potential for localised reduction of adsorbates on surface Au NPs [15, 16]. Evidently good interfacial contact between Au NPs and TiO_2 is critical when synthesising Au/TiO_2 photocatalysts. Au/TiO_2 also potentially has the ability to increase the photo-response of TiO_2 into the visible light region because of the Au absorption band lying at 500–600 nm, which is inferred to be the surface plasmon band. It is proposed that excitation of electrons at the surface plasmon band of Au could allow for electron transfer to the low-energy CB of TiO_2, which is photocatalytically inactive with visible light, therefore also providing charge separation [17–21]. Au NP size, dispersion, loading and NP-support interaction all play critical roles in the photocatalytic activity of Au/TiO_2 with all these parameters dependent upon the catalyst preparation methods [17, 22–24]. Among the many methods of preparing Au, particle deposition–precipitation of Au onto TiO_2 provides the good interfacial contact necessary for photocatalytic activity from Au/TiO_2 as initially observed by Bamwenda et al. [22, 24] when comparing various Au/TiO_2 photocatalyst preparation methods.

Bamwenda et al. [22] used UV light to irradiate suspensions of TiO_2 with various Au loadings and monitored H_2 production from a water–ethanol solution. Below are the proposed chemical equations for H_2 evolution. Following electron-hole formation under UV light, electrons are trapped at Au NPs and reduce protons to produce H_2 gas as shown in Eqs. 1–4:

$$e^-(Au_{tr}) + H^+(sol) \rightarrow H\,(ad) \tag{1}$$

$$2H\,(ad) \rightarrow H_2(ad) \tag{2}$$

$$H_2(ad) \rightarrow H_2(sol) \tag{3}$$

$$H_2(sol) \rightarrow H_2(g) \tag{4}$$

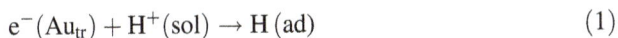

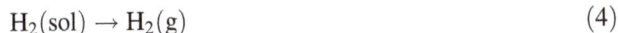

where (Au_{tr}) is an electron trapped at an Au site, (sol) is solution, (ad) is adsorbed and (g) is gas phase.

The corresponding surface trapped holes would interact with either water or ethanol, as shown in Eqs. 5–6, as previously suggested by Sakata and Kawai for H_2 production over Pt/TiO$_2$ from a water–ethanol solution: [27]

$$2h_{tr}^+ + C_2H_5OH \rightarrow 2H^+ + CH_3CHO \tag{5}$$

$$2h_{tr}^+ + CH_3CHO + H_2O \rightarrow 2H^+ + CH_3CHOOH \tag{6}$$

The acetic acid produced in Eq. 6 can then be decomposed into methane and CO_2 as per the photo-Kolbe reaction shown in Eq. 7 ($CH_3^\cdot$ may combine with $H^\cdot$, giving CH_4, or with another $CH_3^\cdot$ to give C_2H_6), with methane also possibly forming via Eq. 8:

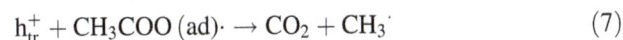

$$h_{tr}^+ + CH_3COO\,(ad)\cdot \rightarrow CO_2 + CH_3^\cdot \tag{7}$$

$$CO_2 + 4H_2 \rightarrow CH_4 + H_2O \tag{8}$$

By depositing precious metal onto the TiO$_2$ surface, it was suggested that the higher E_f of the semiconductor compared to that of the metal allows for smooth electron transfer from TiO$_2$ to the metal and resulted in the formation of a space charge layer at the metal side of the interface [26]. This space charge layer causes charge carriers to be transported in opposite directions and therefore reduces the probability of recombination [27].

Previous work by our group has so far focused on hydrogen production from ethanol over Au-loaded anatase and rutile powdered catalysts. Nadeem et al. [28] investigated the effect of varying the anatase particle size from nanoparticle to microparticle, with Au NPs of the same size, on H_2 rates from ethanol in order to determine the effect of support particle size on the reaction. Upon normalisation of the H_2 rate with respect to the surface areas of the TiO$_2$ nano- and micro-particles, it was found that both particle sizes yielded similar specific reaction rates, indicating that support particle size is not an important parameter for Au/TiO$_2$ photocatalysis. Infrared spectroscopy (IR) of the ethanol exposed Au/TiO$_2$ surface showed deprotonation of ethanol upon ethanol adsorption on the Au/TiO$_2$ surface, similar to most metal/metal oxides systems [29, 30]. Combining the IR data with the desorption profile gained from temperature programmed desorption (TPD), which showed the main reaction to be that of dehydrogenation of ethanol to form acetaldehyde, Nadeem et al. [28] put forward the following reaction pathway for the photoreaction of ethanol over the Au/TiO$_2$ surface as shown in Eqs. 9–11. During the process each ethoxide injects two electrons into the TiO$_2$ CB to produce an α-hydroxyethyl radical (Eq. 10a) and acetaldehyde (Eq. 10b).

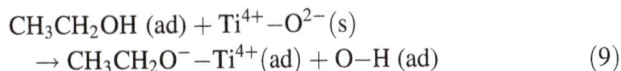

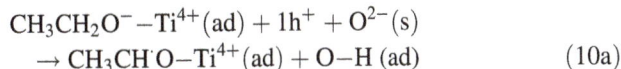

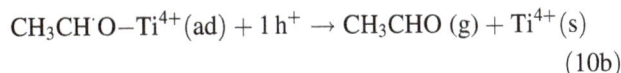

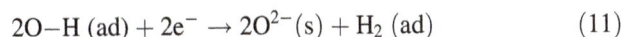

$$CH_3CH_2OH\,(ad) + Ti^{4+}-O^{2-}(s)$$
$$\rightarrow CH_3CH_2O^--Ti^{4+}(ad) + O-H\,(ad) \tag{9}$$

$$CH_3CH_2O^--Ti^{4+}(ad) + 1h^+ + O^{2-}(s)$$
$$\rightarrow CH_3CH^\cdot O-Ti^{4+}(ad) + O-H\,(ad) \tag{10a}$$

$$CH_3CH^\cdot O-Ti^{4+}(ad) + 1\,h^+ \rightarrow CH_3CHO\,(g) + Ti^{4+}(s) \tag{10b}$$

$$2O-H\,(ad) + 2e^- \rightarrow 2O^{2-}(s) + H_2\,(ad) \tag{11}$$

The production of H_2 in this scheme requires some of the photoexcited electrons to migrate from the TiO$_2$ CB to become trapped at Au NPs, which, if in the vicinity of H^+ ions produced from the deprotonation of ethanol upon adsorption, can reduce the protons to form H_2 gas.

This reaction mechanism was further refined by Murdoch et al. [31] when investigating various loadings of Au on both rutile and anatase NPs. The anatase supports had Au particle sizes between $\sim$3 and 20 nm, whilst the rutile supports had Au particle sizes between $\sim$20 and 35 nm. Negligible photoactivity for the production of H_2 from ethanol was observed for the un-doped rutile and anatase photocatalysts alone. The results from loading Au onto the anatase and rutile supports and running the photoreaction are shown in Fig. 2 with the H_2 rate pertaining to anatase on the left axis and the H_2 rate for rutile on the right. From Fig. 2 it can be seen that Au loaded onto anatase NPs led to an increase in the H_2 rate to a maximum of 1.1×10^{-6} mol $m_{cat.}^{-2}$ min^{-1} with increased loading up to 4 wt% Au, after which the H_2 rate decreases. For Au loaded onto rutile hydrogen production is far lower, about two orders of magnitude lower than that on anatase (1.7×10^{-8} mol $m_{cat.}^{-2}$ min^{-1}). The two orders of magnitude

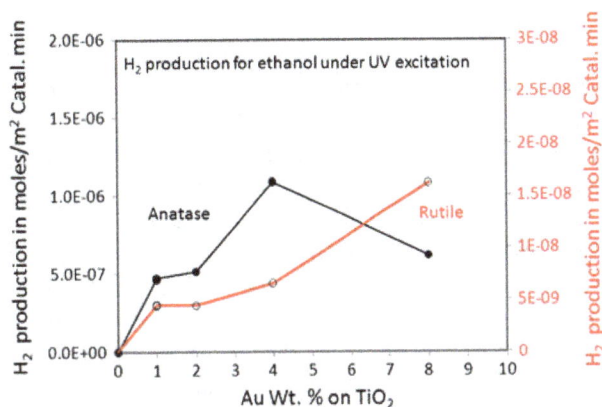

Fig. 2 Photocatalytic H_2 production from ethanol over Au/TiO$_2$ anatase (A) and rutile (R) as a function of wt% Au loading [29]

difference between the H_2 rates, produced over the rutile and anatase catalysts, were explained to be due to the greater charge recombination rate of rutile as compared to anatase. For the 1, 2 and 4 wt% Au/anatase catalysts, Murdoch et al. normalised the H_2 rates to the number of exposed Au atoms on the surface, calculated from XPS data of Au4f. They found the normalised rate to be independent of Au particle size. This is in contrast to results observed for dark reactions where increasing Au particle size above around 4 nm results in a decrease in the catalytic activity [32–35].

Rutile and anatase polymorphs of TiO_2

There are several main differences between the anatase and rutile phases of TiO_2. First, the band gap (BG) of rutile is slightly smaller at 3.0 eV as compared to 3.2 eV for anatase [23]. Anatase can therefore only be excited by UV light whilst rutile has a photo-response to UV light, as well as extending slightly into the visible light region [36]. It is generally accepted that the CB of anatase lies higher than that of rutile [37]; however there is still some debate about the relative positions between the anatase and rutile CBs [38], as well as the exact CB potentials for both of these TiO_2 polymorphs [39–41]. Furthermore anatase has an indirect BG whilst rutile has a direct BG, which may affect electron-hole recombination rates. Many authors have studied the kinetics and mechanisms of charge separation and electron-hole recombination in rutile and anatase. Colbeau-Justin and Kunst [42] used a time-resolved microwave conductivity (TRMC) method to probe charge carrier lifetimes in TiO_2 with the TRMC signal attributable to electron activity because electron mobility is greater than that of the associated holes. From the TRMC signal it was evident that rutile showed faster decay of mobile electrons than observed for anatase. This demonstrates that rutile has a shorter photoexcited electron lifespan and therefore higher recombination rate in relation to anatase, whilst anatase charge carriers are subjected to competition between recombination and hole trapping processes allowing for a longer lifespan of charge carriers. Upon exposure of anatase to ethanol, the TRMC signal showed a much slower decay inferring that ethanol is scavenging holes at the surface and trapping them there. For rutile exposed to ethanol no difference was observed to the signal recorded for unexposed rutile, further indicating that fast recombination is occurring. Using the same method Xu et al. [43] confirmed these findings using single crystals of rutile (110) and anatase (101). As shown in Fig. 3, there is a difference in the observed signal of around one magnitude between both crystals. The signal for rutile rapidly decays, even during the excitation period, and therefore indicates that fast electron-hole recombination has

occurred. Conversely the slower signal decay observed for anatase, as well as the much larger signal amplitude, shows that anatase charge carriers have a lifetime of over 10 ns. This difference in charge carrier lifetimes between these two TiO_2 polymorphs has been attributed by the authors to the differing band structures presumed to arise in anatase and rutile, as shown in the inset in Fig. 3. It is envisioned that in anatase, with an indirect BG as opposed to the direct BG in rutile, following the vertical photoexcitation of electrons into the unoccupied states of the anatase CB electrons then relaxing to the bottom of the CB will be unable to undergo direct recombination with holes because of the indirect BG. This consequently would lead to longer charge carrier lifetimes and an increase in the diffusion length of electron-hole pairs excited within the bulk. Equally important, the CO photo-oxidation was conducted on both surfaces using reflection absorption infrared spectroscopy (RAIRS). $TiO_2(101)$ anatase was found to be about eight times more photo-catalytically active than $TiO_2(110)$ rutile.

Using TiO_2 films Shen et al. [44] directly measured the photoexcited hole activity of both rutile and anatase phases using a lens-free heterodyne detection transient-grating (LF-HD-TG) method. The LF-HD-TG signal is attributed to the change in the amount of photoexcited holes as the effective mass for holes is far smaller than that for electrons. Upon laser excitation the anatase signal quickly decreased (<2 ps) and was followed by a slow decay thereafter, whilst the rutile sample showed only a slow decay. The rapid decrease of the anatase signal was of first-order kinetics and proposed to be due to surface hole trapping whilst the slow decay observed for both samples indicative of recombination and/or trapping processes occurring within the bulk of the particles. Experiments

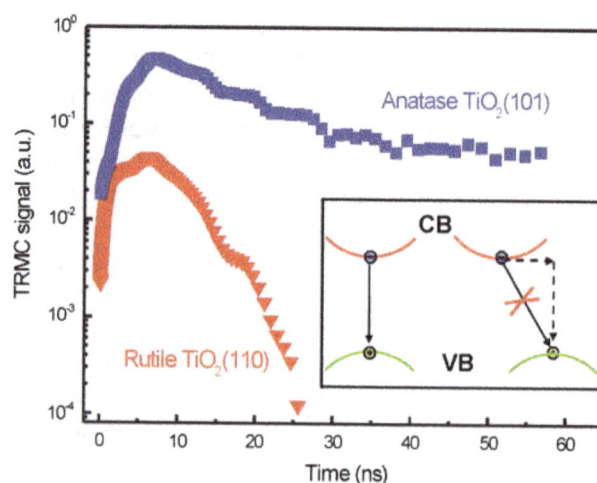

Fig. 3 Transient photoconductance measurements at 30 GHz using 355-nm laser pulses (10-ns FWHM) in rutile (110) and anatase (101) single crystals. The *inset* shows a schematic of the proposed model of the rutile band gap (*left*) and the anatase band gap (*right*) [43]

were also run on anatase and rutile of various NP size, which showed the same LF-HD-TG response regardless of particle size. Based on these findings Shen et al. proposed that the differences in the charge dynamics observed between rutile and anatase could be dependent on the difference in the dielectric constants of the two phases. Surface potential barrier height is inversely proportional to the dielectric constant and so the smaller dielectric constant of anatase gives rise to a larger surface potential barrier, with the potential slope being steeper for anatase than for rutile, as shown in Fig. 4. This results in the faster transport of holes to the surface of anatase than in rutile, with subsequent hole trapping at surface sites, which correlates with the initial fast decay of the signal upon excitation of anatase.

Work was conducted on the photocatalytic reaction of acetic acid over the rutile (011) surface reconstructions {011} and {114} [45, 46]. The {011} surface was found to be far more active for this reaction than the {114} surface, and upon calculating the depletion layer from the quantum yield of the reaction from both surfaces it was found that the more reactive {011} reconstruction had a larger depletion layer and therefore higher potential barrier height than the {114} surface. The increase of reactivity observed as being proportional to the depletion layer was attributed by these authors as being due to the width of the depletion layer affecting the electron-hole recombination rate, which correlates with the work of Shen et al. [44].

TiO$_2$ P25: a case study for the synergistic effect in photocatalysis

TiO$_2$ P25 is the industry standard, mixed phase rutile/anatase TiO$_2$ catalyst that has repeatedly shown enhanced catalytic activity in comparison to using the rutile and anatase phases alone. P25 has an anatase/rutile ratio of $\sim 70/30$ with a surface area of typically 50 ± 15 m^2 g^{-1} [36]. The method used for synthesising P25 must ensure good contact between anatase and rutile in order to observe the synergistic effect between the two phases. The preparation method of the catalyst will also have a direct effect

on the morphology of the interfacial boundaries, in turn affecting the electronic energy levels of electron-hole pairs [45]. Furthermore the calcination temperature of the catalysts must be closely controlled in order to gain the required TiO$_2$ phase composition, with phase transformation of anatase to rutile beginning at around 550 °C [46].

Several authors have monitored H$_2$ production from various reactants over mixed phase TiO$_2$ photocatalysts. Zhang et al. [46] deposited Pt onto almost pure rutile, pure anatase and mixed phase TiO$_2$. Surface phase composition of the TiO$_2$ particles was controlled by increasing the calcination temperature and confirmed using XRD, visible and UV Raman spectroscopy. At 550 °C the anatase phase in the bulk began to transform to rutile with the surface retaining the anatase phase until temperatures above 680 °C. Above 700 °C nearly all bulk anatase is transformed into the rutile phase and around 44 % of surface anatase remains. Complete phase transformation to rutile is achieved upon calcination to 800 °C. From monitoring H$_2$ production from the reaction of methanol and water over these TiO$_2$ photocatalysts, the H$_2$ rate with respect to the surface specific area of the catalyst was calculated. The highest activity observed was for the photocatalyst calcined at 700–750 °C (rutile bulk and mixed phase surface), which yielded a H$_2$ rate around four times greater than that observed for rutile alone. From UV Raman spectra of the 700–750 °C calcined catalysts, the authors confirm that both rutile and anatase coexist at the surface and therefore have surface phase junctions. It has been postulated by many authors [23, 25, 39, 46–48] that it is the presence of surface phase junctions in mixed phase TiO$_2$ that facilitates the electron transfer between the two phases upon photoexcitation and therefore improves charge separation, reduces recombination and subsequently increases photocatalytic activity.

Kho et al. [39] measured the onset potentials versus Ag/AgCl of mixed phase TiO$_2$ NPs and found that above 60 % anatase content there is a steady increase in the onset potential observed. The onset potential can be taken as an evaluation of the quasi-E_f of the photocurrent from the anode and therefore Kho et al. suggested that E_f band bending had occurred, which would indeed require intimate contact between the two TiO$_2$ phases. The authors also monitored H$_2$ production over a TiO$_2$ suspension with an aqueous solution of methanol and H$_2$PtCl$_6$. Within the TiO$_2$ suspension the anatase/rutile content was varied, and an optimum anatase content of 39 % was found to yield the best H$_2$ rate in comparison to pure rutile or anatase as well as other phase composites. Based on the 39 % anatase composition another catalyst was prepared via manual grinding of anatase and rutile samples and tested for H$_2$ production. This time no synergistic effect was observed between anatase and rutile, showing that the grinding

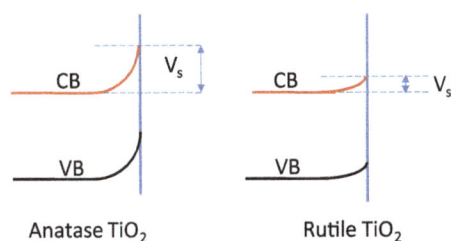

Fig. 4 Schematic of surface band bending in anatase and rutile, V_s is the surface potential barrier height; adapted from [44]

method did not produce sufficient inter-particle mixing, and therefore no good interfacial contact was made between the two phases, which is crucial for the enhanced photocatalytic activity that was previously observed. Kho et al. described the mechanisms for the synergistic effect between anatase and rutile in the presence of methanol as shown in Fig. 5. Electron-hole pairs are generated after photoexcitation in both anatase and rutile with methanol acting as a hole scavenger and oxidised at the VB of both anatase and rutile. Due to the lower potential of the rutile CB, as compared to the anatase CB, Kho et al. proposed that the majority of photoexcited electrons at the anatase CB migrate across the phase junction to the rutile CB. Therefore, the majority of H^+ ion reduction occurs at the rutile CB, with a smaller amount occurring at the anatase CB. Due to this transfer of electrons from the anatase CB to the rutile CB, the likelihood of electron-hole recombination at either the anatase or rutile VBs would be decreased with more H^+ reduction occurring at the rutile CB than would be observed in single-phase rutile. However, due to the rutile CB minimum being closer to the H^+ reduction potential than that of the anatase CB, there is possibly no benefit to be gained from transferring photoexcited electrons from the anatase to rutile CB for H^+ reduction and in fact another mechanism for the observed synergistic effect in P25 could be occurring.

From work on the photooxidation of phenol by various pure and Pt doped TiO_2 P25 and Hombikat UV100 (HK) photocatalysts, Sun et al. [23] found that TiO_2 P25 alone showed the greatest phenol degradation. According to Sun et al. band bending caused by interfacial contact between the two phases resulted in the anatase CB energy increasing within the space charge layer and therefore prohibiting electrons from migrating from anatase to rutile, as shown in Fig. 6. However VB bending of both phases allows for hole migration from the anatase to the rutile VB, explaining P25's enhanced photoactivity in liquid reaction. This

would result in holes mostly populating the rutile VB where the majority of oxidation would occur and electrons concentrated at the anatase CB where mostly reduction reactions would occur. It is thought that rutile's role in the process is purely for charge separation and providing oxidation sites. Photoexcited electrons in rutile are not thought to be directly involved in the reaction and instead undergo recombination with holes.

Other TiO_2 systems studied for the synergistic effect in photocatalysis

Using theoretical studies Deak et al. [38] produced the generic band alignment scheme shown in Fig. 7. It was found that the rutile CB lay higher than the anatase CB by around 0.3–0.4 eV, whilst the anatase VB is around 0.5–0.6 eV lower than for rutile. This is in disagreement with other models discussed already whereby the rutile CB is lower in energy than the anatase CB [23, 39]. Despite this, the authors concluded that this band alignment will result in an accumulation of migrating electrons at the anatase CB and migrating holes at the rutile VB; the latter is in agreement with several other studies [25, 49–52].

Further models that have gained greatest acceptance are commonly referred to as the 'rutile antenna' and the 'rutile sink' models [25]. Bickley et al. [49] proposed the 'rutile sink' model as shown in Fig. 8. In this model Bickley et al. assumed there to be an overlayer of rutile on the anatase and that <1 % of UV radiation was absorbed by the rutile overlayer. On this assumption the majority of incident light is absorbed by, and photoexcited electrons in, the underlying anatase. In agreement with Sun et al. [23], it is thought that when the E_fs of both phases are in contact with each other a space charge layer is created promoting the migration of holes from anatase to rutile and then to the surface, as shown in Fig. 8. Due to the position of the CB

Fig. 5 Schematic of the electron pathway during the photocatalytic production of H_2 over anatase and rutile particles [39]

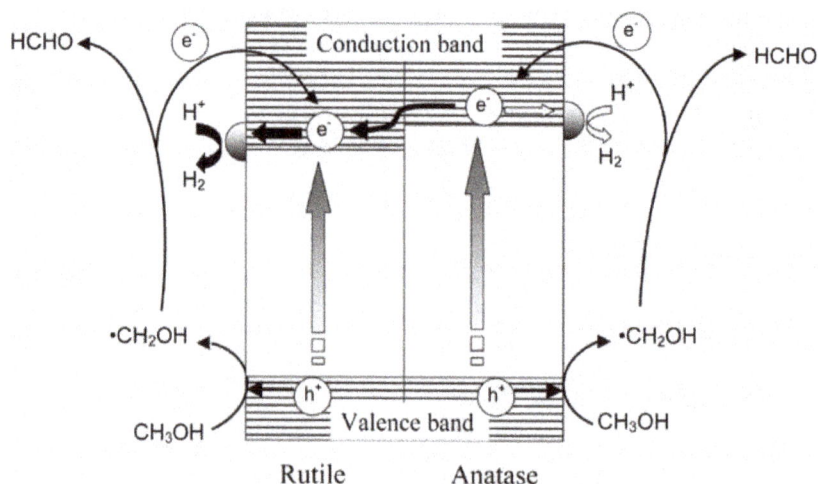

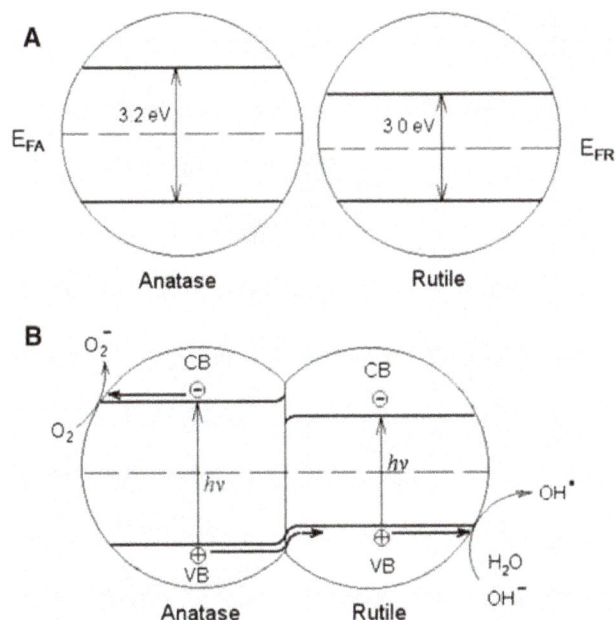

Fig. 6 Mechanism proposed for electron-hole separation in P25 following photo-excitation; adapted from [24]. **a** Before anatase and rutile phases contact each other. **b** After the line up of Fermi levels of anatase and rutile phases

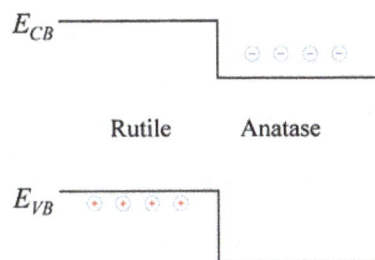

Fig. 7 Schematic of rutile/anatase band alignment [38]

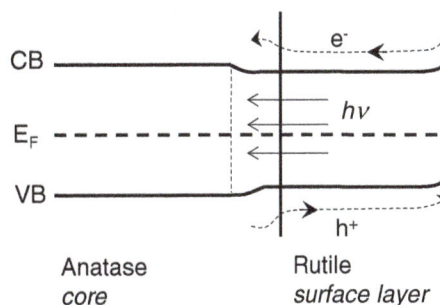

Fig. 8 Mechanism for the synergistic effect of a thin overlayer of rutile on anatase; adapted from [49]

edges and E_f of the two phases, the space charge layer is thought to adversely affect the migration of electrons toward the surface. However, electron migration from the lower energy rutile CB to the higher energy anatase CB, as shown in Fig. 8 by Bickley et al. [49], is energetically unfavourable.

Scott et al. [52] obtained H_2 production rates of the ethanol reaction over pure phase Au/rutile and Au/anatase, as well as for Au/P25. It was found that when the H_2 rates were normalised, in order to take into consideration the number of exposed Au atoms on the surface, that 1.5 wt% Au/P25 had a H_2 rate (7×10^{-5} mol^{-1} m^2 min^{-1}) that was around two times greater than that seen for 2.0 wt% Au/Anatase (4×10^{-5} mol^{-1} m^2 min^{-1}), which was found to have a rate of two orders of magnitude greater than that observed for 2.0 wt% Au/Rutile (6×10^{-7} mol^{-1} m^2 min^{-1}). Scott et al. proposed that the synergistic mechanism operating in P25 that led to the enhancement of the H_2 rate involves electron transport from the rutile phase to the anatase phase. They

propose that the smaller BG of rutile results in the generation of a higher number of charge carriers in rutile than observed in anatase and that these electrons are then transferred to the anatase CB where they benefit from a slower rate of electron-hole recombination; Fig. 9. The enhanced H_2 rate observed for P25 can then be attributed to the reaction occurring mainly at the anatase CB where there is a greater difference in the potentials of the anatase CB and the redox potential for H^+ reduction than is the case for the rutile CB and H^+ reduction potential.

Hurum et al. [50] re-evaluated this 'rutile sink' model using electron paramagnetic resonance (EPR) spectroscopy following visible light photoexcitation in order to take into account any charge transfer trapping on charge carrier pathways. EPR can be used to focus on the electron and hole trap population and recombination behaviour in mixed phase TiO$_2$ and is very useful in this respect as it is able to directly and indirectly detect photo-induced radicals via spin-trapping methods allowing for the identification of electron and hole surface and lattice trapping sites [37]. Upon irradiation of the aqueous P25 sample with visible light, the signal intensity for the population of anatase bulk and surface trapping sites increased, whilst the rutile trapping site signal simultaneously decreased. This evidence directly indicates that electron transfer is occurring from rutile to anatase under visible light. As Hurum et al. used visible light to excite their samples, rather than UV, only rutile has the potential to utilise the highest frequency light available from the visible light source for photoexcitation. Therefore, photoexcitation would only occur in rutile with these electrons transferred to anatase, specifically to trapping sites that lie just below the anatase CB. As shown in Fig. 10, further electron transport fills up surface trapping sites in anatase, resulting in the availability of photogenerated holes in rutile to be available for oxidative reactions.

Nair et al. [27] also investigated the synergistic effect between anatase and rutile using the photodegradation reaction of phenol under both UV and visible light. As expected under UV light the photocatalysts with the

Fig. 9 **a** Production of hydrogen (per unit area) from ethanol over 1.5 Au wt%/TiO$_2$ P25, 2 wt% Au/TiO$_2$ anatase and 2 wt% Au/TiO$_2$ rutile. **b** Model explaining the role of rutile and anatase as synergistic effect. In the model the larger number of charge carriers in rutile associated with the slower rate of charge carriers disappearance in anatase is behind the high activity of P25. **c** HRTEM of the 2 wt% Au/TiO$_2$ P25; the interface between Au, anatase and rutile is clearly seen. The table present results from **a** normalised to the surface Au signal from their XPS Au4f. Still after normalisation the rate of Au/TiO$_2$ P25 is larger than that of Au/TiO$_2$ anatase. The rate of hydrogen production is negligible

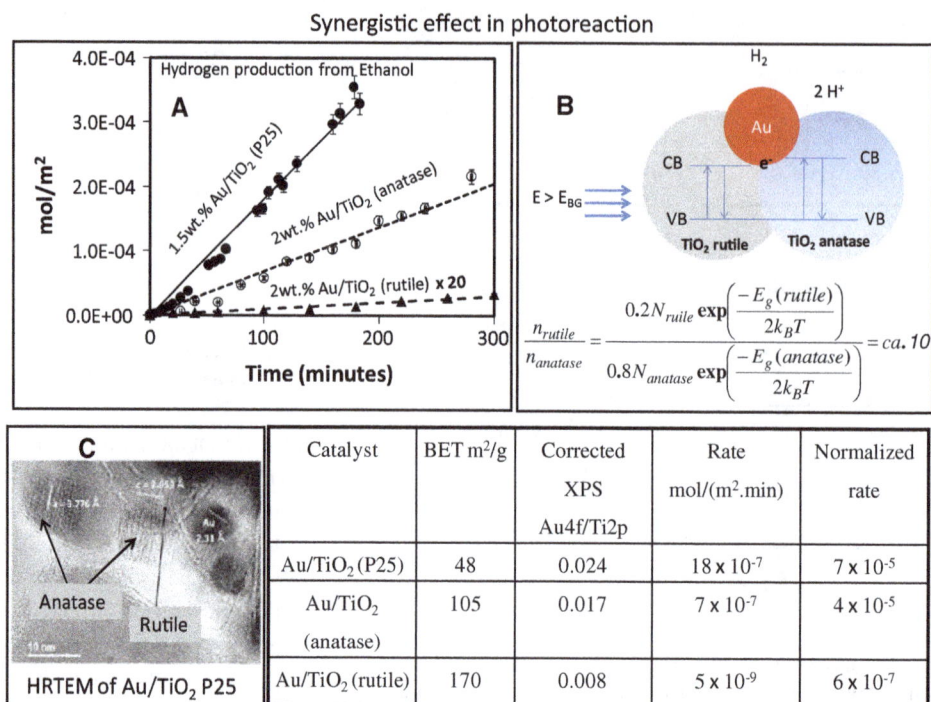

Synergistic effect in photoreaction

$$\frac{n_{rutile}}{n_{anatase}} = \frac{0.2 N_{rutile} \exp\left(\frac{-E_g(rutile)}{2k_BT}\right)}{0.8 N_{anatase} \exp\left(\frac{-E_g(anatase)}{2k_BT}\right)} = ca. 10$$

HRTEM of Au/TiO$_2$ P25

Catalyst	BET m^2/g	Corrected XPS Au4f/Ti2p	Rate mol/(m^2.min)	Normalized rate
Au/TiO$_2$ (P25)	48	0.024	18 x 10^{-7}	7 x 10^{-5}
Au/TiO$_2$ (anatase)	105	0.017	7 x 10^{-7}	4 x 10^{-5}
Au/TiO$_2$ (rutile)	170	0.008	5 x 10^{-9}	6 x 10^{-7}

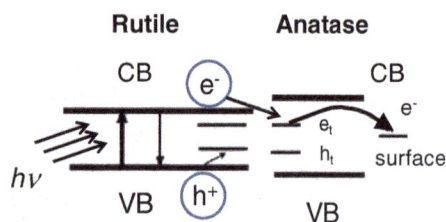

Fig. 10 Rutile antenna model explaining the synergistic effect between rutile and anatase; adapted from [50]

highest anatase content (88 and 92 %) showed the highest activity, whilst under visible light photocatalysts with the highest rutile content (72 and 76 %) showed the highest activity. Nair et al. explained these results by proposing a slightly modified version of the 'rutile sink' and 'rutile antenna' models based on the band structure of the two phases at the phase junction. Figure 11a shows the proposed interfacial band model between rutile with a BG of 3.0 eV and anatase with a BG of 3.2 eV at equilibrium and the potential charge barrier formed at the phase junction. Figure 11b shows the model after UV excitation where excitation of electrons to the CB occurs mainly in anatase. The charge barrier created between the two phases encourages migration of electrons in the interfacial region from the anatase CB to the rutile CB and holes migrate to the anatase VB. Figure 11c shows the model under visible light excitation with a photoresponse only from rutile. This time the charge barrier influences electron transfer in the interfacial region from the rutile CB to the anatase CB with holes remaining at the rutile VB.

Carneiro et al. [51] agree with the work cited in refs. [23, 38, 49] in that hole migration occurs from anatase to rutile in mixed phase TiO$_2$; however they found that the efficiency of photocatalytic reactions carried out in aqueous liquid phase in the presence of rutile was actually reduced because of this hole migration rather than increased. Carneiro et al. monitored the photocatalytic activity of both methylene blue degradation and cyclohexane oxidation over a series of sol–gel TiO$_2$ photocatalysts calcined at various temperatures in order to gain various phase compositions. By increasing the calcination temperature the % of rutile increased, along with the particle size of both anatase and rutile, whilst the presence of surface water and hydroxyl groups was observed to decrease because of the decreased surface area. It is proposed that the lower density of OH groups on the rutile surface than observed for the anatase particles means that holes that have migrated to rutile are not being utilised in surface processes, whereas when no rutile is present photogenerated holes will remain at anatase where they can participate in surface reactions. Carneiro et al. [51] suggested that the observed discrepancy between the photoreactivity of their catalysts, as compared to P25, was due to the improved crystalline quality of the TiO$_2$ particles in P25 and it is therefore anatase particle size and morphology in P25 rather than through charge separation between the two phases that results in the enhanced reactivity observed.

In direct opposition to the majority of previously proposed charge transfer models [38, 49, 50], Scotti et al. [54] used EPR to determine that for hydrothermally synthesised

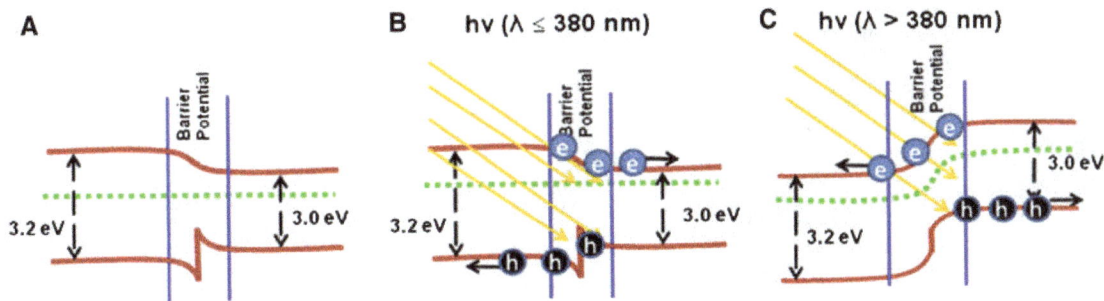

Fig. 11 Proposed interfacial band model, **a** under equilibrium, **b** under UV light and **c** under visible light [28]

mixed phase TiO_2 catalysts UV excited electron transfer was occurring from the anatase to the rutile phase, as previously concluded by Kho et al. [39]. Figure 12 shows the EPR signal for (a) pure rutile, (b) 61 % rutile, (c) 48 % rutile, (d) 20 % rutile and (e) pure anatase at 10 K following UV irradiation for 20 min. From these results it can be seen that the signal obtained for the mixed phase catalysts is similar to that obtained for pure rutile regardless of the anatase % of the catalyst. This indicates that electrons in mixed phase TiO_2 are preferentially trapped at O^- and Ti^{3+} sites on the rutile phase. Scotti et al. therefore conclude that under UV excitation electrons are transferred from the higher energy anatase CB to the lower energy rutile CB, with hole migration simultaneously occurring from the lower energy anatase VB to the higher energy rutile VB.

Komaguchi et al. [47] also show evidence for electron transfer from anatase to rutile due to the presence of trapping sites. Pure phase anatase and rutile as well as P25 catalysts were partially reduced and then subjected to light of wavelength less than the rutile BG. Upon illumination the Ti^{3+} ESR signal for both pure anatase and rutile disappeared and was then restored to full signal intensity upon switching off the light. The photoresponse signal for P25 also disappeared upon illumination; however on turning off the light source the signal exceeded its initial response, reaching 130 % of the initial intensity for the reduced P25 catalyst and 250 % of initial intensity for the air treated P25 catalyst. Furthermore, following restoration of the signal after illumination, the rutile component is observed to be more dominant than previously observed, showing that there is increased electron trapping in rutile. As the light source used in this work is unable to photoexcite electron-hole pairs in TiO_2, Komaguchi et al. proposed that it is trapped electrons that are being excited from trap sites around 0.3–0.8 eV below the CB. Following excitation of trapped electrons into the TiO_2 CB, the ESR signal is unobservable until relaxation of electrons back into Ti^{3+} sites occurs. By taking into account the increase in rutile signal following illumination, the following scheme is suggested, as shown in Eq. 12. First, electrons at anatase trap sites are photoexcited into the anatase CB where the

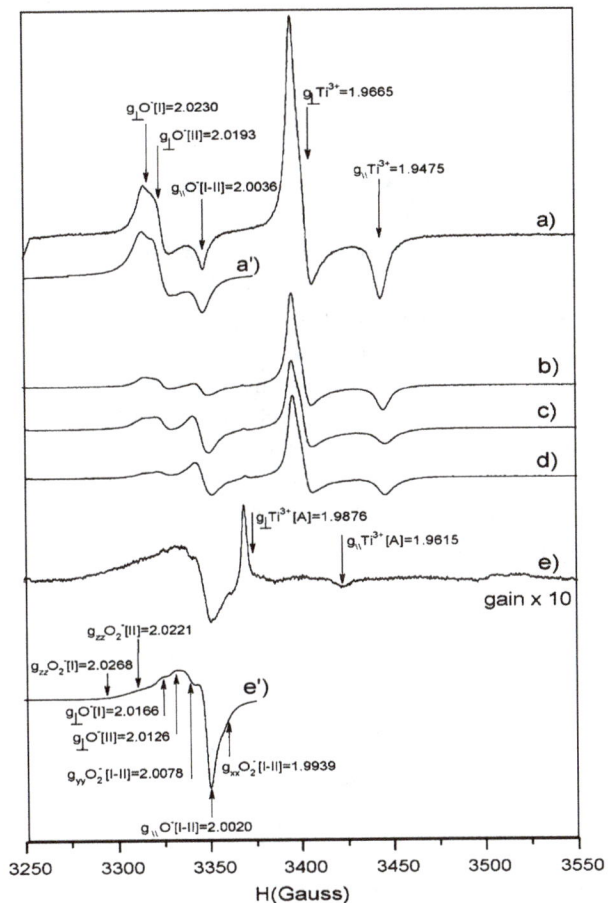

Fig. 12 EPR spectra of hydrothermally synthesised TiO_2 recorded at 10 K following 20 min of UV irradiation showing trapping centres at O^- sites and Ti^{3+} sites. *a* 100 % rutile, *b* 61 % rutile, *c* 48 % rutile, *d* 20 % rutile and *e* 0 % rutile (100 % anatase) [54]

corresponding ESR signal remains silent until the electrons relax and return to trap sites. In this case the majority of excited electrons are transferred to the rutile CB before becoming re-trapped at rutile trapping sites.

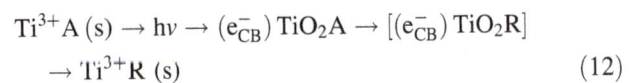

$$Ti^{3+}A\,(s) \rightarrow h\nu \rightarrow (e^-_{CB})\,TiO_2A \rightarrow [(e^-_{CB})\,TiO_2R]$$
$$\rightarrow Ti^{3+}R\,(s) \tag{12}$$

where Ti^{3+}As are anatase electron trapping sites, $[(e^-_{CB})\,TiO_2A]$s are anatase CB electrons, $[(e^-_{CB})\,TiO_2R]$s are

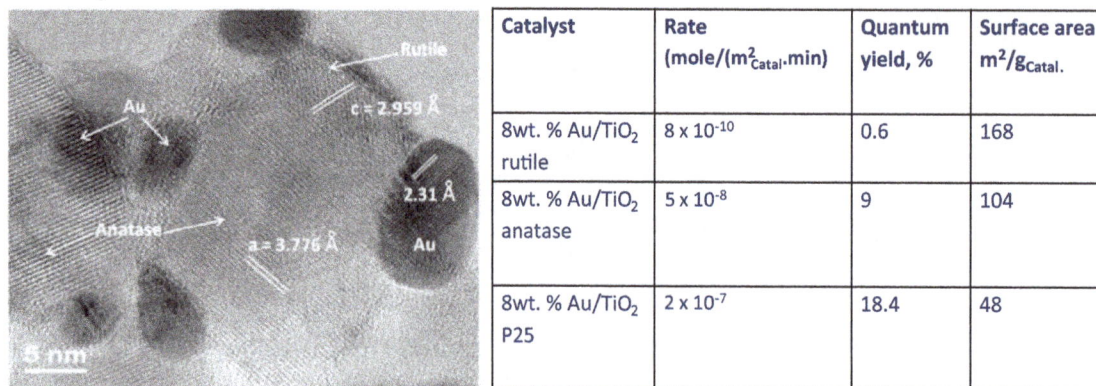

Catalyst	Rate (mole/(m^2_{Catal}.min))	Quantum yield, %	Surface area $m^2/g_{Catal.}$
8wt. % Au/TiO_2 rutile	8×10^{-10}	0.6	168
8wt. % Au/TiO_2 anatase	5×10^{-8}	9	104
8wt. % Au/TiO_2 P25	2×10^{-7}	18.4	48

Fig. 13 HRTEM of 8 wt% Au/TiO_2 P25. Au particles (*dark particles*). The close proximity of anatase, rutile and Au particles ensures electron transfer to reduce hydrogen ions to hydrogen molecules. Despite the high loading of Au the mean particle size is still about 5 nm, evidence of high dispersion. The table presents the rate of reaction, the quantum yield for 8 wt% Au/TiO_2 catalysts and the corresponding surface areas. Flux of UV (360 nm) photons is 1.8×10^{15} photons/s cm^2

rutile CB electrons, Ti^{3+}Rs are rutile electron trapping sites and (s)s are surface ions.

Hurum et al. [53] monitored recombination behaviour of electron-hole pairs at the surface and in the lattice of P25 using EPR. By adding 2,4,6-trichlorophenol (TCP) and then methanol to P25 to act as hole scavengers, the resultant EPR signal should show electrons populating surface trapping sites. Upon illumination with fluorescent light, at room temperature, the EPR signal for anatase surface electron trapping sites is observed to be very broad, indicating that a wide variety of both surface and lattice sites, with varying geometries and therefore energies, have been populated. Also observed is a signal attributed to distorted four-coordinated interfacial sites, as previously observed between silica and anatase coatings. Following illumination of the sample, this time at 10 K, more electron-hole pairs were created because of the lower temperature impeding charge carrier mobility. Subsequently any recombination of trapped species will occur where the electron-hole pair was formed. The resultant EPR signal showed there to be a decrease in both the number of populated surface electron trapping sites and interfacial electron trapping sites, which coincided with an increase in the signal for the anatase and rutile lattice electron trapping sites. These two observations indicate that recombination of electron-hole pairs is dominantly a surface process with photoexcited holes within TiO_2 particles generated on the particle surface and photoexcited electrons quickly trapped at lattice sites. This work directly shows that the surface-to-bulk ratio of particles in photocatalysis influences photoexcitation in particulate systems.

In a very recent work from our group [55], a series of Au/TiO_2 catalysts was studied for TiO_2 anatase, rutile and P25. Au wt% was changed from 1 to 8 %; TEM of Au particles showed a very similar particle size distribution, typically with a mean size between 4 and 5 nm. The quantum yield of the hydrogen production was measured using ethanol (0.5 vol%) as a sacrificial agent under UV excitation of 360 nm with a light flux of about 1 mW/cm^2. Part of the results is presented in Fig. 13. In all cases the rutile quantum yield was found less than 1 % and is therefore of little activity. The quantum yield of the anatase series increased with increasing Au loading. It was found equal to 9 % for 8 wt% anatase. At all Au concentrations the quantum yield of the P25 was found higher than that of the anatase. For example, the quantum yield of 8 wt% Au/TiO_2 P25 was found equal to 18.4 %. This once again is a direct observation of the synergistic effect of the two phases for the reaction.

There are also limited studies on the synergistic effect of TiO_2 phases different from those between anatase and rutile such as between anatase and brookite [56, 57] and rutile-brookite [58] that are treated in this work because of the so far very limited work. Despite the recent increase in studies on the role of each phase, as presented here and elsewhere [58–62], on the possible transfer routes for charge carriers across the mixed phase titania photocatalysts there is still no generally accepted consensus on the exact mechanisms for the synergistic effect between anatase and rutile. The clearest results still indicate that the presence of three phases—nano-particles of the precious metal (such as gold), the rutile and the anatase phase—in intimate contact is needed. Along this interface the transfer of holes from the anatase to the rutile phase appears to be the most plausible [but the transfer of electrons from the rutile to the anatase surface-trapped sites (Figs. 8, 9, 10) or due to a different band bending, interfacial band model (Fig. 11) is still unclear].

References

1. Ni, M., Leung, K.H., Leung, D., Sumathy, K.: A review and recent developments in photocatalytic water-splitting using TiO_2 for hydrogen production. Renew. Sustain. Energy Rev. **11**, 401–425 (2007)

2. Linsebigler, A.L., Lu, G., Yates, J.T. Jr.: Photocatalysis on TiO_2 surfaces: principles, mechanisms, and selected results. Chem. Rev. **95**, 735–758 (1995)

3. Fujishima, A., Honda, K.: Electrochemical photolysis of water at a semiconductor electrode. Nature **238**, 37–38 (1972)

4. Crabtree, G.W., Dresselhaus, M.S., Buchanan, M.V.: The hydrogen economy. Phys. Today **57**, 39–44 (2004)

5. Idriss, H.: Photoreactions of organic compounds with TiO_2 single crystal surfaces. In: Anpo, M., Kamat, P. (eds) Environmentally Benign (Chapter 21), ISBN: 978-0-387-48441-9 (2010)

6. Waterhouse, G.W.N., Idriss, H.: Photoreaction of ethanol and acetic acid over model TiO_2 single crystal surfaces. In: Vayssieres, L (ed) On Solar Hydrogen & Nanotechnology (Chapter 3). Wiley, New York (2009)

7. Henderson, M.A.: A surface science perspective on TiO_2 photocatalysis. Surf. Sci. Rep. **66**, 185–297 (2011)

8. Yates J.T. Jr.: Photochemistry on TiO_2: mechanisms behind the surface chemistry. Surf. Sci. **603**, 1605–1612 (2009)

9. Walter, M.G., Warren, E.L., McKone, J.R., Boettcher, S.W., Mi, Q., Santori, E.A., Lewis, N.S.: Solar water splitting cells. Chem. Rev. **110**, 6446–6473 (2010)

10. Bowker, M.: Sustainable hydrogen production by the application of ambient temperature photocatalysis. Green Chem. **13**, 2235–2246 (2011)

11. Connelly, K.A., Idriss, H.: The photoreaction of TiO_2 and Au/TiO_2 single crystal and powder surfaces with organic adsorbates. Emphasis on hydrogen production from renewables. Green Chem. **14**, 260–280 (2012)

12. Yang, Y.Z., Chang, C.H., Idriss, H.: Photo-catalytic production of hydrogen from ethanol over M/TiO_2 catalysts (M = Pd, Pt or Rh). App. Cat. B Environ. **67**, 217–222 (2006)

13. Mogyorosi, K., Kmetyko, A., Czirbus, N., Vereb, G., Sipos, P., Dombi, A.: Comparison of the substrate dependent performance of Pt-, Au- and Ag-doped TiO_2 photocatalysts in H_2 production and in decomposition of various organics. React. Kin. Cat. Lett. **98**, 215–225 (2009)

14. Subramanian, V., Wolf, E.E., Kamat, P.V.: Catalysis with TiO_2/gold nanocomposites. Effect of metal particle size on the Fermi level equilibration. J. Am. Chem. Soc. **126**, 4943–4950 (2004)

15. Chen, X., Shen, S., Guo, L., Mao, S.S.: Semiconductor-based photocatalytic hydrogen generation. Chem. Rev. **110**, 6503–6570 (2010)

16. Wu, G., Chen, T., Zhou, G., Zong, X., Li, C.: H_2 production with low CO selectivity from photocatalytic reforming of glucose on metal/TiO_2 catalysts. Sci. China Ser. B Chem. **51**, 97–1000 (2008)

17. Primo, A., Corma, A., Garcia, H.: Titania supported gold nanoparticles as photocatalyst. Phys. Chem. Chem. Phys. **13**, 886–910 (2010)

18. Linic, S., Christopher, P., Ingram, D.: Plasmonic-metal nanostructures for efficient conversion of solar to chemical energy. Nat. Mater. **10**, 911–921 (2011)

19. Du, L., Furube, A., Yamamoto, K., Hara, K., Katoh, R., Tachiya, M.: Plasmon-induced charge separation and recombination dynamics in gold–TiO_2 nanoparticle systems: dependence on TiO_2 particle size. J. Phys. Chem. C **113**, 6454–6462 (2009)

20. Seh, Z.W., Liu, S., Low, M., Zhang, S.-Y., Liu, Z., Mlayah, A., Han, M.-Y.: Adv. Mater. **24**, 2310–2314 (2012)

21. Idriss, H., Wahab, A.K.: European Procedure (Patents) (12CHEM0012-EP-EPA) filed at the Patent Office on 03-09-2012 as Serial Number 12006217.9. (2012)

22. Bamwenda, G.R., Tsubota, S., Nakamura, T., Haruta, M.: Photoassisted hydrogen production from a water-ethanol solution: a comparison of activities of Au-TiO_2 and Pt-TiO_2. J. Photochem. Photobiol. A Chem. **89**, 177–189 (1995)

23. Sun, B., Vorontsov, V., Smirniotis, P.G.: Role of platinum deposited on TiO_2 in phenol photocatalytic oxidation. Langmuir **19**, 3151–3156 (2003)

24. Bamwenda, G.R., Tsubota, S., Kobayashi, T., Haruta, M.: Photoinduced hydrogen production from an aqueous solution of ethylene glycol over ultrafine gold supported on TiO_2. J. Photochem. Photobiol. A Chem. **77**, 59–67 (1994)

25. Sakata, T., Kawai, T.: Heterogeneous photocatalytic production of hydrogen and methane from ethanol and water. Chem. Phys. Lett. **80**, 341–344 (1981)

26. Yang, Y., Zhong, H., Tian, C.: Photocatalytic mechanisms of modified titania under visible light. Res. Chem. Intermed. **37**, 91–102 (2011)

27. Nair, R.G., Paul, S., Samdarshi, S.K.: High UV/visible light activity of mixed phase titania: a generic mechanism. Solar Energy Mater. Solar Cells **95**, 1901–1907 (2011)

28. Nadeem, M.A., Murdoch, M., Waterhouse, G.I.N., Metson, J.B., Keane, M.A., Llorca, J., Idriss, H.: Photoreaction of ethanol on Au/TiO_2 anatase: comparing the micro to nanoparticle size activities of the support for hydrogen production. J. Photochem. Photobiol. A Chem. **216**, 250–255 (2010)

29. Scott, M., Idriss, H.: Heterogeneous catalysis for hydrogen production. In: Anstis, P., Crabtree, R.H. (eds) Handbook of Green Chemistry—Green Catalysis, vol. 1, chap. 10, ISBN-10: 3-527-31577-2 (2009)

30. Idriss, H.: Ethanol reactions over the surfaces of transition metals/cerium oxide catalysts. Platin. Met. Rev. **48**, 105–115 (2004)

31. Murdoch, M., Waterhouse, G.I.N., Nadeem, M.A., Metson, J.B., Keane, M.A., Howe, R.F., Llorca, J., Idriss, H.: The effect of gold loading and particle size on photocatalytic hydrogen production from ethanol over Au/TiO_2 nanoparticles. Nat. Chem. **3**, 489–492 (2011)

32. Haruta, M.: Size- and support-dependency in the catalysis of gold. Catal. Today **36**, 153–166 (1997)

33. Valden, M., Lai, X., Goodman, D.W.: Onset of catalytic activity of gold clusters on titania with the appearance of non-metallic properties. Science **281**, 1647–1650 (1998)

34. Lai, X., Goodman, D.W.: Structure-reactivity correlations for oxide-supported metal catalysts: new perspectives from STM. J. Mol. Cat. A **162**, 33–50 (2000)

35. Cárdenas-Lizana, F., Gómez-Quero, S., Idriss, H., Keane, M.A.: Gold particle size effects in the gas phase hydrogenation of dinitrobenzene over Au/TiO_2. J. Catal. **268**, 223–234 (2009)

36. Li, G., Chen, L., Graham, M.E., Gray, K.A.: A comparison of mixed phase titania photocatalysts prepared by physical and chemical methods: the importance of the solid–solid interface. J. Mol. Cat. A. **275**, 30–35 (2007)

37. Kudo, A., Miseki, Y.: Heterogeneous photocatalyst materials for water splitting. Chem. Soc. Rev. **38**, 25–278 (2009)

38. Deak, P., Aradi, B., Frauenheim, T.: Band lineup and charge carrier separation in mixed rutile-anatase systems. J. Phys. Chem. C **115**, 3443–3446 (2011)

39. Kho, Y.K., Iwase, A., Teoh, W.Y., Madler, L., Kudo, A., Amal, R.: Photocatalytic H_2 evolution over TiO_2 nanoparticles. The synergistic effect of anatase and rutile. J. Phys. Chem. C. **114**, 2821–2829 (2010)

40. Long, R., Dai, Y., Huang, B.: Structural and electronic properties of iodine-doped anatase and rutile TiO_2. Comp. Mat. Sci. **45**, 223–228 (2009)

41. Lee, H.S., Woo, C.S., Youn, B.K., Kim, S.Y., Oh, S.T., Sung, Y.E., Lee, H.I.: Bandgap modulation of TiO_2 and its effects on

the activity in photocatalytic oxidation of 2-isopropyl-6-methyl-4-pyrimidinol. Top. Catal. **35**, 255–260 (2005)

42. Colbeau-Justin, C., Kunst, M.: Structural influence on charge-carrier lifetimes in TiO$_2$ powders studied by microwave absorption. J. Mater. Sci. **38**, 2429–2437 (2003)

43. Xu, M., Gao, Y., Moreno, E.M., Kunst, M., Muhler, M., Wang, Y., Idriss, H., Wöll, C.: Photocatalytic activity of bulk TiO$_2$ anatase and rutile single crystals using infrared absorption spectroscopy. Phys. Rev. Lett. **106**, 138302 (2011)

44. Shen, Q., Katayama, K., Sawada, T., Yamaguchi, M., Kumagai, Y., Toyoda, T.: Photoexcited hole dynamics in TiO$_2$ nanocrystalline films characterised using a lens-free heterodyne detection transient grating technique. Chem. Phys. Lett. **419**, 464–468 (2006)

45. Wilson, J.N., Idriss, H.: Structure sensitivity and photocatalytic reactions of semiconductors. Effect of the last layer atomic arrangement. J. Am Chem. Soc. **124**, 11284–11285 (2002)

46. Wilson, J.N., Idriss, H.: Effect of surface reconstruction of TiO$_2$(001) single crystal on the photoreaction of acetic acid. J. Catal. **214**, 46–52 (2003)

47. Komaguchi, K., Nakano, H., Araki, A., Harima, Y.: Photoinduced electron transfer from anatase to rutile in partially reduced TiO$_2$ (P-25) nanoparticles: an ESR study. Chem. Phys. Lett. **428**, 338–342 (2006)

48. Zhang, J., Xu, Q., Feng, M.L., Li, C.: Importance of the relationship between surface phases and photocatalytic activity of TiO$_2$. Agnew. Chem. Int. Ed. **47**, 1766–1769 (2008)

49. Bickley, R.I., Gonzalez-Carreno, T., Lees, J.S., Palmisano, L., Tilley, R.J.D.: A structural investigation of titanium dioxide photocatalysts. J. Sol. State Chem. **92**, 178–190 (1991)

50. Hurum, D.C., Agrios, A.G., Gray, K.A., Rajh, T., Thurnauer, M.C.: Explaining the enhanced photocatalytic activity of Degussa P25 mixed-phase TiO$_2$ using EPR. J. Phys. Chem. B **107**, 4545–4549 (2003)

51. Carneiro, J.T., Savenije, T.J., Moulijn, J.A., Mul, G.: How phase composition influences optoelectronic and photocatalytic properties of TiO$_2$. J. Phys. Chem. C **115**, 2211–2217 (2011)

52. Scott, M., Nadeem, A.M., Waterhouse, G.I.W., Idriss, H.: Hydrogen production from ethanol. Comparing thermal catalytic reactions to photo-catalytic reactions. MRS Proceed. **1326**, 1764–1769 (2011)

53. Hurum, D.C., Gray, K.A., Rajh, T., Thurnauer, M.C.: Recombination pathways in the Degussa P25 formulation of TiO$_2$: surface versus lattice mechanisms. J. Phys. Chem. B **109**, 977–980 (2005)

54. Scotti, R., D'Arienzo, M., Testino, A., Morazzoni, F.: Photocatalytic mineralization of phenol catalyzed by pure and mixed phase hydrothermal titanium dioxide. Appl. Catal. B **88**, 497–504 (2009)

55. Wahab, A.K., Waterhouse G.I.N., Hedhili, M., Anjum, D., Al-Hazza, A., Llorca, J., Idriss, H.: Hydrogen production from water over semiconductor materials: the role of synergism between different phases. American Chemical Society 244th meeting, August 2012 (USA) and Topics in Catalysis (in press)

56. Kandiel, T.A., Feldhoff, A., Robben, L., Dillert, R., Bahnemann, D.W.: Tailored titanium dioxide nanomaterials: anatase nanoparticles and brookite nanorods as highly active photocatalysts. Chem. Mater. **22**, 2050–2060 (2009)

57. Ismail, A.A., Kandiel, T.A., Bahnemann, D.W.: Novel (and better?) titania-based photocatalysts: brookite nanorods and mesoporous structures. J. Photochem. Photobiol. A Chem. **216**, 183–193 (2010)

58. Štengl, V., Králová, D.: Photoactivity of brookite–rutile TiO$_2$ nanocrystalline mixtures obtained by heat treatment of hydrothermally prepared brookite. Mater. Chem. Phys. **129**, 794–801 (2011)

59. Wang, C., Zhang, X., Shao, C., Zhang, Y., Yang, J., Sun, P., Liu, X., Liu, H., Liu, Y., Xie, T., Wang, D.: Rutile TiO$_2$ nanowires on anatase TiO$_2$ nanofibers: a branched heterostructured photocatalysts via interface-assisted fabrication approach. J. Coll. Interf. Sci. **363**, 157–164 (2011)

60. Xu, Q., Ma, Y., Zhang, J., Wang, X., Feng, Z., Li, C.: Enhancing hydrogen production activity and suppressing CO formation from photocatalytic biomass reforming on Pt/TiO$_2$ by optimizing anatase–rutile phase structure. J. Catal. **278**, 329–335 (2011)

61. Kim, S., Ehrman, S.H.: Photocatalytic activity of a surface-modified anatase and rutile titania nanoparticle mixture. J. Coll. Interf. Sci. **338**, 304–307 (2009)

62. Zhang, P., Yin, S., Sato, T.: A low-temperature process to synthesize rutile phase TiO$_2$ and mixed phase TiO$_2$ composites. Mater. Res. Bull. **45**, 275–278 (2010)

2

Thermochemical hydrogen production from water using reducible oxide materials: a critical review

Lawrence D'Souza

Abstract This review mainly focuses on summarizing the different metal oxide systems utilized for water-splitting reaction using concentrated solar energy. Only two or three cyclic redox processes are considered. Particle size effect on redox reactions and economic aspect of hydrogen production via concentrated solar energy are also briefly discussed. Among various metal oxides system CeO_2 system is emerging as a promising candidate and researchers have demonstrated workability of this system in the solar cavity-receiver reactor for over 500 cycles. The highest solar thermal process efficiency obtained so far is about 0.4 %, which needs to be increased for real commercial applications. Among traditionally studied oxides, thin-film ferrites looks more promising and could meet US Department of energy target of \$2.42/kg H_2 by 2025. The cost is mainly driven by high heliostat cost which needs to reduced significantly for economic feasibility. Overall, more work needs to be done in terms of redox material engineering, reactor technology, heliostat cost reduction and gas separation technologies before commercialization of this technology.

Keywords Hydrogen · Water splitting · Solar thermal

Introduction

Hydrogen is considered as next generation fuel to propel airplanes, automotive vehicles and virtually any stationary power system using fuel cells. The non-polluting by-product 'water' upon hydrogen combustion has attracted world attention to save ever polluting earth environment for sustainable future. Currently, the hydrogen is derived from fossil fuels. The smallest molecule of universe sees highest demand due to its non-polluting end product as well as its remarkable chemical and physical properties. There are number of chemical transformation in which hydrogen is used as hydrogenating or reducing agent. Moreover, present trend to harvest CO_2 into useful chemicals demands hydrogen. Many scientists around the world are pessimistic about CO_2 hydrogenation since they see raising demand for hydrogen and currently there are no real alternatives to fulfill other than fossil fuels. Researchers have been looking at different possibilities to generate hydrogen by biological and chemical means. Electrolysis of water is one of the easy and greener route to generate hydrogen only if electricity comes from wind, tidal, photovoltaics, geothermal or hydropower. The other greener routes are photoelectrochemical water splitting [1], by direct splitting of water [2, 3] and solar thermochemical cycles. It is hoped that combination of several technologies can fulfill future hydrogen demand.

Water splitting by low valent metal oxides at high temperature is one of the clean way of hydrogen production since the temperature needed to perform chemical reaction comes from concentrated solar thermal heat. Though the technology is known since more than three decades commercial realization is yet to happen due to numerous challenges in this technology. The off-sun hours, cloudy and rainy seasons are main drawbacks for commercial realization. Moreover, technology cannot be implemented in geographically poor sun receiving regions.

This review summarizes the work done in high-temperature hydrogen production via two-step redox processes

L. D'Souza (✉)
SABIC Corporate Research and Innovation Center (CRI)
at KAUST, Saudi Basic Industries Corporation, P.O. Box
4545-4700, Thuwal 23955-6900, Saudi Arabia
e-mail: dsouzal@sabic.com

using various metal oxides (Table 1). Only the high-temperature experiments demonstrated either in solar furnace or in laboratory fixed-bed reactor have been considered. This review does not cover the hybrid technologies or other forms of hydrogen production technologies.

Bilgen et al. [4] have demonstrated the possibility of splitting water directly at high temperature. The theoretical calculation for the said reaction is depicted in Fig. 1. It was found that the amount of hydrogen produced decreases with increase in H_2O partial pressure. Figure 1 gives the compounded results for partial pressure of water equal to 0.1 bar between 1,500 and 5,000 K. Bilgen [4] experimentally demonstrated that between 2,273 and 2,773 K formation of about 2–3 % H_2 when mixture of steam and argon was passed in the crucible at the focus of the solar furnace.

The dissociation of metal oxide to their respective metal is written as follows [14]:

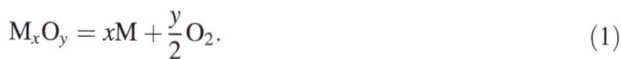

$$M_xO_y = xM + \frac{y}{2}O_2. \tag{1}$$

The temperature required for few metal oxides conversion to their metallic form is given in Table 2. Except for ZnO, achieving temperature needed to reduce metal oxide to their metallic form is practically impossible due to the high temperature required. Concentration ratios of up to 10,000 suns have been achieved by researchers which translate to 3,800 K. But high-temperature operation, reactor material thermal stability and radiation heat losses makes the process almost impossible. The temperature required to attain $\Delta G°$ of the reaction (1) equals zero can be substantially brought down by the use of hydrocarbon as reducing agents, for example graphite or methane, which can be written as follows:

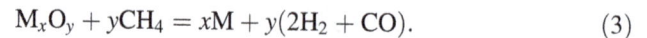

$$M_xO_y + yC(gr) = xM + yCO \tag{2}$$

$$M_xO_y + yCH_4 = xM + y(2H_2 + CO). \tag{3}$$

Two-step cyclic redox processes are simplest way of producing hydrogen by utilizing metal oxide. The solar reduction step is endothermic and can be written as shown in (4),

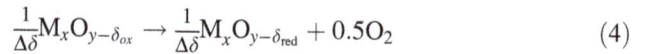

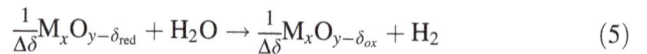

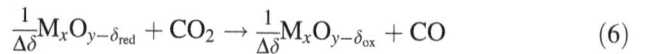

$$\frac{1}{\Delta\delta}M_xO_{y-\delta_{ox}} \rightarrow \frac{1}{\Delta\delta}M_xO_{y-\delta_{red}} + 0.5O_2 \tag{4}$$

$$\frac{1}{\Delta\delta}M_xO_{y-\delta_{red}} + H_2O \rightarrow \frac{1}{\Delta\delta}M_xO_{y-\delta_{ox}} + H_2 \tag{5}$$

$$\frac{1}{\Delta\delta}M_xO_{y-\delta_{red}} + CO_2 \rightarrow \frac{1}{\Delta\delta}M_xO_{y-\delta_{ox}} + CO \tag{6}$$

(where δ is non-stoichiometric coefficient and $\Delta\delta$ is change in non-stoichiometric coefficient).

The reaction (4) takes place at temperature above 1,000 K and many metal oxide systems have been studied over the past four decades. Several two- and three-step H_2O-splitting thermochemical cycles based on metal oxides redox reactions have been reported in the literature. Nakamura [5] first proposed the two-step redox cycle in 1977 for Fe_3O_4/FeO redox cycle; interest then diminished for the next two decades and thereafter a spurt of interest resulted in investigation of several other oxide systems for thermochemical redox cycle for hydrogen generation. The high temperature required for reduction reaction can be supplied by either concentrated solar energy or fossil fuels. The solar reduction is usually carried out in the presence of an inert gas, if a reducing gas is used the reduction temperature can be brought down substantially. The reduced metal oxide can be oxidized back to the original state by oxidants like H_2O or CO_2. If H_2O is used H_2 can be produced and if CO_2 is used CO can be produced as shown in Eqs. (5) and (6), respectively. If H_2O and CO_2 are used to oxidize the redox material alternatively or together one can produce the synthesis gas (CO + H_2) from totally renewable sources (CO_2 and H_2O) [30].

Sibieude et al. [10] demonstrated reduction of Fe_3O_4 to FeO in a solar furnace by heating the material 300 °C above its melting point. They could obtain up to 40 % conversion in air and 100 % conversion in argon atmosphere. Figure 2 gives the conversion rate of Fe_3O_4 to FeO as a function of temperature under argon flow (20 l/h).

As observed by many researchers, they also experienced that quenching the reduced oxides is very important. Table 3 summarized the FeO yield with various quenching rates. It can be seen that in presence of air up to 40 % conversion can be obtained with 373 K/s cooling rate.

As per literature the reduction of Mn_3O_4 to MnO occurs above 1,773 K [31]. Sibieude et al. [10] have studied reduction of Mn_3O_4 to MnO in a solar furnace. They could

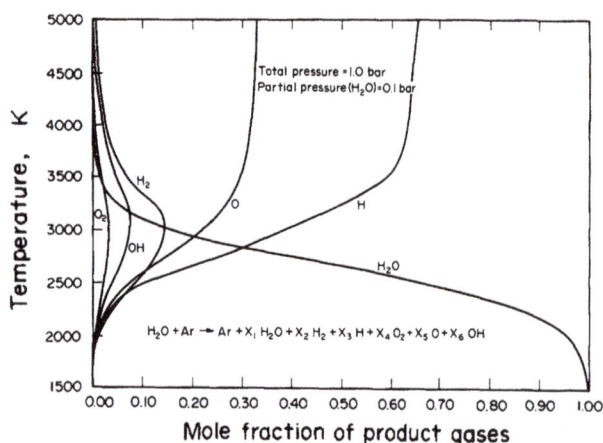

Fig. 1 Theoretical composition of the different products in the dissociation of water at high temperature; total pressure is 1 bar and partial pressure of H_2O is 0.1 bar

Table 1 Summary of potential two-step water-splitting reaction systems reported in the literature

Main theme	T (K) for $\Delta G^0 = 0$	T (K) for reduction	H_2 yield (%)	T (K) for oxidation	References
$Fe_3O_4 = 3FeO + 1/2\ O_2$		2,500		<1,000	[5]
$2Mn_3O_4 = 6MnO + O_2$	2,000	1,810	0.002	900	[6]
$2\ Co_3O_4 = 6CoO + O_2$	1,000	1,175	4×10^{-7}	900	[6]
$2\ Nb_2O_5 = 4NbO_2 + O_2$	4,000	3,600	99.7	900	[6]
$ZnO = Zn + 1/2\ O_2$	2,350	2,300	Na	Na	[7–9]
Mn_3O_4 to MnO	Na	1,773	Na	Na	[10]
$(Fe_{1-x}M_x)_3O_4 = (Fe_{1-x}Mn_x)O$	Na		Very low	773–1,173	[11, 12]
$(Fe_{1-x}M_x)_3O_4 = (FeM)O$, M = Ni, Mn, Zn	Na	>1,073	Na	<1,073	[13]
$(Fe_{1-x}Mx)_3O_4 = (FeM)O$, M = Mn, Co, Mg	Na		Na	Na	[14]
$2CdO = 2Cd + O_2$	Na	>1,473	Na	Na	[10]
$SnO_2 = Sn + O_2$	Na	>1,873	90 %	773–873	[4, 15]
$ZnFe_2O_4 = Zn_xFe_{3-x}O_4$, Zn(g), O_2	Na	1,173			[16]
$x/3Fe_3O_4 + Y_yZr_{1-y}O_{2-y/2} = Fe_xY_yZr_{1-y}O_{2-y/2+x} + x/6\ O_2$	Na	1,673	Na	<1,273	[17, 18]
$Fe_xY_yZr_{1-y}O_{2-y/2+x} + x/3H_2O = x/3Fe_3O_4 + Y_yZr_{1-y}O_{2-y/2} + x/3H_2$					
$2CeO_2(s) = Ce_2O_3(s) + 1/2O_2(g); Ce_2O_3(s) + H_2O(g) = 2CeO_2(s) + H_2(g)$	Na	2,273	Na	673–873	[19–21]
$Ce_{1-x}Zr_xO_2\ (0 \leq x \leq 0.3)$	Na	1,773	Na	Na	[22–24]
$TiO_2 = TiO_{x,}\ x = 1.83–1.96$	Na	2,573–3,073	Na	Na	[25]
$2\ SiO_2 = 2\ SiO + O_2$	Na	$3,250^a$	Na	2,729	[4, 26]
$SiO\ (g) + H_2O + SiO_2 + H_2$					
$WO_3\ (s) = W + 3/2\ O_2$	Na	$4,183^a$	Na	1,157	[4, 26]
$W + 3H_2O = WO_3\ (s) + 3H_2$					
$MoO_2 = Mo + O_2$	Na	$3,986^a$	Na	1,816	[4, 26]
$Mo + 2H_2O = MoO_2\ (s) + 2H_2$					
$3\ In_2O_3 = In_2O_3 + 4\ In$	Na	$>2,780^a$	Na	1,000	[4]
Solid acids viz. SiO_2, Al_2O_3, TiO_2, ZnO, $CaCO_3$	Na	Na	Na	Na	[27]
$In_2O_3 = In_2O + O_2\ In_2O + 2H_2O = In_2O_3 + H_2$	Na	2,473	Na	1,073	[28]
$MnFe_2O_4 + 3CaO + (1 - y)H_2O = Ca_3(Fe,Mn)_3O_{8-y} + (1 - y)H_2$	Na	1,273	Na	873	[29]
$\quad Ca_3(Fe,Mn)_3O_{8-y} = MnFe_2O_4 + 3CaO + (1 - y)/2\ O_2$					

Na data not available

[a] Process is practically not feasible

Table 2 Approximate temperature required for which ΔG° of the reaction (1) equals zero

Metal oxide	T (K) for $\Delta G^0 = 0$
$Fe_2O_3^a$	3,700
Al_2O_3	>4,000
MgO	3,700
ZnO	2,335
TiO_2^a	>4,000
SiO_2^a	4,500
CaO	4,400

[a] Fe_2O_3, TiO_2 and SiO_2 decompose to lower valence oxides before complete dissociation to the final

obtain about 80 % conversion at 2,173 K under atmospheric pressure of air with a cooling rate of 373 K/s. The quenching of the MnO is very important to stop backward reaction, i.e., formation of Mn_3O_4 again. They have not performed more extensive work on this system.

Ehrensberger et al. [11] have studied non-stoichiometric FeMn oxides for two-step water-splitting reaction. They calculated ΔG_R values for two-step Nakamura cycles FeO–Fe_3O_4 and MnO–Mn_3O_4 and the plotted results are shown in the Figs. 3, 4. Figure 3 indicates that ΔG^0 equals to zero for the reduction of Mn_3O_4 to MnO is at least 500 K less than that of Fe_3O_4 to FeO system. However, Fig. 4 indicates that FeO can produce hydrogen between 873 and 1,073 K but MnO system is unable to produce hydrogen in significant levels. This led the authors to think of the possibility of combining Fe and Mn oxides to reduce spinel at lower temperature as well as produce hydrogen in significant amount in oxidation step. Authors demonstrated the oxidation of $Fe_{1-y}O$ and $(Fe_{1-x}Mn_x)_{1-y}O$ ($x \leq 0.3$) in a laboratory tubular furnace and monitored gaseous

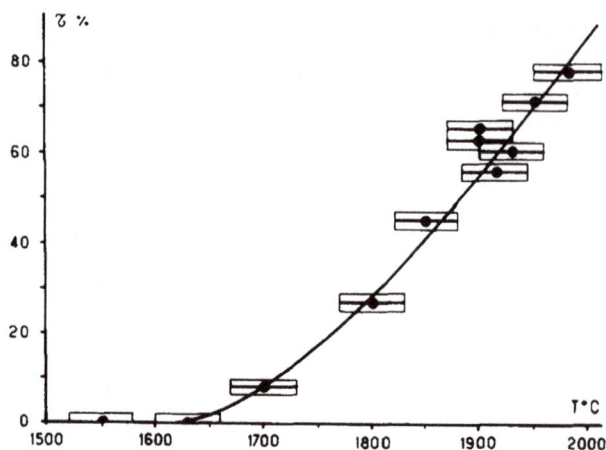

Fig. 2 Conversion rate of magnetite versus temperature

Table 3 Influence of cooling rate on FeO yield in air and argon atmosphere

Atmosphere	Quenching speed (K s^{-1})	%mol FeO
Air[a]	278	0
	293	25
	373	40
	1,273	50
Argon[a]	278	40
	293	45
	373	55
	1,273	60

[a] Residence time = 1 min, temperature = 2,173 K, flow rate = 20 l/h

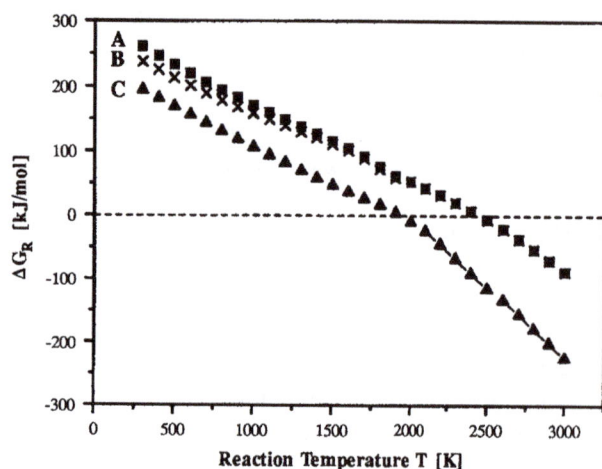

Fig. 3 Gibbs free enthalpy ΔG_R for the decomposition of Fe_3O_4 (A, B) and Mn_3O_4 (C) to FeO (A), $Fe_{0.947}O$ (B) and MnO (C) as a function of temperature

products using mass spectrometer. At atmospheric pressure, water with a partial pressure of about 4,200 Pa in nitrogen was able to oxidize $(Fe_{1-x}Mn_x)_{1-y}O$ (x = 0.0,

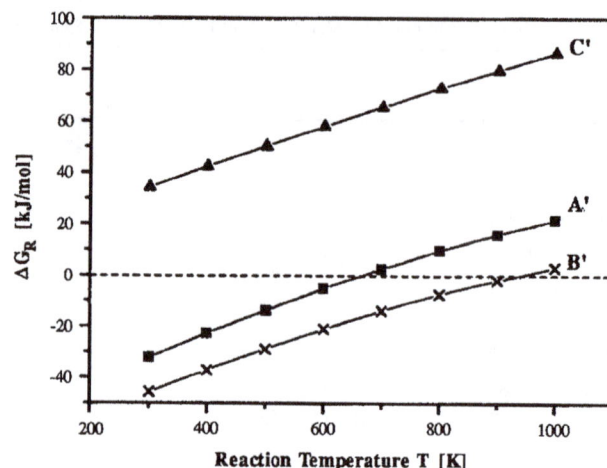

Fig. 4 Gibbs free enthalpy ΔG_R for the water splitting reaction of FeO (A'), $Fe_{0.947}O$ (B') and MnO (C') as a function of temperature

0.1, 0.3) to $(Fe_{1-x'}Mn_{x'})_3O_4$ with $x' < x$ forming molecular hydrogen. The substitution of iron with 10–30 % Mn in the wuestite phase did not lower the total amount of hydrogen formed, but it changed the kinetics of the process significantly. It was also found that the process is thermodynamically controlled at high temperature. The rate of water splitting decreased with increase in manganese concentration.

They also found that during water-splitting reaction $(Fe_{1-x}Mn_x)_{1-y}O$ forms manganese-rich rock salt phase and an iron-rich spine phase due to phase segregation processes [12].

Tamura et al. [13] extended the work to 'NiMnFe' system, as shown in the reaction schemes 7 and 8.

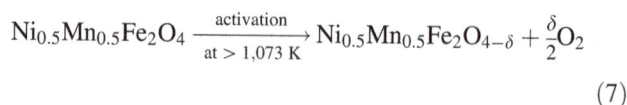

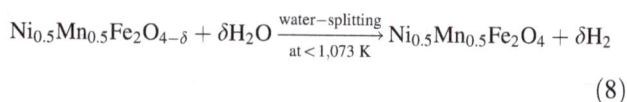

$$Ni_{0.5}Mn_{0.5}Fe_2O_4 \xrightarrow[\text{at} > 1,073 \text{ K}]{\text{activation}} Ni_{0.5}Mn_{0.5}Fe_2O_{4-\delta} + \frac{\delta}{2}O_2$$
(7)

$$Ni_{0.5}Mn_{0.5}Fe_2O_{4-\delta} + \delta H_2O \xrightarrow[\text{at} < 1,073 \text{ K}]{\text{water-splitting}} Ni_{0.5}Mn_{0.5}Fe_2O_4 + \delta H_2$$
(8)

They performed the above two-step reaction in a solar reactor at 1,073 K. In the first endothermic step, $Ni_{0.5}Mn_{0.5}Fe_2O_4$ was thermally activated to get oxygen-deficient compound, in the second step the oxygen-deficient compound was oxidized using H_2O to produce H_2. Since, O_2 and H_2 were produced in two different steps, high-temperature separation of those gases can be eliminated in the proposed method. They have demonstrated the workability of two-step water-splitting reaction with $NiFe_2O_4$, $Ni_{0.5}Mn_{0.5}Fe_2O_4$ and $Ni_{0.5}Zn_{0.5}Zn_2O_4$ systems using thermogravimetric experiments. They found that $NiFe_2O_4$ system needs lower reactivation rate (conducted after the water-splitting reaction) compared to

$Ni_{0.5}Mn_{0.5}Fe_2O_4$ system. The oxygen released during reduction step in $NiFe_2O_4$, $Ni_{0.5}Mn_{0.5}Fe_2O_4$ and $Ni_{0.5}Zn_{0.5}Zn_2O_4$ systems were 0.2, 0.3 and 0.4 %, respectively. They also demonstrated the workability of the two-step hydrogen production in solar reactors. They performed two redox cycles to prove the oxygen and hydrogen evolution in activation (reduction) and reactivation (oxidation) processes. The activation was conducted at 1,373 K in presence of Ar and reactivation was conducted in presence of (steam + Ar) flow at 573 K. In the case of $ZnFe_2O_4$, reduction follows two pathways [16] as shown in Eqs. 9 and 10.

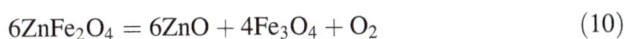

$$3ZnFe_2O_4 = 3Zn + 2Fe_3O_4 + 2O_2 \qquad (9)$$

$$6ZnFe_2O_4 = 6ZnO + 4Fe_3O_4 + O_2 \qquad (10)$$

The reduction and oxidation steps have been demonstrated using Xe beam experiment and solar furnace experiments. It took less than 60 s for the Zn-ferrite to release the expected amount of O_2 from the lattice at 1,750 K. Authors have seen deposition of Zn on the reactor walls during reduction step and have measured O_2 released using mass spectrometer.

Abanades et al. [19] examined CeO_2/Ce_2O_3 redox pairs and demonstrated the feasibility in a solar reactor featuring an inert gas atmosphere at $T = 2,273$ K, $P = 100$–200 mbar. It consists of two chemical steps: (1) reduction, $2CeO_2 \rightarrow Ce_2O_3 + 0.5O_2$; (2) hydrolysis, $Ce_2O_3 + H_2O \rightarrow 2CeO_2 + H_2$. The reduction step is endothermic and takes place at $T = 2,273$ K, $P = 100$–200 mbar; however, oxidation step takes place at 673–873 K resulting in pure hydrogen which can be directly used in fuel cells application. The main advantages of the process are low cost material which is abundantly available in nature and the process uses non-corrosive chemicals. The reduced phase is very stable at ambient temperature and nonreactive to moisture and oxygen which makes this material ideal for on-site hydrogen generation which in turn overcomes problem associated with transportation. However, this technology has few drawbacks, a maximum heat input temperature slightly higher than 2,273 K, the cycle working temperature of the endothermic step must be optimized to be compatible with dish or tower technologies, and to reduce sample vaporization. High molecular weight of cerium oxides poses problem in the flow of solids in the process.

Chueh et al. [20] have extended the work on CeO_2 system. They demonstrated the O_2 evolution during reduction step, CO and H_2 generation during oxidation step using the solar cavity-receiver reactor over 500 cycles. They could achieve solar-to-fuel efficiencies of 0.7–0.8 % and concluded that efficiency is limited by the system scale and design rather than by chemistry. However, Rager [32] pointed out that the efficiency 0.7–0.8 % efficiency refers to 'peak instantaneous efficiency' but after averaging the efficiency over 80 % of the fuel production, the actual efficiency is just 0.4 %. He recalculated the solar thermal process efficiency and found that the value is still lower than that of reported by Chueh et al. [20], mainly because the later authors did not consider the energy need for large amount of purge gas used in redox processes. Purge gas takes up lot of solar heat and hence results in lower solar thermal efficiency.

Kang et al. [23] have extended the work on CeO_2 system. They synthesized $Ce_xZr_{1-x}O_2$ ($x = 0.6, 0.7, 0.8, 1.0$) solid solutions and tested for redox reactions. They found that the reduced $Ce_xZr_{1-x}O_{2-\delta}$ ($x = 0.5, 0.6, 0.7, 0.8, 1.0$) samples exhibited higher hydrogen production ability for water splitting due to improved oxygen diffusion through the bulk. Kaneko et al. [22] have extended the work on $Ce_xZr_{1-x}O_{2-\delta}$ solid solution system. They introduced Zr^{4+} on various ratios in CeO_2 lattices and found that the oxygen releasing capacity or extent of CeO_2 reduction increases with the increase of Zr^{4+} ions similar to Kang et al.'s [23] observations. The highest oxygen release was found at $x = 2$ ($Ce_{0.8}Zr_{0.2}O_2$) at 1,773 K in air and the amount of reduced cerium was found to be about 11 % which is seven times higher than just with bare CeO_2. The enhancement of the O_2-releasing reaction with CeO_2–ZrO_2 oxide is found to be caused by an introduction of Zr^{4+}, which has smaller ionic radius than Ce^{3+} or Ce^{4+} in the fluorite structure.

Le Gal and Abanades [24] doped trivalent lanthanides, viz. La, Sm and Gd in CeO_2 to form binary oxides and used in hydrogen production by solar thermal redox cycles. They found that trivalent lanthanide-doped material improves the thermal stability of the material during consecutive redox cycles, but hydrogen production remains the same as ceria. They also doped trivalent lanthanides in CeO_2–ZrO_2 to form ternary oxides. They found that with 1 % gadolinium to ceria–zirconia solid solutions nearly 338.2 μmol of hydrogen per gram during one cycle with the O_2-releasing step at 1,400 °C and the H_2-generation step at 1,050 °C. This quantity of hydrogen is more than with CeO_2–ZrO_2 system. They also found that the addition of lanthanum enhances the thermal stability of ceria–zirconia solid solution similar to as observed in cases of lanthanum-doped CeO_2 binary oxides.

Lipinski et al. [21] applied first and second laws of thermodynamics to analyze the potential of applying heat recovery for realizing high efficiency in solar-driven CeO_2-based non-stoichiometric redox cycles to split H_2O or CO_2. They found that at 2,000 K, with 80 % solid phase heat recovery, advanced materials can only increase efficiency from 16 to 20 %, while, at 1,850 K, advanced materials can improve efficiency from 14 to 23 %, a higher maximum value because of decreased re-radiation and gas heating at the lower value of T_{red}.

Inoue et al. [33] demonstrated effectiveness of a ZnO/MnFe$_2$O$_4$ system in a lab furnace at 1,273 K. When H$_2$O was contacted with ZnO/MnFe$_2$O$_4$ at 1,273 K H$_2$ formation happens with the expense of oxidation of ZnO/MnFe$_2$O$_4$. The later forms spinel kind of material containing Zn^{II}, Mn^{II}, Mn^{III} and Fe^{III} ions. The reaction happens by incorporation of Zn^{II} ions into MnFe$_2$O$_4$ crystal structure, accompanied by the partial oxidation of Mn^{II} in MnFe$_2$O$_4$ to Mn^{III}. The second step, oxygen releasing can be carried out using solar thermal route but this is not demonstrated experimentally by authors. Similarly, they have also demonstrated H$_2$ production using CaO (or Na$_2$CO$_3$) and MnFe$_2$O$_4$ by passing steam at 1,273 K [29]. The mechanism of H$_2$ formation is similar to that explained earlier, i.e., oxidation of Mn^{II} to Mn^{III} to form spinel kind of material (Ca$_3^{2+}$Fe$_{2.02}^{3+}$Mn$_{0.96}^{2+}$Mn$_{0.02}^{3+}$O$_{7.02}$).

Roeb et al. [34] used monolith coatings for redox system. They noticed that the potential of the monolith coatings to absorb oxygen from steam and to release hydrogen decreased with the number of completed cycles which is due to sintering of the material which increases with the redox cycles.

Lundberg [6] performed computer model calculation for various systems for two-step solar hydrogen productions, the systems considered were CoO/Co$_3$O$_4$, MnO/Mn$_3$O$_4$, FeO/Fe$_3$O$_4$, NbO$_3$/Nb$_2$O$_5$ and the halide systems FeX$_2$/Fe$_3$O$_4$ where X = F, Cl, Br and I. In his calculation he found that the ratio of H$_2$/H$_2$O is controlled by the temperature and oxygen partial pressure generated by the redox system. The yield of the hydrogen is defined as follows:

$$Y\,(\%) = \frac{H_2(\text{formed})}{H_2(\text{max})} \times 100 \tag{11}$$

where H$_2$max is the maximum amount of hydrogen that can be formed as per the formula:

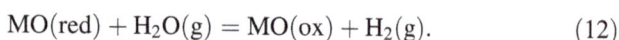

$$MO(\text{red}) + H_2O(g) = MO(\text{ox}) + H_2(g). \tag{12}$$

Calculations showed that FeO–Fe$_3$O$_4$ and NbO$_2$–Nb$_2$O$_5$ systems give more H$_2$ yield at lower temperature and that of MnO–Mn$_3$O$_4$ and CoO–Co$_3$O$_4$ systems give >1 % H$_2$ yield at any temperature. In reduction step, in order to reduce thermally oxidized metal oxide needs to be heated

up to a temperature where its oxygen partial pressure is higher than in atmosphere (0.21 atm). It was found that though FeO–Fe$_3$O$_4$ and NbO$_2$–Nb$_2$O$_5$ give higher yield they need to be heated above their melting point to reduce them. On the other hand, MnO–Mn$_3$O$_4$ and CoO–Co$_3$O$_4$ systems can be reduced below their melting point but hydrogen yield in these systems are very low (Table 4). Therefore, none of the systems studied are suitable to fulfill both desired conditions for the redox reactions.

It was also tried to combine metal oxide which yields higher H$_2$ with metal oxide which can be reduced below its melting point to find out whether this fulfills the need of redox cycle. Considering the spinel phase composition of (Fe$_{0.85}$Co$_{0.15}$)$_3$O$_4$ the H$_2$ yield obtained was 45 %, but during the oxidation of the (Fe$_{0.85}$Co$_{0.15}$)O system the equilibrium oxygen pressure of the redox system will successfully increases and the yield of the H$_2$ will gradually decreases down to about 3 %. The opposite effect was found during the reduction step, the spinel phase with composition (Fe$_{0.85}$Co$_{0.15}$)$_3$O$_4$ will start to be reduced at 2,020 K, but while reduction of the spinel the wuestite phase will become rich with iron and the oxygen partial pressure will decrease leading to gradual increase in the reduction temperature of 2,135 K by the time the initial composition is reached.

An yttrium-stabilized cubic zirconia material coated with iron oxide was proposed to split water in the temperature range 1,273–1,673 K [35, 36]. Kodam et al. [17] studied supported Fe$_3$O$_4$–FeO system. Various amount of iron oxide was supported on yttrium-supported ZrO$_2$ for cyclic redox study. It was found that the Fe$_3$O$_4$ reacts with YSZ to produce Fe^{2+}-containing ZrO$_2$ phase by releasing oxygen molecules in the first step. It was also found that the Fe^{2+} ions enters into the cubic YSZ lattice. In the second step, the Fe^{2+}-containing YSZ generated hydrogen via steam splitting to reproduce Fe$_3$O$_4$ on the cubic YSZ support. The system showed good reproducibility. It was found that when the Fe$_3$O$_4$ content was increased up to 30 wt% on the Fe$_3$O$_4$/YSZ sample [17], the sample became denser and harder mass after the thermal reduction step, similar to the unsupported Fe$_3$O$_4$. This is due to the fact that the limitation of Fe^{2+} solubility in the YSZ exists close

Table 4 The yield of H$_2$ at 900 K for the different metal oxide systems together with the enthalpy of the reaction, the reduction temperature in air and the melting points of the system

System	Yield H$_2$ at 900 K (%)	ΔH_r/H$_2$ at 900 K (kJ)	Reduction temperature in air	Melting point (K)	
				Reduced phase	Oxidized phase
NbO$_2$/Nb$_2$O$_5$	99.7	−62.7	3,600	2,175	1,785
FeO/Fe$_3$O$_4$	63	−49.5	2,685	1,650	1,870
MnO/Mn$_3$O$_4$	0.002	17	1,810	2,115	1,835
CoO/Co$_3$O$_4$	4×10^{-7}	251.2	1,175	2,080	Decomposes at 1,175

to the 25 wt% Fe_3O_4 content in the Fe_3O_4/YSZ. When raising the Fe_3O_4 content above 25 wt%, excess Fe^{2+} ions would form FeO crystals on the ZrO_2 surface, which in turn melts at 1,713 K. Therefore, the Fe_3O_4 contents should be limited to <25 % to avoid sintering of redox material and its cyclic reproducibility.

The disadvantage of mixed iron oxide cycles where oxides are partially reduced and oxidized is their low molar ratio of released oxygen to the total oxygen present in the system. The major drawback of all systems using reactive coatings is their low ratio of hydrogen mass generated to support structure mass. Considering the properties of the above problems, the cycle based on the ZnO/Zn redox pair [7–9] is of special interest since no cyclic heating and cooling is required and a pure metal state is achieved. It consists of the solar endothermal dissociation of ZnO(s) into its elements; and the non-solar exothermal steam-hydrolysis of Zn into H_2 and ZnO(s), and represented by Eqs. 13 and 15.

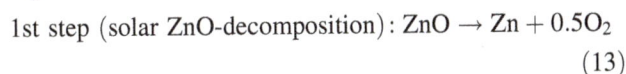

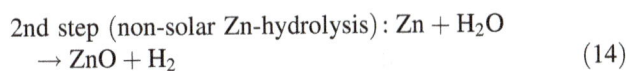

1st step (solar ZnO-decomposition): $ZnO \rightarrow Zn + 0.5O_2$
(13)

2nd step (non-solar Zn-hydrolysis): $Zn + H_2O \rightarrow ZnO + H_2$
(14)

H_2 and O_2 are derived in different steps, thereby eliminating the need for high-temperature gas separation. This cycle has been proposed to be a promising route for solar H_2 production from H_2O because of its potential of reaching high-energy conversion efficiencies and thereby its economic competitiveness [37, 38].

The first step of the two-step ZnO/Zn water-splitting cycle was first demonstrated in a solar furnace in 1977 by Bilgen et al. [4]. They have demonstrated the decomposition of ZnO in a solar furnace. They also found that Zn yield increases if ZnO is diluted with other refractory materials like ZrO_2 and Y_2O_3 and if the reaction is carried

out in the presence of argon inert gas. Table 5 summarized the Zn yield found in the different experiments and different conditions.

Recently, the solar thermal ZnO dissociation was demonstrated by Lédé et al. [39] in a quartz vessel containing sintered ZnO, by Haueter et al. [40] in a rotating cavity reactor type, and by Perkins et al. [41] in an aerosol reactor type. Perkins et al. reported the O_2 measurement, which is the only clear indicator of the ongoing thermal ZnO dissociation. The maximum net Zn yield was 17 % [41]. However, to-date there is no report in the literature which claims continuous dissociation of ZnO monitored by product gas analysis for more than few minutes.

Palumbo et al. [25] have studied TiO_2 system for two-step solar production of Zn from ZnO, the primary reaction schemes can be written as shown in reactions (15) and (16).

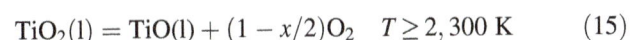

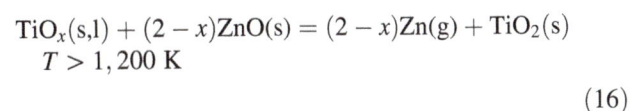

$$TiO_2(l) = TiO(l) + (1 - x/2)O_2 \quad T \geq 2,300 \text{ K} \quad (15)$$

$$TiO_x(s,l) + (2 - x)ZnO(s) = (2 - x)Zn(g) + TiO_2(s) \\ T > 1,200 \text{ K} \quad (16)$$

But the authors have not tried water splitting using partially reduced TiO_x. The minimum values of x that the authors obtained experimentally were 1.91, 1.86 and 1.83 for temperatures of 2,300, 2,500, and 2,700 K, respectively, in an Ar atmosphere at 1 bar. They used the latter material to reduce ZnO to produce Zn as indicated in reaction (16). It is to be noted that the higher the degree of decomposition, the greater the vaporization of TiO_2, this limits the efficiency of the water-splitting cycle using TiO_2 system.

Sibieude et al. [10] have used CdO for two-step water-splitting reaction. They demonstrated reduction of CdO to Cd in a solar furnace at high temperature. The reaction scheme is shown in Eqs. 17–19.

Table 5 Mol% zinc content of condensed vapors from ZnO and mixed oxides ZnO–Y_2O_3, ZnO–ZrO_2 samples heated at the focus of 2 kW solar concentrator

	Air p (bar)		Argon atmosphere p (bar)					
	<0.001	1	<0.001	0.034	0.092	0.263	0.789	
ZnO	Between 20 and 30 mol% of Zn	No Zn formation	Difficulties exist in obtaining due to strong volatilization of ZnO sample; the results are poorly reproducible					Static atmosphere
			68 mol% was obtained for $p < 0.001$ bar Ar					
			45 mol% was obtained for $p = 0.263$ bar Ar (in a flow of gas)					
10 mol% ZnO				70 %	62 %	60 %	25 %	Static atmosphere
90 mol% Y_2O_3			71 %	76 %	66 %	68 %	65 %	Gas circulation
10 mol% ZnO				67 %	60 %	60 %	30 %	Static atmosphere
90 mol% ZrO_2			75 %	74 %	65 %	70 %	67 %	Gas circulation

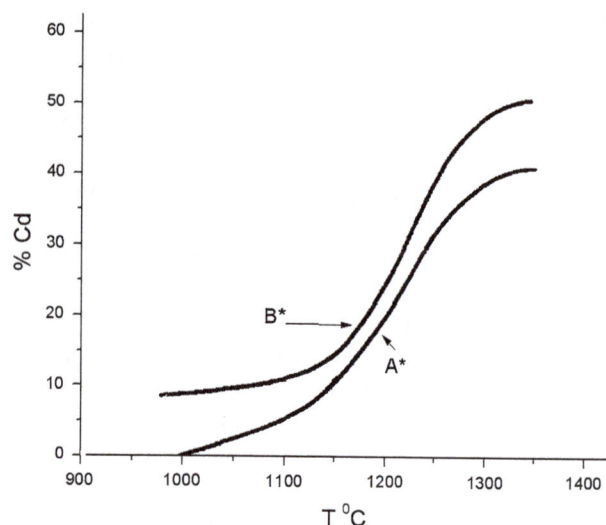

Fig. 5 Cd (metal) content of condensates versus temperature after thermal decomposition of CdO, flow rate of argon was $A*$ 3.4 cm^3/s and $B*$ 10 cm^3/s

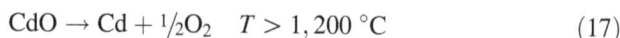

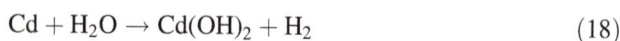

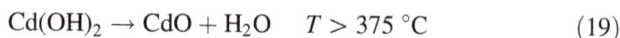

$$CdO \rightarrow Cd + \tfrac{1}{2}O_2 \quad T > 1,200\ °C \tag{17}$$

$$Cd + H_2O \rightarrow Cd(OH)_2 + H_2 \tag{18}$$

$$Cd(OH)_2 \rightarrow CdO + H_2O \quad T > 375\ °C \tag{19}$$

They observed that subjecting CdO alone to solar radiation did not reduce the oxide, but when CdO was mixed with refractory material, in their case 20 %mol ZrO$_2$, resulted in the formation of Cd metal in the stream of Ar. The amount of Cd metal in the deposited condensate at different temperatures is shown in Fig. 5.

Quenching of evaporated metal was very important in this reaction. When CdO was dissociated into Cd(g) and O(g) the recombination will also takes place very fast.

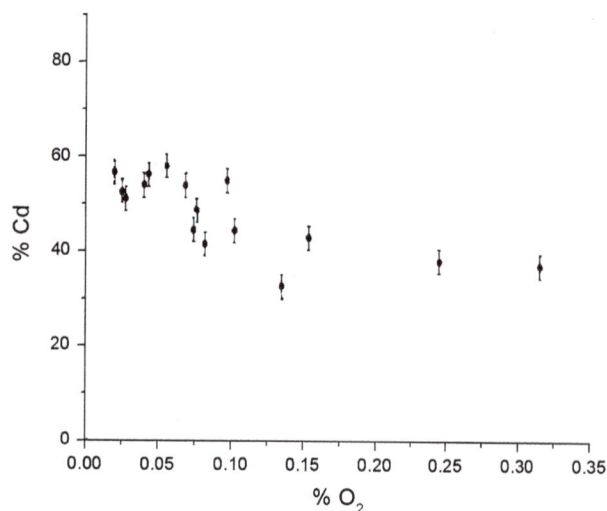

Fig. 6 Dependence of Cd (metal) content of condensates on the oxygen concentration of the argon flow

When CdO alone was heated strong vaporization produces large amount of dissociated vapors which is insufficiently quenched by the argon flow on a water cooled wall of the condenser. The problem was overcome by mixing the CdO with ZrO$_2$; in this case vaporization rate of Cd metal was lowered by its dispersion in the refractory metal oxide matrix which permits the effective quenching of vaporized metal. It is to be noted that partial pressure of oxygen plays a main role in the Cd yield. Figure 6 gives the %Cd metal recovered in various O$_2$ partial pressures.

Abanades et al. [15] have studied SnO$_2$ = Sn + 1/2O$_2$ cycle which consists of a solar endothermic reduction of SnO$_2$ into SnO(g) and O$_2$ followed by a non-solar exothermic hydrolysis of SnO(s) to form H$_2$ and SnO$_2$(s). The thermal reduction occurs under atmospheric pressure at about 1,873 K and over. The solar step encompasses the formation of SnO nanoparticles that can be hydrolyzed efficiently in the temperature range of 500–600 °C with a H$_2$ yield over 90 %. A preliminary process design is also proposed for cycle integration in solar chemical plants. They also compared their system with literature reported 'Sn-Souriau' [42] three-step cycles and inferred that the reaction (22) producing hydrogen from the Sn/SnO$_2$ mixture produced from reaction (21) is slow and partial at 600 °C which results in low H$_2$ yield of <45 %. The three-step cycling process proposed by them is as follows:

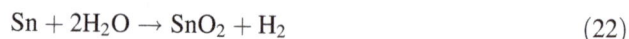

$$SnO_2 \rightarrow SnO + 1/2O_2 \tag{20}$$

$$2SnO \rightarrow Sn + SnO_2 \tag{21}$$

$$Sn + 2H_2O \rightarrow SnO_2 + H_2 \tag{22}$$

Fan et al. [43] have studied steam to hydrogen conversion using six different metals. It is interesting to note that only Fe and Sn are found to generate reasonable hydrogen at 873 K as shown in Table 6. Other metals did not show a good amount of hydrogen production at 873 K. Considering melting point of different metallic and their oxides states (as shown in Table 7) of Fe and Sn it can be inferred that Fe is very suitable for given application unless there is a provision to handle liquid metal in the solar reactor similar to Zn–ZnO$_2$ case. The steam to H$_2$ value (γ^a) of 40.82 % is lower compared to one reported by Abanades et al. [15] which is equivalent to 90 % at similar conditions. If solar reactor is designed to handle liquid metals, then both Zn and Sn seems to be better candidates for two-step redox reactions with good hydrogen yield and at low-temperature operation.

Recently, Cho and Kim [27] reported production of H$_2$ using solid acids such as silica gel, activated Al$_2$O$_3$, CaCO$_3$, TiO$_2$ and ZnO. This is very interesting study as it reports on liberation of hydrogen gas at very low temperature. They have demonstrated the possibility of H$_2$ production using a laboratory plug flow reactor. Figure 7 gives

Table 6 Maximum per-pass conversion of H_2O to H_2 in the regeneration reactor and the stable phase obtained at 873 K for countercurrent gas–solid operation

Metal phase	γ^a (%)	Oxidized phase
Ni	0.4	NiO
Cd	1.83	CdO
Cu	0	Cu_2O
Co	2.27	CoO
Sn	40.82	SnO_2
MnO	0	Mn_3O_4
Fe	74.79	Fe_3O_4
Fe	74.79	Fe_2O_3

γ^a conversion of H_2O to H_2

Table 7 Melting points [44] of various phases of Fe and Sn

Material	Melting point (K)
Fe (cast)	1,548
Fe (pure)	1,808
FeO	1,693
Fe_3O_4	1,538
Fe_2O_3	1,811
Sn	504
SnO	1,353
SnO_2	1,400

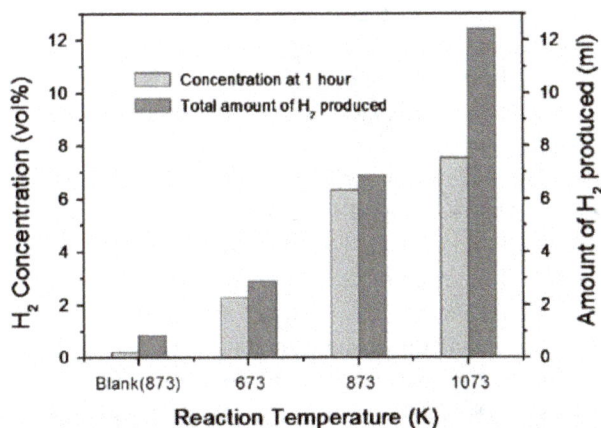

Fig. 7 H_2 concentration in the product gas stream at a reaction time of 1 h and total amount of H_2 produced versus reaction temperature resulted from using a wetted Al_2O_3. For the experiments, 60 g of Al_2O_3 (5.5 wt% H_2O) in a stainless steel reactor and 2 ml/min of CO_2 carrier gas were used at atmospheric pressure

the concentration and total amount of H_2 liberated at various temperatures using activated Al_2O_3.

The concentration of hydrogen produced in product gas stream using five different oxides at ≈ 610 K is shown in the Table 8. Though $CaCO_3$ shows highest concentration

of H_2, i.e., 1,590 ppm in 1 h reaction time the activated Al_2O_3 produces highest amount of total H_2 at 1,073 K.

Though the authors demonstrated the workability of the producing hydrogen from solid acids on a laboratory-scale fixed-bed reactor, replication of the results in a solar reactor needs to be performed to know the feasibility of the process.

One of the main problem to tackle is overcoming sintering of redox material. Agglomeration due to sintering brings down the recyclability over multiple redox cycles. The key properties of the redox material should include a high oxygen carrying capacity, good mechanical properties and cheap and easy synthetic procedures. If redox material do not fulfill any one of these key properties it would not be a suitable material for commercial-scale operation.

Particle size or grain size effect on rate of oxidation

It is generally accepted that smaller the particle size easier is to oxidize or reduce. In case of two-phase alloys the rate of oxidation may significantly improve with grain size reduction because both mutual solubility and diffusivity among the system will enhance [45, 46]. But this is not always the case, during the oxidation if the top layer acts as a protective layer then the further oxidation of the metals will be hampered.

Figure 8 shows the oxidation kinetics of three different alloy systems with two different grain sizes at 1,073 K. The grain size reduced Cu–Cr alloy showed very slow oxidation kinetics compared to As-cast alloy. But in case of Cu–Fe and Cu–Co oxidation kinetics found to be much faster when nano-crystals (20–30 nm) were used compared to the As-cast alloy. This is because Cr_2O_3 scale formed on alloy prevents further oxidation. This is similar to in case aluminum where external layer forms Al_2O_3 and prevents further oxidation or corrosion of the aluminum metal.

The reduction kinetics of metal oxides depends on many factors such as whether they are supported or unsupported, particle size, gas atmosphere, kind of metal oxides and whether single or mixed metal oxides. There are not many reports available on high-temperature reduction of metal oxides in an inert atmosphere as in the case of solar thermal reduction, but there are plenty of studies available in the literature on reduction of metal oxides using H_2 or CO as reducing agents. For example, Syed-Hassan and Chun-Zhu [47] have studied the particle size effect on reduction of NiO in H_2 atmosphere. The reduction profiles for NiO particles of size 20 and 24 nm are very different from that of particle size of 3.3 nm (please refer Fig. 5a in Ref. [47]). The profiles for 20 and 24 nm are almost similar and very much resemble to that of 55 nm NiO supported on SiO_2 substrate. Authors concluded that reduction kinetics is independent of supported or unsupported NiO, but merely

Table 8 H_2 concentration in the product gas stream at a reaction time of 1 h

Solid acid (amount)	Al_2O_3 (50 g)		SiO_2 (30 g)		TiO_2 (50 g)		ZnO (50 g)		$CaCO_3$ (50 g)	
H_2O (wt%)	0	5.2	0	5.3	0	5.1	0	5.1	0	5.0
H_2 (ppm)	0	870	0	200	0	260	0	630	0	1,590

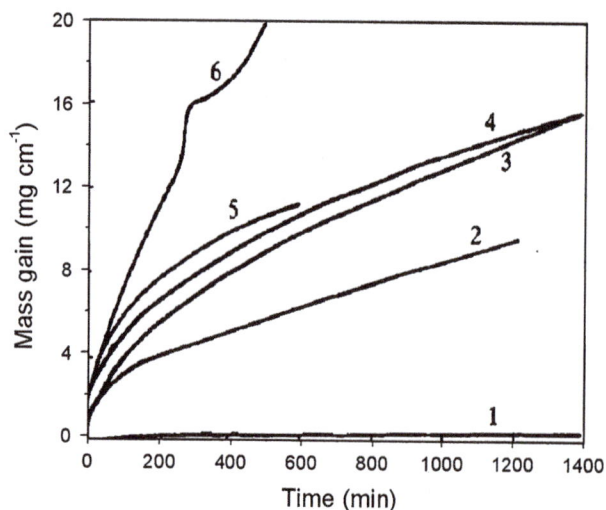

Fig. 8 Oxidation kinetics of As-cast and grain size reduced Cu–M alloys at 1,073 K. *1* Grain size reduced Cu–Cr alloy, *2* As-cast Cu–Co alloy, *3* As-cast Cu–Cr alloy, *4* grain size reduced Cu–Fe alloy, *4* As-cast Cu–Fe alloy, *6* grain size reduced Cu–Co alloy. [45]

Table 9 Hydrogen cost estimation (per kg H_2) in different processes

Years	Hy-S	CuCl	Ferrite	S–A	ZnO	CdO	MnO	S–I
2015	$5.68	$6.83	$4.06	$7.78	$6.07	NA	$	$
2025	$3.85	$5.39	$2.42	$4.71	$4.18	NA	$4.63	$4.68

the particle) takes place. In general, solid-state diffusion requires higher activation energy [48]. The number of steps of diffusion in the solid state would appear to increase with NiO conversion, resulting in continuous increases in the activation energy. This is the reason in general the reduction of bigger particles crystallites needs higher activation energy than smaller one.

Economic evaluation

The US DOE has established a target of $2 to $3 per kg hydrogen by 2025 to make it economically affordable. The short term, i.e., 2015, DOE target is $6/kg hydrogen. Any competitive technology to produce hydrogen considers this figure as a reference for their process efficiency and economic evaluation.

DOE in collaboration with TIAX (TIAX is a laboratory-based technology development company with a focus on clean energy) led the effort of cost calculation for solar thermochemical hydrogen (STCH) in many US national laboratories. They considered eight promising technologies for cost calculations, viz., hybrid-sulfur (HyS), copper chloride (CuCl), thin-film nickel ferrite ("ferrite"), sulfur-ammonia (S–A), zinc oxide (ZnO), manganese oxide (MnO), sulfur-iodine (S–I), and cadmium oxide (CdO).

Five out of eight technologies mentioned in Table 9 appear to meet DOE's short-term target (by 2015) of $6/kg hydrogen but meeting long-term target seems quite difficult. Only thin film ferrite is very close to DOE's long-term requirement. Even in this case achieving the long-term targets require significant technological advances in multiple dimensions. The primary cost driver for all the processes that were analyzed was heliostats costs. Reducing the heliostats cost or increase in plant efficiency will bring down the CAPEX and OPEX. Most probably mass production of heliostats and improving its efficiency will help to reduce the CAPEX and help to meet DOE hydrogen cost targets.

depends on the particle/crystal size. The E_a versus %NiO converted trends are very different for first reduction (particle size 3.3 nm) to that of second (particle size = 20 nm) and third reduction (particle size = 24 nm) (please refer Fig. 5b in Ref. [47]). The E_a for first reduction remains almost unchanged (indicating single rate-limiting step) throughout the whole reduction process, but for second and third reduction steps the E_a profiles continuously increased till the complete reduction of NiO. The main reason for the difference could be due to the difference in surface to bulk atoms in different particle size crystals. The size of the metallic island which forms during initial stages of reduction is bigger than particle size, i.e., 3.3 nm, or the whole particle surface can be instantaneously covered by the metallic layer without requiring a significant growth of islands; therefore, the reduction happens immediately in those particles size crystals. But for a big particle with significant atoms in the interior (e.g., 20 nm) it might take quite some time for the growth of island to cover the whole particle's surface. Once the surface has been covered completely by the metal product, the Ni–NiO boundary would then progressively advance inward. As reduction continues, the hydrogen radicals moves towards Ni–NiO boundary on the surface and the diffusion of atoms (i.e., movement and rearrangements of atoms in the interior of

Conclusions

Several metal oxides have been proposed to apprehend redox cycles. To-date, the solar-to-fuel efficiencies of prototype reactors are low, on the order of <1 %. The main problem in large-scale solar application would be an efficient fast quenching system to hinder the reoxidation of the reduced metal in liquid or in vapor phase. When molten metal is made to react with steam, an oxide layer will form on the surface and it floats on top of the melt, which prevents further oxidation reaction.

So far, there are no reports demonstrating good repeatability of the cyclic two-step reaction to satisfy the practical use of the process. This would be one of the most difficult achievements in this technology because the high-temperature process will cause significant sintering of the metal oxide, which severely deactivates metal oxide for repeated cyclic reactions.

Though there are hundreds of publication available in the literature on metal oxide redox cycle for hydrogen production, only few of them are practically demonstrated in prototype solar reactor. The challenges while conducting redox experiments in a solar reactor are very different than in a laboratory-scale plug flow reactor. The commercial realization of redox technology for hydrogen production seems still far away.

Acknowledgments I thank SABIC CRI for giving me an opportunity to work on this review and publish; Dr. Hicham Idriss for his valuable input on the manuscript and Dr. Sandro Gambarotta for his constant support, guidance and encouragement.

References

1. Licht, S.: Solar water splitting to generate hydrogen fuel—a photothermal electrochemical analysis. Int. J. Hydrog. Energy **30**(5), 459–470 (2005)
2. Kogan, A.: Direct solar thermal splitting of water and on-site separation of the products—II. Experimental feasibility study. Int. J. Hydrog. Energy **23**(2), 89 (1998)
3. Lede, J., Villermaux, J., Ouzane, R., Hossain, M.A., Ouahes, R.: Production of hydrogen by simple impingement of a turbulent jet of steam upon a high temperature zirconia surface. Int. J. Hydrog. Energy **12**(1), 3–11 (1987)
4. Bilgen, E., Ducarroir, M., Foex, M., Sibieude, F., Trombe, F.: Use of solar energy for direct and two-step water decomposition cycles. Int. J. Hydrog. Energy **2**(3), 251–257 (1977)
5. Nakamura, T.. Hydrogen production from water utilizing solar heat at high temperatures. Sol. Energy **19**(5), 467 (1977)
6. Lundberg, M.: Model calculations on some feasible two-step water splitting processes. Int. J. Hydrog. Energy **18**(5), 369 (1993)
7. Palumbo, R., Léde, J., Boutin, O., Elorza Ricart, E., Steinfeld, A., Möller, S., Weidenkaff, A., Fletcher, E.A., Bielicki, J.: The production of Zn from ZnO in a high-temperature solar decomposition quench process—I. The scientific framework for the process. Chem Eng Sci **53**(14), 2503–2517 (1998)
8. Palumbo, R.: Solar thermal chemical processing: Challenges and changes. J Phys IV Fr **9**, Pr3-35–Pr33-40 (1999)
9. Weidenkaff, A., Brack, M., Möller, S., Palumbo, R., Steinfeld, A.: Solar thermal production of zinc: Program strategy and status of research. J Phys IV Fr **9**, Pr3-313–Pr3-318 (1999)
10. Sibieude, F., Ducarroir, M., Tofighi, A., Ambriz, J.: High temperature experiments with a solar furnace: The decomposition of Fe_3O_4, Mn_3O_4, CdO. Int. J. Hydrog. Energy **7**(1), 79–88 (1982)
11. Ehrensberger, K., Frei, A., Kuhn, P., Oswald, H.R., Hug, P.: Comparative experimental investigations of the water-splitting reaction with iron oxide $Fe_{1-y}O$ and iron manganese oxides $(Fe_{1-x}Mn_x)_{1-y}O$. Solid State Ionics **78**, 151–160 (1995)
12. Ehrensberger, K., Kuhn, P., Shklover, V., Oswald, H.R.: Temporary phase segregation processes during the oxidation of $(Fe_{0.7}Mn_{0.3})_{0.99}O$ in $N_2 + H_2O$ atmosphere. Solid State Ionics **90**, 75–81 (1996)
13. Tamaura, Y., Steinfeld, A., Kuhn, P., Ehrensberger, K.: Production of solar hydrogen by a novel, 2-step, water-splitting thermochemical cycle. Energy **20**(4), 325 (1995)
14. Steinfeld, A., Kuhn, P., Reller, A., Palumbo, R., Murray, J., Tamaura, Y.: Solar-processed metals as clean energy carriers and water-splitters. Int. J. Hydrog. Energy **23**(9), 767–774 (1998)
15. Abanades, S., Charvin, P., Lemont, F., Flamant, G.: Novel two-step SnO_2/SnO water-splitting cycle for solar thermochemical production of hydrogen. Int. J. Hydrog. Energy **33**(21), 6021 (2008)
16. Kaneko, H., Kodama, T., Gokon, N., Tamaura, Y., Lovegrove, K., Luzzi, A.: Decomposition of Zn-ferrite for O_2 generation by concentrated solar radiation. Sol. Energy **76**, 317 (2004)
17. Kodama, T., Nakamuro, Y., Mizuno, T.J.: A two-step thermochemical water splitting by iron-oxide on stabilized zirconia. J. Sol. Energy Eng. **128**, 3 (2006)
18. Kodama, T.: High-temperature solar chemistry for converting solar heat to chemical fuels. Prog. Energy Combust. Sci. **29**(6), 567 (2003)
19. Abanades, S., Flamant, G.: Thermochemical hydrogen production from a two-step solar-driven water-splitting cycle based on cerium oxides. Sol. Energy **80**(12), 1611 (2006)
20. Chueh, W.C., Falter, C., Abbott, M., Scipio, D., Furler, P., Haile, S.M., Steinfeld, A.: High-flux solar-driven thermochemical dissociation of CO_2 and H_2O using nonstoichiometric ceria. Science **330**(6012), 1797–1801 (2010)
21. Lapp, J., Davidson, J.H., Lipinski, W.: Efficiency of two-step solar thermochemical non-stoichiometric redox cycles with heat recovery. Energy **37**, 591–600 (2012)
22. Kaneko, H., Taku, S., Tamaura, Y.: Reduction reactivity of CeO_2–ZrO_2 oxide under high O_2 partial pressure in two-step water splitting process Sol. Energy **85**(9), 2321–2330 (2011)
23. Kang, K.-S., Kim, C.-H., Park, C.-S., Kim, J.-W.: Hydrogen reduction and subsequent water splitting of Zr-added CeO_2. J. Ind. Eng. Chem. **13**(4), 657–663 (2007)
24. SP Gal, A., Abanades, A.: Dopant incorporation in Ceria for enhanced water-splitting activity during solar thermochemical hydrogen generation. J. Phys. Chem. C **116**, 13516–13523 (2012)
25. Palumbo, R., Rouanet, A., Pichelin, G.: The solar thermal decomposition of TiO_2 at temperatures above 2200 K and its use in the production of Zn from ZnO. Energy **20**(9), 857–868 (1995)
26. Abanades, S., Charvin, P., Flamant, G., Neveu, P.: Screening of water-splitting thermochemical cycles potentially attractive for hydrogen production by concentrated solar energy. Energy **31**(14), 2805–2822 (2006)

27. Cho, Y.S., Kim, J.H.: Hydrogen production by splitting water on solid acid materials by thermal dissociation. Int. J. Hydrog. Energy **36**(14), 8192–8202 (2011)

28. Bilgen, E., Bilgen, C., Beghi, GE., Ducarroir, M.: Thermochemical hydrogen producing processes. Contract file No. 08SX.31155-8-6602. Prepared by Exergy Research Corporation, prepared for NRC of Canada, Montreal Road, Ottawa, KlA OR6, J.J. Murray (1979)

29. Tamaura, Y., Ueda, Y., Matsunami, J., Hasegawa, N., Nezuka, M., Sano, T., Tsuji, M.: Solar hydrogen production by using ferrites sol. Energy **65**(1), 55–57 (1999)

30. Loutzenhiser, P., Meier, A., Steinfeld, A.: Review of the two-Step H_2O/CO_2-splitting solar thermochemical cycle based on Zn/ZnO redox reactions. Materials **3**, 4922–4938 (2010)

31. Hed, A.Z., Tannhauser, D.S.: Contribution to the Mn-O phase diagram at high temperature. J. Electrochem. Soc. **4**, 314–318 (1967)

32. Rager, T.: Re-evaluation of the efficiency of a ceria-based thermochemical cycle for solar fuel generation. Chem. Commun. **48**, 10520–10522 (2012)

33. Inoue, M., Asegawa, N., Uehara, R., Gokon, N., Kaneko, H., Tamaura, Y.: Solar hydrogen generation with $H_2O/ZnO/MnFe_2O_4$ system. Sol. Energy **76**(1–3), 309–315 (2004)

34. Roeb, M., Sattle, C., Klüser, R., Monnerie, N., Oliveira, L.D., Al, E.: Solar hydrogen production by a two-step cycle based on mixed iron oxides. J. Sol. Energy Eng. Trans. ASME **128**(2), 125–133 (2006)

35. Kodama, T., Nakamuro, Y., Mizuno, T.: A two-step thermochemical water splitting by iron-oxide on stabilized zirconia. J. Sol. Energy Eng. Trans. ASME **128**(1), 3–7 (2006)

36. Gokon, N., Mizuno, T., Nakamuro, Y., Tamaura, K., Kodama, T.: Iron-containing yttria-stabilized zirconia system for two-step thermochemical water splitting. J. Sol. Energy Eng. Trans. ASME **130**(1), 011018 (2008)

37. Perkins, C., Weimer, A.W.: Likely near-term solar-thermal water splitting technologies. Int. J. Hydrog. Energy **29**(15), 1587–1599 (2004)

38. Steinfeld, A.: Solar hydrogen production via a two-step water-splitting thermochemical cycle based on Zn/ZnO redox reactions. Int. J. Hydrog. Energy **27**(6), 611 (2002)

39. Lédé, J., Elorza-Ricart, E., Ferrer, M.: Solar thermal splitting of zinc oxide: A review of some of the rate controlling factors. J. Sol. Energy Eng. Trans. ASME **123**(2), 91–97 (2001)

40. Haueter, P., Moeller, S., Palumbo, R., Steinfeld, A.: The production of zinc by thermal dissociation of zinc oxide—Solar chemical reactor design. Sol. Energy **67**(1–3), 161–167 (1999)

41. Perkins, C., Lichty, P.R., Weimer, A.W.: Thermal ZnO dissociation in a rapid aerosol reactor as part of a solar hydrogen production cycle. Int. J. Hydrog. Energy **33**(2), 499–510 (2008)

42. Souriau, D.: Procédé et dispositif pour l'utilisation d'énergie thermique à haute température, en particulier d'origine nucléaire. Device and method for the use of high temperature heat energy, in particular of nuclear origin. France Patent FR2135421

43. Gupta, P., Velazquez-Vargas, L.G., Fan, L.S.: Syngas redox (SGR) process to produce hydrogen from coal derived syngas. Energy Fuels **21**(5), 2900–2908 (2007)

44. Perry, R.H., Green, D.: Perry's Chemical Engineers Handbook, 6th edn. McGraw-Hill, New York (1984)

45. Yuan-shi, L., Yan, N., Guang-yan, F., Wei-tao, W., Gesmundo, F.: Effect of grain size reduction on high temperature oxidation of binary two-phase alloys. Trans. Nonferrous Met. Soc. China **11**(5), 644–648 (2001)

46. Zhong-qiu, C., Yan, N., Li-Jie, C., Wei-tao, W.: Effect of grain size reduction on high temperature oxidation of behaviour of Cu-80Ni alloy. Trans. Nonferrous Met. Soc. China **13**(4), 908–911 (2003)

47. Syed-Hassan, S.S.A., Li, C.-Z.: Effects of crystallite size on the kinetics and mechanism of NiO reduction with H_2. Int. J. Chem. Kinet. **43**(12), 667–676 (2011)

48. Mrowec, S.: Defects and Diffusion in Solids: An Introduction. Elsevier, Amsterdam (1980)

3

Preparation and electrochemical characterization of $(100 - x)(0.7Li_2S \cdot 0.3P_2S_5) \cdot xLiBr$ glass–ceramic electrolytes

Satoshi Ujiie · Akitoshi Hayashi · Masahiro Tatsumisago

Abstract Glass and glass–ceramic electrolytes of $(100 - x)(0.7Li_2S \cdot 0.3P_2S_5) \cdot xLiBr$ ($x = 0$, 5, 10, 12.5, 15 and 20) (mol %) were synthesized by mechanical milling and subsequent heat treatment. Glass powders with no crystal phase were obtained by mechanical milling. The conductivities of the glasses increased concomitantly with increasing LiBr content. The conductivity at $x = 20$ was 3.1×10^{-4} S cm^{-1} at room temperature. The $Li_7P_3S_{11}$ crystal with high Li ion conductivity was precipitated in the glass–ceramics obtained by heat treatment. LiBr crystal was also precipitated in the glass–ceramics containing LiBr as a starting material. The glass–ceramic at $x = 10$ showed conductivity of 6.4×10^{-3} S cm^{-1}. It increased to 8.4×10^{-3} S cm^{-1} by control of the milling period to prepare the precursor glass. $90(0.7Li_2S \cdot 0.3P_2S_5) \cdot 10LiBr$ glass–ceramic with high Li ion conductivity had a wide electrochemical window of 10 V. An all-solid-state lithium secondary battery using $90(0.7Li_2S \cdot 0.3P_2S_5) \cdot 10LiBr$ glass–ceramic as an electrolyte was charged and discharged successfully.

Keywords Solid electrolyte · Glass–ceramic · Sulfide · Lithium ion conductivity · Lithium battery

S. Ujiie · A. Hayashi (✉) · M. Tatsumisago
Department of Applied Chemistry, Graduate School of
Engineering, Osaka Prefecture University, 1 - 1 Gakuencho,
Naka-ku, Sakai, Osaka 599-8531, Japan
e-mail: tatsu@chem.osakafu-u.ac.jp

S. Ujiie
Energy Use R&D Center, The Kansai Electric Power Co., Inc,
11-20-3 Nakoji Amagasaki, Hyogo 661-0974, Japan

Introduction

Electric power generation using renewable energy such as sunlight and wind is increasing. Electrical storage will be an important function of the power system because power is not supplied continuously by such renewable energy resources. Batteries are suitable for power storage because of their high energy efficiencies. Lithium ion batteries with high energy densities are expected to be used in large-scale power storage systems. Safety and reliability are strongly required for large storage systems, while lithium ion batteries present the risk of burning because of their flammable liquid electrolytes. All-solid-state batteries using non-flammable solid electrolytes are expected to improve the safety and reliability of lithium ion batteries. However, the conductivities of the solid electrolytes are not sufficiently high. Solid electrolytes with high lithium ion conductivities and wide electrochemical windows are necessary to put all-solid-state batteries to practical use.

Sulfide-based electrolytes as lithium ion conductors have been investigated [1–12]. The crystals in the Li_2S–GeS_2–P_2S_5 system called thio-LISICON have conductivities of 10^{-4}–10^{-3} S cm^{-1} at room temperature [5]. Recently, Kanno et al. [13] reported a new crystal of $Li_{10}GeP_2S_{12}$ showing extremely high conductivity of 1.2×10^{-2} S cm^{-1}. The glass electrolytes in the Li_2S–P_2S_5 system have conductivities of 10^{-4} S cm^{-1} at room temperature. They are enhanced by the addition of a lithium halide [1]. The lithium ion conductivity of $70Li_2S \cdot 30P_2S_5$ glass increased from 1.3×10^{-4} to 5.6×10^{-4} S cm^{-1} by the addition of 20 mol % LiI [14]. Presumably, the enhancement of conductivity resulted from the increase of the total lithium ion concentration in the glass. The glass–ceramic electrolytes in the Li_2S–P_2S_5 system also show high lithium ion conductivities [6–9].

Above all, $70Li_2S\cdot30P_2S_5$ glass–ceramic containing the $Li_7P_3S_{11}$ crystal has high conductivity of 4.2×10^{-3} S cm^{-1} [6]. We added LiI to the $70Li_2S\cdot30P_2S_5$ glass–ceramic in expectation of increased conductivity, but the conductivity decreased sharply with increasing LiI contents [14]. Unknown crystal phase was precipitated in the glass–ceramic added with LiI, although the fraction of the $Li_7P_3S_{11}$ crystal with high conductivity decreased. A solid solution of LiI in the $Li_7P_3S_{11}$ crystal was not confirmed. The conductivity of the $70Li_2S\cdot30P_2S_5$ glass–ceramic would be enhanced if the lithium ion concentration could be increased without decreasing the $Li_7P_3S_{11}$ crystal phase.

In this study, LiBr was added to the $70Li_2S\cdot30P_2S_5$ glass–ceramic. The effective ionic radii of I^-, Br^- and S^{2-} are, respectively, 220, 196 and 184 pm [15]. LiBr might be dissolved in the $Li_7P_3S_{11}$ crystal because the radius of Br^- is smaller than that of I^-; moreover, it is close to that of S^{2-}. The solid solution is expected to reduce the precipitation of other crystal phases that would decrease the glass–ceramic conductivity. The addition of LiBr increases the Li content in the $70Li_2S\cdot30P_2S_5$ glass–ceramic. It is therefore expected that the addition of LiBr to the $70Li_2S\cdot30P_2S_5$ glass–ceramic enhances the conductivity. We synthesized the glass–ceramic electrolytes by mechanical milling and subsequent heat treatment. The crystalline phases and the conductivities of the glass–ceramics were examined. An all-solid-state cell was fabricated. Then its charge–discharge performance was investigated.

Experimental

$(100 - x)(0.7Li_2S\cdot0.3P_2S_5)\cdot x$LiBr (mol %) ($x = 0$, 5, 10, 12.5, 15 and 20) glasses were synthesized by mechanical milling. Reagent-grade Li_2S (99.9 %; Nippon Chemical Ind. Co. Ltd.), P_2S_5 (99 %; Aldrich Chemical Co. Inc.) and LiBr (99.999 %; Aldrich Chemical Co. Inc.) were used as starting materials. They were mixed in an agate mortar for 10 min and put into a 45 ml ZrO_2 pot with 500 ZrO_2 balls of 4 mm diameter. The pot was mounted on a planetary ball mill apparatus (Pulverisette 7; Fritsch GmbH). Then the materials were milled at 500 rpm for 10 or 20 h. Some samples labeled as "$(A + B) + C$" were synthesized using two-step milling: A and B were mixed and milled for 10 h, then the obtained compound and C were mixed and milled for 10 h. All processes were conducted in a dry Ar atmosphere. Glass–ceramic samples were obtained by heating the milled samples.

Differential thermal analyses (DTA) were conducted using a thermal analyzer (Thermo-plus 8120; Rigaku Corp.) to observe crystallization temperatures. The glass samples were sealed in Al pans in an Ar-filled glove box

and heated at 10 °C min^{-1} under N_2 gas flow up to 400 °C. X-ray diffraction (XRD) measurements (CuKα) were performed using a diffractometer (SmartLab; Rigaku Corp.) to identify crystal phases of the glass samples and the glass–ceramic samples. Ionic conductivities were measured for the pelletized samples pressed under 360 MPa. The pellet diameter and thickness were, respectively, 10 mm and about 1.5 mm. Carbon paste was painted as electrodes on both faces of the pellets and stainless steel disks were attached to the pellets as current collectors. The prepared two-electrode cell was packed in a silica glass tube. Then AC impedances were measured under dry Ar gas flow using an impedance analyzer (1260; Solartron Analytical). The frequency range and the applied voltage were, respectively, 10 Hz to 8 MHz and 50 mV. The electrochemical stability of the glass–ceramic sample was evaluated using cyclic voltammetry at room temperature using a potentiostat (1287; Solartron Analytical). A stainless steel disk as a working electrode and a lithium metal foil as a counter and reference electrode were attached to each face of the pelletized sample. The potential was swept between -0.1 and $+10$ V with a scanning rate of 5 mV s^{-1}. An all-solid-state two-electrode cell was assembled using the glass–ceramic sample as an electrolyte. The positive electrode was prepared by mixing two powders of $LiCoO_2$ as an active material and the glass–ceramic as an electrolyte. The $LiCoO_2$ particles were coated with $LiNbO_3$ thin film in advance because the $LiNbO_3$-coated $LiCoO_2$ shows good charge–discharge performance in all-solid-state batteries using a sulfide-based electrolyte [16]. An indium foil was used as a negative electrode.

Results and discussion

The XRD patterns of the $(100 - x)(0.7Li_2S\cdot0.3P_2S_5)\cdot x$LiBr samples prepared by mechanical milling for 10 h are presented in Fig. 1. Halo patterns were observed for all the samples, indicating that amorphous powders with no crystal phase were obtained in the composition of $0 \leq x \leq 20$.

Figure 2 shows the DTA curves of the milled samples. Glass transition phenomena occurred between 180 and 220 °C, which suggested that the amorphous powders obtained by milling were glasses. Marked exothermic peaks were observed between 230 and 260 °C. These peaks were regarded as resulting from crystallization of the glasses because the samples heated to temperatures just above these temperatures exhibited crystalline XRD patterns as depicted in Fig. 3. The glass transition temperatures and the crystallization temperatures were shifted to a lower temperature with increasing LiBr content. The shifts of the temperatures would result from the change of the

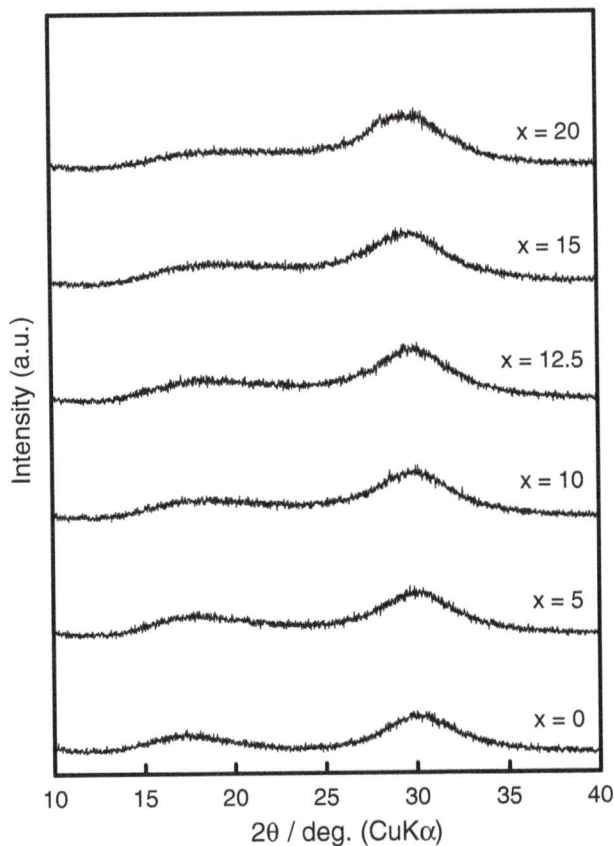

Fig. 1 XRD patterns of the $(100 - x)(0.7Li_2S \cdot 0.3P_2S_5) \cdot xLiBr$ samples prepared by milling for 10 h

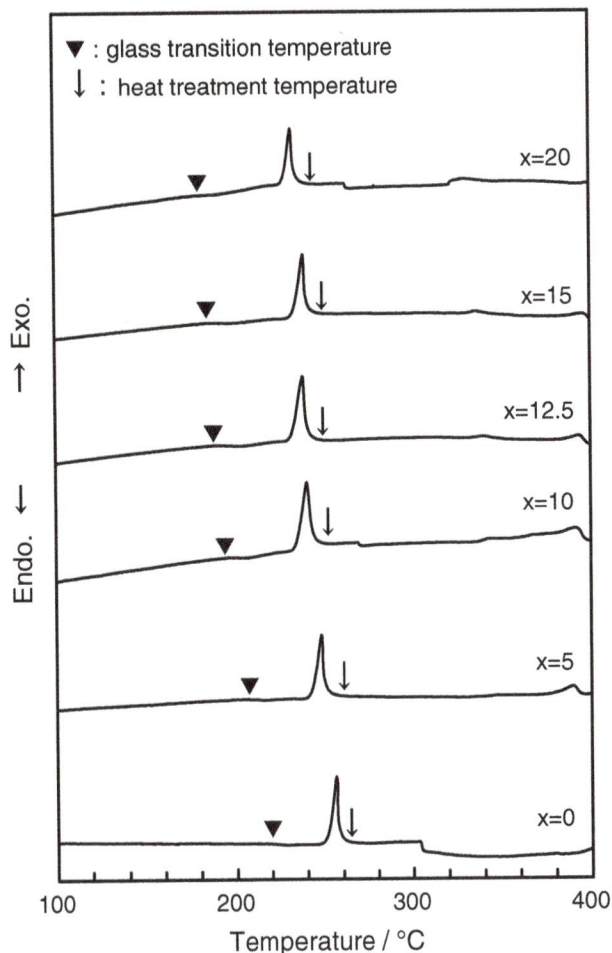

Fig. 2 DTA curves of the $(100 - x)(0.7Li_2S \cdot 0.3P_2S_5) \cdot xLiBr$ samples prepared by milling for 10 h

glass composition and the glass structure. Results suggest that the milled LiBr was not simply amorphous, but mainly entered into the binary Li_2S–P_2S_5 glass.

The XRD patterns of the $(100 - x)(0.7Li_2S \cdot 0.3P_2S_5) \cdot xLiBr$ samples prepared using heat treatment are depicted in Fig. 3. Several crystalline peaks were observed, indicating that glass–ceramics were obtained. The $Li_7P_3S_{11}$ crystals with high lithium ion conductivity were precipitated in all samples. The $Li_7P_3S_{11}$ crystal peaks maintained almost identical intensity in the composition range from $x = 0$ to 15. At $x = 20$, the intensity of the peaks of the $Li_7P_3S_{11}$ crystal was weakened slightly, suggesting that the $Li_7P_3S_{11}$ crystallinity was degraded. No marked shift of the peaks caused by the $Li_7P_3S_{11}$ crystal was observed. However, the formation of a solid solution of LiBr in the $Li_7P_3S_{11}$ crystal cannot be denied. Br^- ions would be partially substituted for S^{2-} ions to form a $Li_{7 - x}P_3S_{11 - x}Br_x$ crystal if LiBr is dissolved in the $Li_7P_3S_{11}$ crystal. The shift of the diffraction peaks caused by the solid solution must be slight because the ionic radii of Br^- and S^{2-} are close [15]. The little shift of the diffraction peaks would not be clearly observed using our laboratory XRD instruments. Further structural analyses are necessary to confirm the

solid solution. The peaks attributable to LiBr crystal were also observed in samples containing LiBr as a starting material. The intensity of the peaks increased concomitantly with increasing LiBr content. The LiBr crystals in the glass–ceramics were not caused by starting crystals, but were precipitated from the glasses because the crystal phase of LiBr disappeared completely during mechanical milling, as shown in Fig. 1. The peaks at 26.5° were attributable to carbon paste (graphite) painted on the pellets to facilitate AC impedance measurements.

The composition dependence of electrical conductivities at 25 °C for the glasses and glass–ceramics are portrayed in Fig. 4. The glass conductivities increased concomitantly with increasing LiBr content. The conductivity at $x = 20$ was 3.1×10^{-4} S cm^{-1}. The increase of the conductivity is expected to result from the increased Li^+ ion content in the glasses. The glass–ceramic conductivities also increased concomitantly with increasing LiBr content between $x = 0$–10. The conductivity of the glass–ceramic at $x = 0$ was 3.9×10^{-3} S cm^{-1}. It increased to

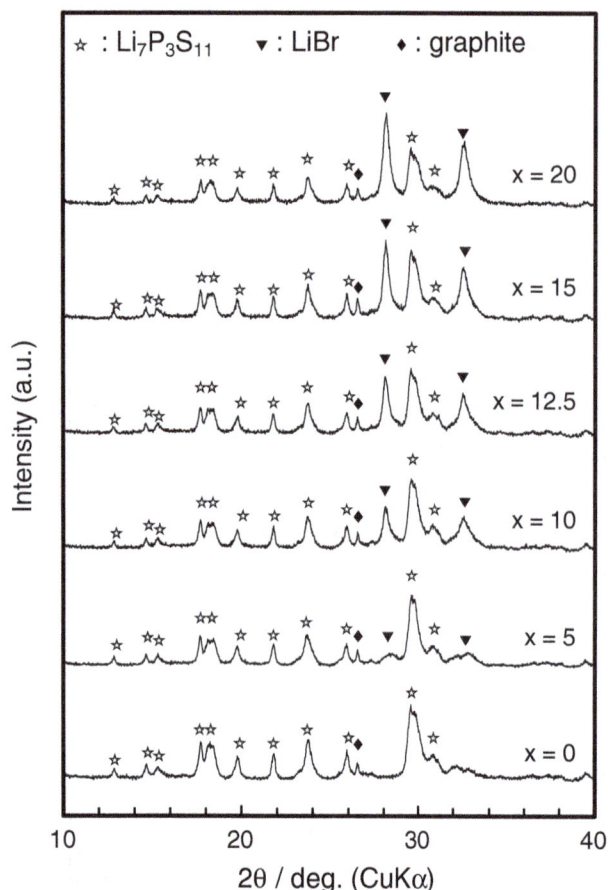

Fig. 3 XRD patterns of the $(100 - x)(0.7Li_2S \cdot 0.3P_2S_5) \cdot x$LiBr glass–ceramics prepared by milling for 10 h and subsequent heat treatment

6.5×10^{-3} S cm^{-1} at $x = 10$. In the composition range of $x = 12.5\text{–}20$, the glass–ceramic conductivities were decreased concomitantly with increasing LiBr contents.

The change in the conductivity of glass–ceramic was inferred to result from the increased Li$^+$ ion content, the precipitation of the Li$_7$P$_3$S$_{11}$ crystal, and the precipitation of LiBr crystal. The increase in the Li$^+$ ion content would increase the conductivity of glass–ceramic because Li$^+$ ions are only charge carriers. The precipitation of Li$_7$P$_3$S$_{11}$ crystal provides high conductivity to glass–ceramic because the Li$_7$P$_3$S$_{11}$ crystal has extremely high conductivity [6]. The LiBr crystal precipitation would decrease the conductivity of glass–ceramic because of its low conductivity [17]. In the range of $x = 0\text{–}10$, the Li$_7$P$_3$S$_{11}$ crystal caused high conductivity of glass–ceramic. The Li$^+$ ions as charge carriers were increased by the non-crystallized LiBr component. Therefore, the conductivity of glass–ceramic was enhanced in spite of the precipitation of LiBr crystal. In addition, the possibility exists that a solid solution of LiBr in the Li$_7$P$_3$S$_{11}$ crystal improved the conductivity. In the range of $x = 12.5\text{–}20$, the Li$_7$P$_3$S$_{11}$ crystal provides

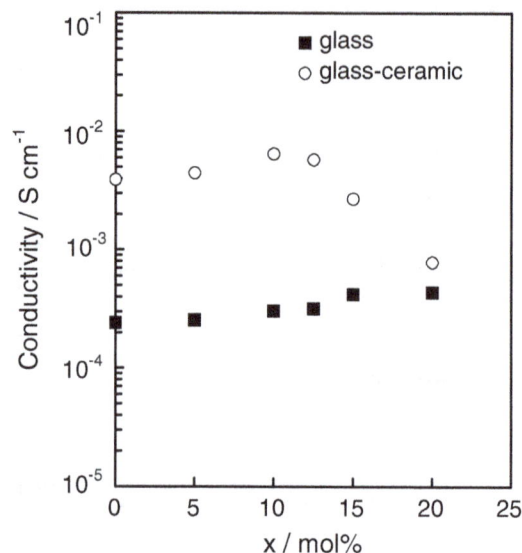

Fig. 4 Composition dependence of conductivities at 25 °C for the $(100 - x)(0.7Li_2S \cdot 0.3P_2S_5) \cdot x$LiBr glasses and glass–ceramics prepared by milling for 10 h and subsequent heat treatment

high conductivity to glass–ceramic in the same manner between $x = 0\text{–}10$. However, the excessive increase in LiBr crystal decreased glass–ceramic conductivity despite the increase in the Li$^+$ ion content. In addition, at $x = 20$, the decrease in the conductivity was accelerated by the degradation of the Li$_7$P$_3$S$_{11}$ crystal crystallinity.

As portrayed in Fig. 4, the highest conductivity of 6.5×10^{-3} S cm^{-1} was obtained for the $90(0.7Li_2S \cdot 0.3P_2S_5) \cdot 10$LiBr glass–ceramic. We sought to enhance the conductivity of the $90(0.7Li_2S \cdot 0.3P_2S_5) \cdot 10$LiBr glass–ceramic by changing the milling process. Figure 5 shows the XRD patterns of the $90(0.7Li_2S \cdot 0.3P_2S_5) \cdot 10$LiBr glasses prepared using various milling processes. The samples labeled as "10hMM", "20hMM", and "30hMM" were milled, respectively, for 10, 20, and 30 h. The samples labeled as "(Li$_2$S + P$_2$S$_5$) + LiBr", "(LiBr + P$_2$S$_5$) + Li$_2$S", and "(Li$_2$S + LiBr) + P$_2$S$_5$" were milled using two-step milling defined in "Experimental". Halo patterns were observed for all samples. Glasses were obtained despite the difference in milling processes. An unknown peak was observed for the (Li$_2$S + LiBr) + P$_2$S$_5$ sample, and also for the "(Li$_2$S + LiBr)" sample, suggesting that the peak resulted from a compound of Li$_2$S and LiBr. The peak assignment has not been clarified yet.

The DTA curves of the $90(0.7Li_2S \cdot 0.3P_2S_5) \cdot 10$LiBr glasses prepared using various milling processes are depicted in Fig. 6. The exothermic peaks attributable to crystallization were observed between 220 and 230 °C. Glass transition phenomena were observed at about 195 °C in all samples. The glasses prepared by milling for over 20 h except for the (Li$_2$S + LiBr) + P$_2$S$_5$ glass showed a

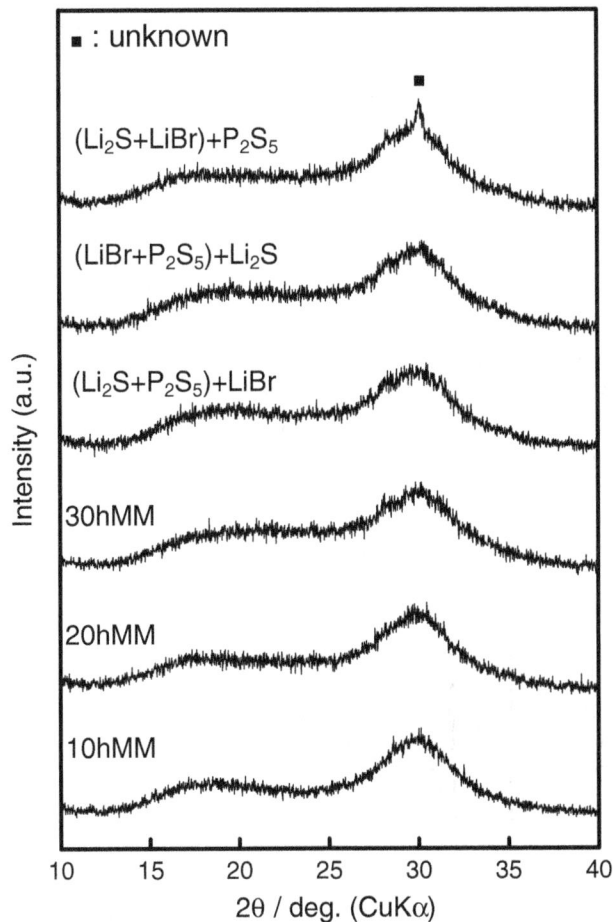

Fig. 5 XRD patterns of the 90(0.7Li$_2$S·0.3P$_2$S$_5$)·10LiBr glasses prepared using various milling processes. The samples labeled as "10hMM", "20hMM" and "30hMM" were milled for, respectively, 10, 20 and 30 h. The samples labeled as "(Li$_2$S + P$_2$S$_5$) + LiBr", "(LiBr + P$_2$S$_5$) + Li$_2$S" and "(Li$_2$S + LiBr) + P$_2$S$_5$" were milled by the two-step milling

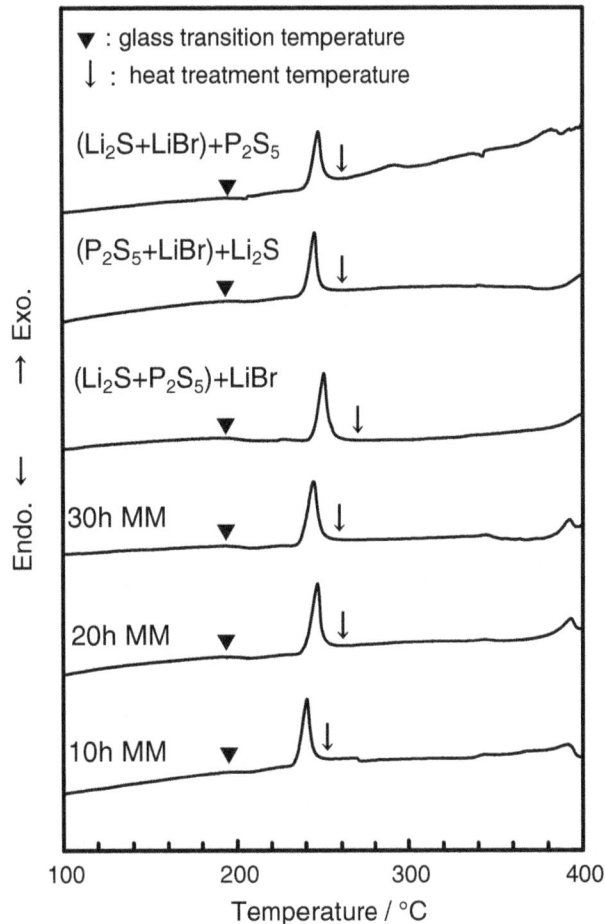

Fig. 6 DTA curves of the 90(0.7Li$_2$S·0.3P$_2$S$_5$)·10LiBr glasses prepared using various milling processes

clearer glass transition than the 10hMM glass. Milling for 20 or 30 h was believed to be effective for obtaining homogeneous glasses [8].

Figure 7 shows XRD patterns of the 90(0.7Li$_2$S·0.3P$_2$S$_5$)·10LiBr glass–ceramics prepared using various milling processes and subsequent heat treatment. Each sample was heated at a temperature between 250 and 270 °C. The peaks attributable to the Li$_7$P$_3$S$_{11}$ crystal and LiBr crystal were observed for all samples. The intensities of the peaks attributable to LiBr crystal in the 10hMM and (Li$_2$S + P$_2$S$_5$) + LiBr samples were slightly larger than those in the other samples. Results suggest that the milling period for the compounds containing LiBr affected the LiBr crystal precipitation in the glass–ceramic. In the (Li$_2$S + LiBr) + P$_2$S$_5$ sample, the peak intensity in the Li$_7$P$_3$S$_{11}$ crystal was weakened. A peak attributable to Li$_2$S crystal was observed.

Table 1 shows the conductivities at 25 °C for the 90(0.7Li$_2$S·0.3P$_2$S$_5$)·10LiBr glasses and glass–ceramics

prepared using various milling processes and subsequent heat treatment. The 90(0.7Li$_2$S·0.3P$_2$S$_5$)·10LiBr glass–ceramics milled for over 20 h showed extremely high conductivity of about 8 × 10^{-3} S cm^{-1}, except for the (Li$_2$S + LiBr) + P$_2$S$_5$ glass–ceramic. In the two-step milling samples, glass–ceramics with high conductivities tended to be obtained from glasses with high conductivities. The highest conductivity of 8.4 × 10^{-3} S cm^{-1} was obtained for the (LiBr + P$_2$S$_5$) + Li$_2$S glass–ceramic. Results suggest that enhancement of the conductivity of the glass–ceramic resulted from improvement of the glass homogeneity. A highly homogeneous glass would decrease the precipitation of LiBr crystal with low conductivity. Low conductivity was obtained for the (Li$_2$S + P$_2$S$_5$) + LiBr glass–ceramic. The (Li$_2$S + P$_2$S$_5$) + LiBr glass–ceramic showed XRD peaks of LiBr crystal with a large intensity as portrayed in Fig. 7. Therefore, the growth of LiBr crystal would decrease the conductivity of the (Li$_2$S + P$_2$S$_5$) + LiBr glass–ceramic. The lowest conductivity of the (Li$_2$S + LiBr) + P$_2$S$_5$ glass–ceramic would result from the degradation of the crystallinity of the

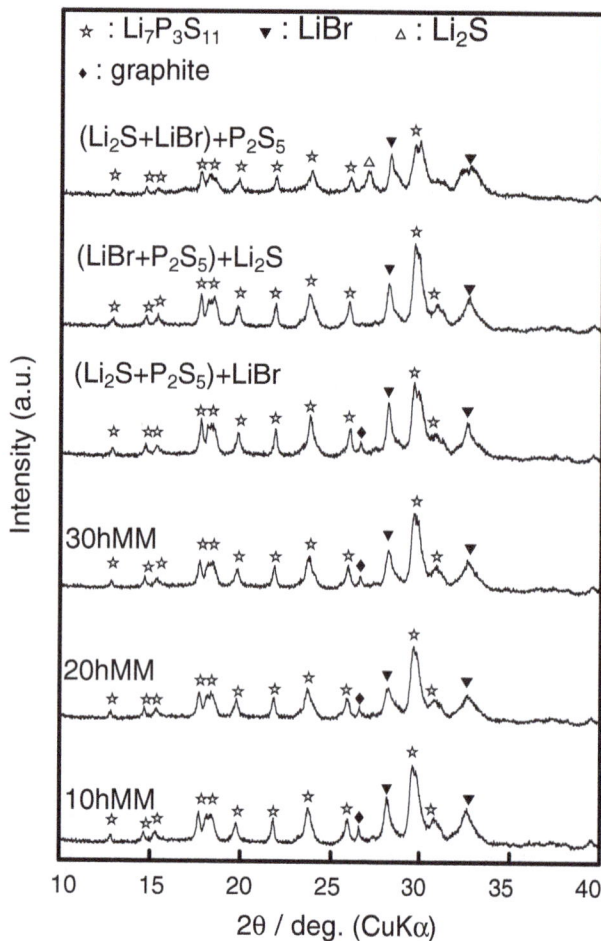

Fig. 7 XRD patterns of the $90(0.7Li_2S \cdot 0.3P_2S_5) \cdot 10LiBr$ glass-ceramics prepared using various milling processes and subsequent heat treatment

Table 1 Conductivities at 25 °C for the $90(0.7Li_2S \cdot 0.3P_2S_5) \cdot 10LiBr$ glasses and glass-ceramics prepared using various milling processes and subsequent heat treatment

Preparation conditions for $90(0.7Li_2S \cdot 0.3P_2S_5) \cdot 10LiBr$ (mol %) samples		Conductivity (S cm^{-1})	
Sample	Milling time (h)	Glass	Glass-ceramic
$(Li_2S + LiBr) + P_2S_5$	20	2.4×10^{-4}	1.9×10^{-3}
$(LiBr + P_2S_5) + Li_2S$	20	3.4×10^{-4}	8.4×10^{-3}
$(Li_2S + P_2S_5) + LiBr$	20	2.9×10^{-4}	7.4×10^{-3}
30hMM	30	3.5×10^{-4}	7.7×10^{-3}
20hMM	20	3.4×10^{-4}	8.0×10^{-3}
10hMM	10	3.1×10^{-4}	6.5×10^{-3}

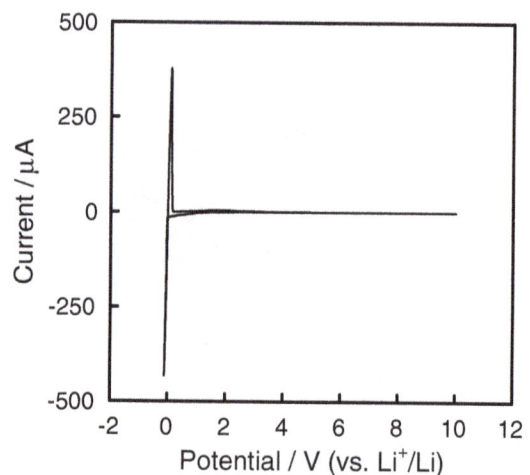

Fig. 8 Cyclic voltammogram of the $90(0.7Li_2S \cdot 0.3P_2S_5) \cdot 10LiBr$ glass-ceramic electrolyte. The scan rate was 5 mV s^{-1}. The potential was swept between -0.1 and $+10$ V

$Li_7P_3S_{11}$ crystal and the precipitation of Li_2S crystal, as portrayed in Fig. 7. To obtain high conductivity for the $90(0.7Li_2S \cdot 0.3P_2S_5) \cdot 10LiBr$ glass-ceramic, P_2S_5 as a network former of the glass should be mixed from the beginning of the vitrification process. The milling period of time for the vitrification should also be controlled to enhance the glass-ceramic conductivity, because the precipitation of LiBr crystal by heat treatment is affected by the milling period of time for the preparation of glasses.

The $90(0.7Li_2S \cdot 0.3P_2S_5) \cdot 10LiBr$ glass-ceramic (20hMM) was used in the cells for cyclic voltammetry and charge-discharge measurement. Figure 8 shows the cyclic voltammogram of glass-ceramic. Large cathodic and anodic currents attributable to deposition and dissolution of lithium metal were observed in the potential range from -0.1 to $+0.1$ V vs. Li^+/Li. The glass-ceramic electrolyte exhibited a wide electrochemical window, because no obvious cathodic or anodic current was observed up to 10 V. Reportedly, the electrochemical stability range of $67Li_2S \cdot 33P_2S_5$ glass added with LiI was about 2.9 V vs.

Li^+/Li [18]. On the other hand, we reported a wide electrochemical window of $80(0.7Li_2S \cdot 0.3P_2S_5) \cdot 20LiI$ glass up to 10 V vs. Li^+/Li [14]. The present study confirmed the high electrochemical stability of $90(0.7Li_2S \cdot 0.3P_2S_5) \cdot 10LiBr$ glass-ceramic.

An all-solid-state cell using the $90(0.7Li_2S \cdot 0.3P_2S_5) \cdot 10LiBr$ glass-ceramic electrolyte was fabricated. The charge-discharge curves of the all-solid-state cell are shown in Fig. 9. The cell was charged up to 3.6 V vs. Li–In and was discharged to 2.0 V vs. Li–In at the current density of 0.064 mA cm^{-2}. The cell showed an average voltage plateau at 3.3 V in both charge and discharge processes. The cell voltage of 3.3 V vs. Li–In corresponds to the potential of 3.9 V vs. Li^+/Li, which is the charge-discharge potential of $LiCoO_2$. The discharge capacity of 90 mAh g^{-1} was obtained at the first cycle. Coulombic efficiencies of greater than 95 % were obtained except for the first cycle, which indicated that the cell using the

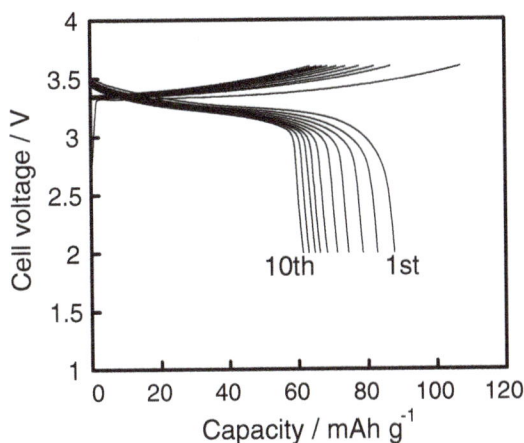

Fig. 9 Charge–discharge curves of the all-solid-state cell In/90(0.7Li$_2$S·0.3P$_2$S$_5$)·10LiBr glass–ceramic/LiCoO$_2$ at 25 °C. The cell was charged and discharged between 2.0 and 3.6 V at the current density of 0.064 mA cm^{-2}

90(0.7Li$_2$S·0.3P$_2$S$_5$)·10LiBr glass–ceramic electrolyte operated as a lithium secondary battery at room temperature. The capacity was decreased gradually with charge–discharge cycling. Decrease of the capacity with cycling was also observed for the cell using the 70Li$_2$S·30P$_2$S$_5$ glass–ceramic electrolyte without LiBr. Consequently, the added LiBr would not decrease the capacity. Results of TEM-EDX analysis indicated that sulfur and phosphorus in the electrolyte added to the composite electrode diffused to the LiCoO$_2$ active material [19]. The diffusion increased the interfacial resistances between the electrolyte and LiCoO$_2$ with cycling. The cycle performance of the cell would be improved using adequate active materials for the electrolyte [20]. The 90(0.7Li$_2$S·0.3P$_2$S$_5$)·10LiBr glass–ceramic material with a high lithium ion conductivity is attractive as a solid electrolyte for a separator layer in all-solid-state batteries.

Conclusion

New electrolytes of (100 − x)(0.7Li$_2$S·0.3P$_2$S$_5$)·xLiBr (x = 0, 5, 10, 12.5, 15 and 20) were synthesized. Glass electrolytes were obtained by mechanical milling in the composition range of x = 0–20. The glass conductivity increased concomitantly with increasing LiBr content. The glass at x = 20 showed conductivity of 3.1 × 10^{-4} S cm^{-1}. Glass–ceramic electrolytes were prepared using heat treatment for the glasses. The Li$_7$P$_3$S$_{11}$ crystals with high lithium ion conductivity and LiBr crystals were precipitated in all glass–ceramic electrolytes with added LiBr. High conductivity of 6.4 × 10^{-3} S cm^{-1} was obtained at x = 10. Furthermore, the conductivity of the 90(0.7Li$_2$S·0.3P$_2$S$_5$)·10LiBr glass–ceramic electrolyte was improved

to 8.4 × 10^{-3} S cm^{-1} by controlling of the milling period used to prepare the precursor glass. The 90(0.7Li$_2$S·0.3P$_2$S$_5$)·10LiBr glass–ceramic electrolyte showed a wide electrochemical window up to 10 V and functioned as an electrolyte for an all-solid-state lithium secondary battery.

References

1. Malugani, J.P., Robert, G.: Preparation and electrical properties of the 0,37 Li$_2$S-0,18P$_2$S$_5$-0,45LiI glass. Solid State Ion. **1**, 519–523 (1980)
2. Paradel, A., Ribes, M.: Electrical properties of lithium conductive silicon sulfide glasses prepared by twin roller quenching. Solid State Ion. **18–19**, 351–355 (1986)
3. Kennedy, J.H., Sahami, S., Shea, S.W., Zhang, Z.: Preparation and conductivity measurement of SiS$_2$-Li$_2$S glasses doped with LiBr and LiCl. Solid State Ion. **18–19**, 368–371(1986)
4. Zhang, Z., Kennedy, J.H.: Synthesis and characterization of the B$_2$S$_3$–Li$_2$S, the P$_2$S$_5$–Li$_2$S and the B$_2$S$_3$–P$_2$S$_5$–Li2S glass systems. Solid State Ion. **38**, 217–224 (1990)
5. Kanno, R., Murayama, M.: Lithium ionic conductor thio-LIS-ICON. J. Electrochem. Soc. **148**, A742–A746 (2001)
6. Mizuno, F., Hayashi, A., Tadanaga, K., Tatsumisago, M.: High lithium ion conducting glass–ceramics in the system Li$_2$S–P$_2$S$_5$. Solid State Ion. **177**, 2721–2725 (2006)
7. Minami, K., Hayashi, A., Tatsumisago, M.: Electrical and electrochemical properties of the 70Li$_2$S·(30−x)P$_2$S$_5$·xP$_2$O$_5$ glass–ceramic electrolytes. Solid State Ion. **179**, 1282–1285 (2008)
8. Hayashi, A., Minami, K., Ujiie, S., Tatsumisago, M.: Preparation and ionic conductivity of Li$_7$P$_3$S$_{11-z}$ glass–ceramic electrolytes. J. Non-Cryst. Solids. **356**, 2670–2673 (2010)
9. Minami, K., Hayashi, A., Tatsumisago, M.: Preparation and characterization of lithium ion conducting Li$_2$S–P$_2$S$_5$–GeS$_2$ glasses and glass–ceramics. J. Non-Cryst. Solids. **356**, 2666–2669 (2010)
10. Rao, R.P., Adams, S.: Studies of lithium argyrodite solid electrolytes for all-solid-state batteries. Phys. Status Solidi A. **208**, 1804–1807 (2011)
11. Homma, K., Yonemura, M., Kobayashi, T., Nagao, M., Hirayama, M., Kanno, R.: Crystal structure and phase transitions of the lithium ionic conductor Li$_3$PS$_4$. Solid State Ion. **182**, 53–58 (2011)
12. Boulineau, S., Courty, M., Tarascon, J.M., Viallet, V.: Mechanochemical synthesis of Li-argyrodite Li$_6$PS$_5$X (X = Cl, Br, I) as sulfur-based solid electrolytes for all solid state batteries application. Solid State Ion. **221**, 1–5 (2011)
13. Kamaya, N., Homma, K., Yamakawa, Y., Hirayama, M., Kanno, R., Yonemura, M., Kamiyama, T., Kato, Y., Hara, S., Kawamoto, K., Mitsui, A.: A lithium superionic conductor. Nat. Mater. **10**, 682–686 (2011)
14. Ujiie, S., Hayashi, A., Tatsumisago, M.: Structure, ionic conductivity and electrochemical stability of Li$_2$S–P$_2$S$_5$–LiI glass and glass–ceramic electrolytes. Solid State Ion. **211**, 42–45 (2012)
15. Shannon, R.D.: Revised effective ionic radii and systematic studies of interatomic distances in halides and chalcogenides. Acta. Cryst. **A32**, 751–767 (1976)
16. Ohta, N., Takada, K., Sakaguchi, I., Zhang, L., Ma, R., Fukuda, K., Osada, M., Sasaki, T.: LiNbO$_3$-coated LiCoO$_2$ as cathode material for all solid-state lithium secondary batteries. Electrochem. Commun. **9**, 1486–1490 (2007)

17. Mercier, R., Tachez, M., Malugani, J.P., Robert, G.: Effect of homovalent(I^-–Br^-)ion substitution on the ionic conductivity of $LiI_{1-x}Br_x$ system. Solid State Ion. **15**, 109–112 (1985)

18. Mercier, R., Malugani, J.P., Fahys, B., Robert, G.: Superionic conduction in Li_2S–P_2S_5–LiI–glass. Solid State Ion. **5**, 663–666 (1981)

19. Ohtomo, T., Hayashi, A., Tatsumisago, M., Tsuchida, Y., Hama, S., Kawamoto, K.: All-solid-state lithium secondary batteries using the $75Li_2S{\cdot}25P_2S_5$ glass and the $70Li_2S{\cdot}30P_2S_5$ glass–ceramic as solid electrolytes. J. Power Sour. **233**, 231–235 (2013)

20. Minami, K., Hayashi, A., Ujiie, S., Tatsumisago, M.: Electrical and electrochemical properties of glass-ceramic electrolytes in the systems Li_2S–P_2S_5–P_2S_3 and Li_2S–P_2S_5–P_2O_5. Solid State Ion. **192**, 122–125 (2011)

4

Growth mechanisms of ceria- and zirconia-based epitaxial thin films and hetero-structures grown by pulsed laser deposition

Daniele Pergolesi · Marco Fronzi · Emiliana Fabbri · Antonello Tebano · Enrico Traversa

Abstract Thin films and epitaxial hetero-structures of doped and undoped CeO_2, and 8 mol% Y_2O_3 stabilized ZrO_2 (YSZ), were fabricated by pulsed laser deposition on different single crystal substrates. Reflection high energy electron diffraction was used to monitor in situ the growth mechanism of the films. Two distinct growth mechanisms were identified along the (001) growth direction for the Ce- and Zr-based materials, respectively. While the doped or undoped ceria films showed a 3-dimensional growth mechanism typically characterized by a pronounced surface roughness, YSZ films showed an almost ideal layer-by-layer 2-dimensional growth. Moreover, when the two materials were stacked together in epitaxial hetero-structures, the two different growth mechanisms were preserved. As a result, a 2-dimensional reconstruction of the ceria-based layers determined by the YSZ film growing above was observed. The experimental results are explained in terms of the thermodynamic stability of the low-index surfaces of the two materials using computational analysis performed by density functional theory.

Keywords Pulsed laser deposition · Hetero-structure · Oxygen-ion conducting oxides · Reflection high energy electron diffraction · Density functional theory

Introduction

Thin films of doped CeO_2 and Y_2O_3-stabilized ZrO_2 (YSZ) fabricated by different thin film deposition methods have been widely investigated as high temperature oxygen-ion conductors. Among the most important applications of these materials is the fabrication of electrolyte membranes for solid oxide fuel cells (SOFCs). Particularly, large oxygen ion conductivity characterizes doped CeO_2. Typical dopants are Gd and Sm with concentration ranging from 10 to 20 % [1]. The bulk oxygen-ion conductivity of 15 % Sm-doped CeO_2 (SDC) is as large as 0.02 S cm^{-1} at about 600 °C, making this material one of the most performing solid state electrolyte in the so-called intermediate temperature range (500–800 °C) [2]. YSZ is also widely used as an electrolyte, for SOFCs but mostly for oxygen sensors used to control the air-to-fuel ratio in vehicles, as well as for the fabrication of thermal barrier coatings due to its low thermal conductivity [3].

Pure zirconia (ZrO_2) shows a complicated phase diagram having a monoclinic crystal structure at temperatures below about 1,000 °C, with transitions to tetragonal and cubic structures with increasing the temperature. Such phase transformations induce very large stresses that cause pure zirconia to crack, limiting its practical application. On the contrary, pure ceria (CeO_2) has a stable cubic phase that can easily become non-stoichiometric in oxygen content, showing important catalytic activity in oxygen reduction reaction processes. This material finds many practical applications, of which the most important are in

D. Pergolesi (✉) · M. Fronzi · E. Fabbri
International Research Center for Materials Nanoarchitectonics (WPI-MANA), National Institute for Materials Science (NIMS), 1-1 Namiki, Tsukuba, Ibaraki 305-0044, Japan
e-mail: pergolesi.daniele@nims.go.jp

A. Tebano
CNR-SPIN and Dipartimento di Informatica Sistemi e Produzione, University of Roma Tor Vergata, Rome, Italy

E. Traversa
International Research Center for Renewable Energy, State Key Laboratory of Multiphase Flow in Power Engineering, Xi'an Jiaotong University, Xi'an 710049, Shaanxi, China

catalysis [4]. Thin films of CeO_2 have been also widely used as buffer layers for the growth of thin film hetero-structures, such as insulating layers for microelectronic devices or as diffusion barriers to avoid chemical reactions at interfaces.

The increasing miniaturization of solid state electro-chemical devices leads to an increasing importance of the thin film deposition technology for their fabrication. Besides, it is well known that the microstructural and morphological characteristics of thin films can strongly affect their physical and chemical properties. In particular, especially for doped ceria, the degree of crystallinity, the average grain size and the relative grain boundary extent can significantly modify the charge transport properties and the chemical stability in operating environment [5–7].

More recently, multilayered thin film hetero-structures comprising ceria-based oxides and/or YSZ have been fabricated in order to study the conducting properties of hetero-phase interfaces. A large increase in ionic conduc-tivity was observed in case of incoherent interfaces due to the faster conduction pathways along dislocation lines [8, 9]. More ordered oxygen-ion conducting hetero-inter-faces have been also fabricated to investigate the effect on the conducting properties of the compressive or ten-sile strain occurring at quasi-coherent hetero-interfaces [10–12]. The fabrication of samples appropriately designed to allow isolating the effect of the interfacial strain field requires a careful control of the deposition process and a deep understanding of the growth mechanisms.

For this work, we used reflection high energy electron diffraction (RHEED) and X-ray diffraction (XRD) analyses to investigate the growth mechanism of doped and undoped ceria and YSZ thin films grown onto different single crystal substrates. Different growth mechanisms were observed for the two materials depending on the crystallographic growth direction.

Density functional theory calculations were used to theoretically investigate the thermodynamic stability of the (100) and (111) surfaces of the two materials in equilib-rium with the gas phase, and a very good agreement with the experimental observations was found.

Experimental

Thin films of CeO_2, 15 % Sm-doped CeO_2 (SDC) and 8 mol% Y_2O_3 stabilized ZrO_2 (YSZ) were grown by pulsed laser deposition (PLD) on different single crystal substrates listed in Table 1. Sintered ceramic pellets prepared in our laboratory were used as target materials.

The substrates were ultrasonically cleaned in de-ionized water, acetone and methanol, and dried with pure nitrogen prior to insertion into the deposition system. Before starting

Table 1 Single crystal substrates used in this work, their crystallo-graphic structures, lattice parameters and crystallographic orientations

Material	Crystal structure	Lattice parameters (nm)	Crystal orientation
MgO	Cubic rock salt	$c = 0.421$	(001)
C-cut Al_2O_3	Hexagonal	$a = 0.476$	(0001)
		$b = 0.476$	
		$c = 1.300$	
9.5 mol % YSZ	Cubic fluorite	$c = 0.512$	(001)
$YAlO_3$ (YAO)	Orthorhombic perovskite	$a = 0.518$	(110)
		$b = 0.531$	
		$c = 0.735$	
$SrTiO_3$ (STO)	Cubic perovskite	$c = 0.3905$	(001)

the deposition of the films, the substrates were kept at about 800 °C in about 40 Pa of high purity oxygen partial pressure for about 20 min. It was observed that such a thermal treatment often resulted in an evident enhancement of the crystalline quality of the substrate surfaces.

The custom made PLD system (AOV Ltd) consisted of a vacuum chamber with a base pressure of about 10^{-6} Pa, equipped with a load-lock chamber. The target carousel can hold up to six targets, and a stainless steel shield reduces cross contamination during the ablation process. Each target can simultaneously rotate and oscillate allow-ing a uniform ablation of the target surface. A KrF excimer laser (Coherent Lambda Physik GmbH) with a wavelength of 248 nm and a pulse width of 25 ns was focused on the target material in a spot area of about 5 mm^2. The energy of the laser shots onto the target surface was set at about 160 mJ. A laser repetition rate between 2 and 5 Hz was used. The deposition of the films was carried out in an oxygen background pressure ranging from 0.5 to 5 Pa, at a substrate temperature of about 700 °C. The target-to-substrate distance was 75 mm.

The samples were cooled from the deposition tempera-ture down to room temperature at 10 degrees min^{-1} in an oxygen background pressure of about 40 Pa.

A high pressure reflection high energy electron diffrac-tion (RHEED) system (AOV Ltd), equipped with a dif-ferential pumping system was used to monitor in situ the surface evolution of the films. An accelerating voltage of 28 kV and an emission current of about 100 μA were used. The RHEED patterns were recorded using a CCD camera.

X-ray analysis (PANalytical X'pert Pro MPD) was used to calibrate the deposition rate by X-ray reflectometry (XRR) and to investigate the out-of-plane crystalline structure of the films by X-ray diffraction (XRD).

To confirm that the PLD process actually pro-vided samples with the expected electrical properties, the electrical conductivities of the highly textured films of SDC and YSZ were measured by electrochemical impedance spectroscopy (EIS). Two Ti–Pt electrodes were deposited onto the film surfaces by electron beam deposition. The electrical characterization was performed in air, using a multichannel potentiostat VMP3 (Bio-Logic), in the frequency range between 1 MHz and 100 mHz, varying the temperature between 400 and 700 °C.

Results and discussion

The growth mechanism of ceria-based epitaxial thin films

Among the substrates used in this work, and listed in Table 1, (001)-oriented STO single crystals are particularly suitable for the epitaxial growth of doped and undoped CeO_2 films [5]. CeO_2 and SDC have a cubic fluorite crystal structure with a lattice parameter of about 5.41 and 5.44 Å, respectively. Epitaxial films can be obtained on STO with the in-plane orientation $(100)CeO_2/(110)STO$. Owing to this 45° in-plane rotation of the CeO_2 unit cell with respect to the STO unit cell, the resulting lattice misfit with the STO substrate is about 1.4 % for SDC and about 2 % for pure ceria. Figure 1a shows the 2θ-θ scan of a 250 Å-thick film of SDC epitaxially oriented with the STO substrate. Figure 1b shows the size effect interference fringes around the (002) SDC reflection, indicating the good crystallographic quality of the film. The red curve in Fig. 1b represents a simulation for a 45 unit cells thick SDC film (about 245 Å), which is in very good agreement with the expected thickness, as derived from the calibration of the deposition rate performed by XRR (Fig. 2b).

Figure 1c shows the typical RHEED patterns acquired for a film of SDC grown on STO, relative to the (100) in-plane orientation of the substrate. A spotty pattern associated with a 3D growth mechanism appeared after few laser shots implying that this island-like growth arose immediately at the early stage of the formation of the first layers.

Analogous results were obtained using (001)-oriented 9.5 YSZ single crystal substrates for the growth of epitaxial CeO_2 films, as shown in Fig. 2. In this case, the two materials have the same crystalline structure and the CeO_2 has a lattice misfit of about −5.6 % with respect to the substrate. Such a relatively large lattice misfit can be accommodated by the introduction of a regular network of misfit dislocations at the interface [13] allowing a well-ordered cube-on-cube growth driven by the substrate along the (001) direction. A mosaic spread of about 0.5° was evaluated by measuring the FWHM of the Gaussian fit of the rocking curve acquired along the (002) reflection peak of the film.

Figure 2b shows the XRR plot used for the calibration of the deposition rate of doped and undoped ceria. With the selected deposition parameters, a deposition rate of about 0.24 Å shot^{-1} was measured. Figure 2c shows that, also in this case, the RHEED patterns of the growing films showed the typical features of a 3D growth mechanism.

The spotty RHEED patterns characterizing the growth of pure and doped ceria films were observed over a relatively wide range of deposition parameters (substrate temperature of 700 ± 100 °C, oxygen background pressure from 0.1 up to 5 Pa, deposition rate from 0.3 up to about 2 Å shot^{-1}). Analogous RHEED patterns were also observed for (001)-oriented CeO_2 films grown by means of different thin film deposition techniques [14].

To check whether such a 3D growth mechanism depends on the growth crystallographic axis, C-cut Al_2O_3 (0001)-oriented single crystalline substrates were used. C-cut (0001) and R-cut (1102) sapphire crystals have been widely used for the growth of crystalline thin films of doped and undoped ceria. The R-cut surface is expected to favour the (001) orientation, while the C-cut should favour the (111)

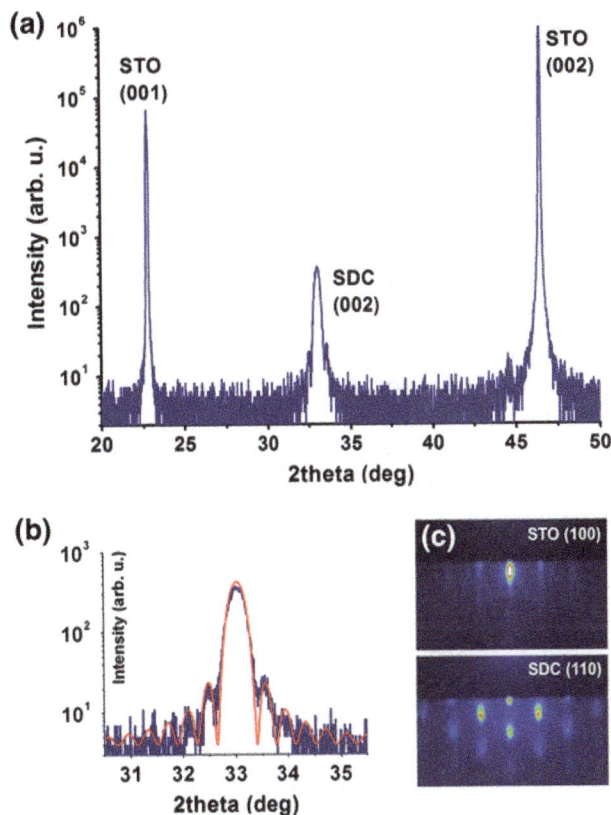

Fig. 1 a XRD analysis of an SDC film grown on STO. b XRD plot of the (002) SDC reflection with superimposed simulation (*red curve*) of the size effect interference fringes for an SDC film thickness of 45 unit cells. c RHEED pattern recorded along the in-plane (100) orientation of the substrate and RHEED pattern of the SDC film at the end of the deposition

(a)

(b)

(c)

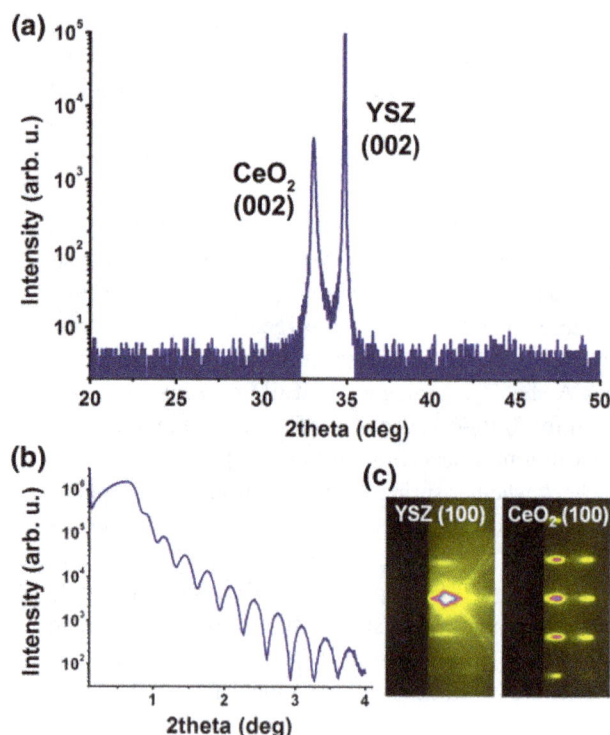

Fig. 2 a XRD analysis of a CeO$_2$ film grown on a (001)-oriented 9.5 YSZ substrate. **b** XRR measurement performed for the calibration of the deposition rate of the film. **c** RHEED patterns of the substrate and of the film recorded along the in-plane (100) orientation of the substrate

Fig. 3 XRD analysis of an SDC film grown on C-cut Al$_2$O$_3$ single crystal substrate. The *inset* shows the RHEED patterns recorded for the substrate and for the SDC film at the end of the deposition

film orientation [15, 16]. Nevertheless, both orientations were obtained on both surfaces, as well as films showing mixed (001)/(111) orientation, depending on process parameters and deposition technique [17, 18]. Using PLD, the fabrication of highly (111)-oriented CeO$_2$ films on C-cut sapphire has been reported for example in [6].

Figure 3 shows that preferentially (111)-oriented SDC films were grown with a minor (001) orientation. We observed that, in our experimental condition, (111)-oriented films were obtained at relatively low oxygen partial pressure (in the order of 0.1–0.5 Pa), while for larger values of oxygen partial pressure (few Pa), the films showed well-defined mixed (001)/(111) orientation.

The RHEED patterns acquired during the growth at low oxygen partial pressure (inset in Fig. 3) showed the typical features that characterize a quasi-2D layer-by-layer growth mechanism. Such streaky features were never observed for CeO$_2$ or SDC film grown onto STO or YSZ substrates where the growth was along the (001) axis.

To summarize, these measurements showed that doped or undoped ceria presents two different growth mechanisms depending on the growth directions; the film grows predominantly quasi-2D along the (111) direction, while an evident 3D growth mechanism was observed along the (001) direction.

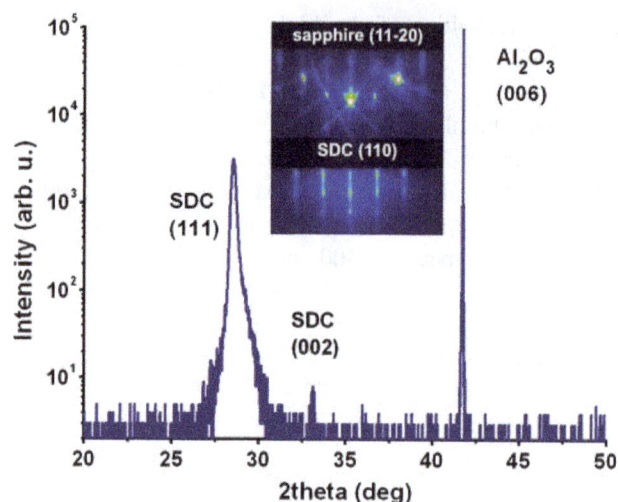

Finally, the electrical conductivity of an SDC film grown on sapphire was measured by EIS in air. The measured conductivity ranged from 0.03 S cm^{-1} at about 680 °C down to 0.001 S cm^{-1} at about 400 °C, with activation energy of about 0.70 eV. This result is in very good agreement with the literature data relative to the bulk conductivity of doped CeO$_2$ films [6]. The electrical characterization of the films grown on STO and YSZ single crystal wafers cannot give reliable results due to the conductive properties of the deposition substrates at high temperatures.

The growth mechanism of YSZ epitaxial thin films

The growth of YSZ films was studied using (110)-oriented YAO and (001)-oriented MgO single crystal substrates. YSZ has a cubic fluorite structure with lattice parameter of about 5.14 Å, which results in a lattice misfit of about 1.5 % with (110)-oriented YAO (Table 1). Figure 4a shows the out-of-plane XRD analysis of a YSZ film grown on YAO. The presence of well-defined size effect interference fringes around the (002) reflection line of the film (Fig. 4b) suggests a very good crystallographic quality. The black curve in Fig. 4b shows a simulation for a 35 unit cells thick YSZ film (about 180 Å). From this simulation, we could estimate a deposition rate of about 0.059 Å shot^{-1}.

Opposite to what observed in the case of (001)-oriented ceria-based films, the RHEED patterns of (001)-oriented YSZ films clearly showed a different growth mechanism. Figure 3b shows the RHEED patterns of the substrate and of the YSZ film acquired toward the (100) in-plane direction of YAO. The RHEED patterns consisted of

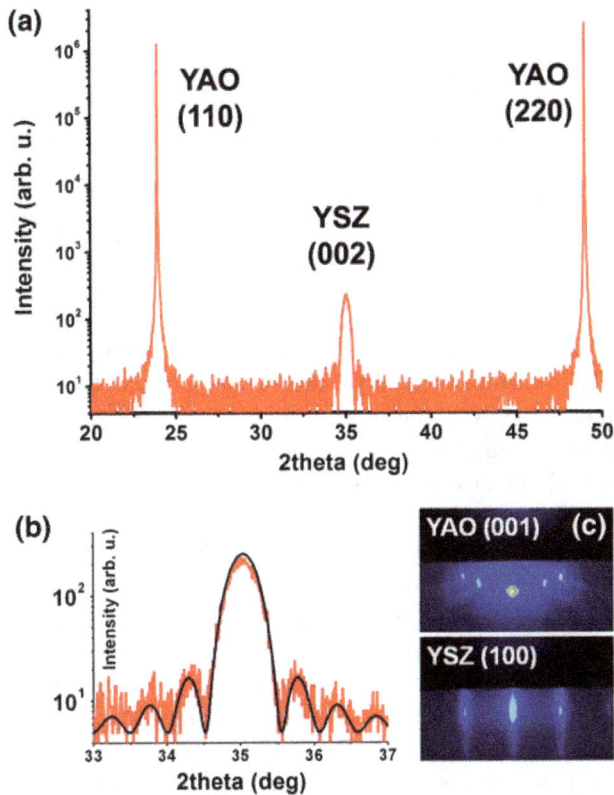

Fig. 4 **a** XRD analysis of a YSZ film grown on (110)-oriented YAO. **b** XRD plot of the (002) YSZ reflection with superimposed simulation (*black curve*) of the size effect interference fringes for a YSZ film thickness of 35 unit cells. **c** RHEED patterns of the substrate and the YSZ film recorded along the in-plane (100) orientation of the substrate

Fig. 5 **a** XRD analysis of a YSZ film grown on (001)-oriented MgO substrate. The inset shows the XRR measurement performed for the calibration of the deposition rate of the film. **b** RHEED patterns of the substrate and of the film recorded along the in-plane (100) orientation of the substrate

well-defined streaks revealing a 2D layer-by-layer growth mechanism.

Very similar result was obtained analysing a 300 Å thick film of YSZ grown on (001)-oriented MgO substrate. In spite of a lattice misfit as large as −18 %, a highly textured growth was observed by XRD analysis (Fig. 5). A relatively large value of 0.64° was found for the FWHM of the (002) ω-scan of the film. Figure 5a shows the XRR plot used to measure the deposition rate of the film with the selected deposition parameters. A value of about 0.053 Å shot^{-1} was estimated in very good agreement with the value obtained from the simulation of the size effect interference fringes showed in Fig. 4b. The final RHEED pattern showed the typical features associated with a 2D layer-by-layer growth (Fig. 5b). However, in this case, the RHEED pattern disappeared during the deposition of the first layers, suggesting the formation of a disordered interface probably characterized by a large density of misfit dislocations introduced to release the excess interfacial strain [8].

According to the measured deposition rate, a streaky RHEED pattern originated from the surface of the YSZ film growing on MgO was not observed below a minimum

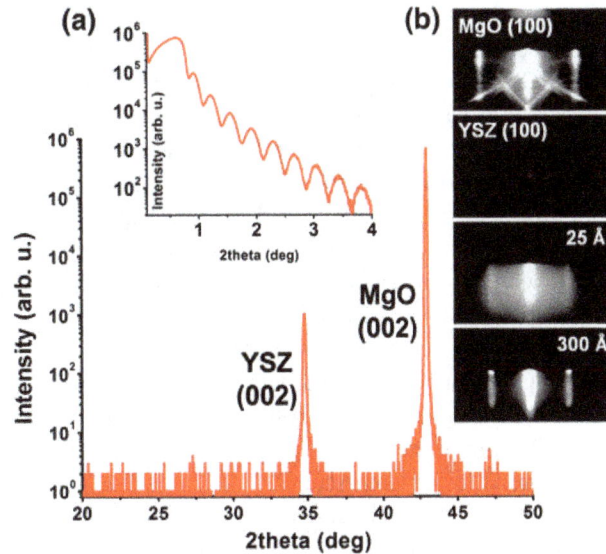

thickness of about 20 Å. However, even in the case of a particularly unfavourable crystalline matching with the deposition substrate, YSZ showed a clear tendency to a 2D growth toward the (001) direction driven by the deposition substrate.

The electrical conductivity of a YSZ film grown on MgO was measured in air by EIS and the conductivity was found to range from 9.6×10^{-3} down to 8.5×10^{-5} S cm^{-1} at 700 and 400 °C, respectively, showing an activation energy of about 0.98 eV. The measured conductivity was in excellent agreement with the conductivity of YSZ single crystals [19].

Calculation of the low index surface energies

The comparison between the results obtained with YSZ and those obtained with doped and undoped ceria films strongly suggests a significant difference in the thermodynamic stability of the (001) surfaces of the two materials. To understand the driving mechanism for the different behaviours experimentally observed, a theoretical evaluation of the surface energy of the low-index surface orientation of CeO_2 and cubic zirconia (c-ZrO_2) was computed using first-principles density functional theory (DFT).

The low-index surface energies of CeO_2 nano-particles have been analysed in a previous work and the stable surfaces were identified as a function of the temperature, the chemical potential and the oxygen background pressure [20]. It was found that unless the chemical potential of the oxygen is extremely low (i.e. very high temperature and very low oxygen partial pressure), the (111) surface orientation is stable if compared with the (100) and (110). By minimization of the Gibbs free energy as indicated by the Gibbs–Wulff theorem, it was possible then to calculate the equilibrium crystal shape of a CeO_2 nano-particles and show that under oxygen-rich conditions, the (111) surface is the only orientation present in the particle and the resulting morphology is described as the octahedron shown in Fig. 6a.

For this work, an analogous analysis has been carried out to calculate the low-index surface energies of c-ZrO_2 nano-particles. As previously mentioned, the cubic fluorite zirconia is stable only at very high temperatures. In YSZ, the effect of the dopant (Y_2O_3), besides creating oxygen vacancies allowing oxygen-ion diffusion, is to stabilize the cubic structure at lower temperatures.

Although YSZ is the material under investigation in this work, for the calculation of the low-index surface energies, we only considered the cubic undoped crystal, since it has been shown that the trend of the low-index surface relative stability for c-ZrO_2 and YSZ does not change [21]. Therefore, the computed surface energies obtained considering undoped cubic zirconia are expected to bring, at least qualitatively, to the same conclusions.

The surface energies of (111) and (001) surfaces of c-ZrO_2 nano-particles under oxygen-rich condition were calculated. Table 2 compares the low-index surface energies calculated for CeO_2 and c-ZrO_2 nano-particles, showing that both crystalline structures favour the (111) surface that strongly minimize the Gibbs free energy, but the difference between the (001) and (111) surface energies in the case of c-ZrO_2 turned out to be about a factor of two smaller than in the case of CeO_2.

The equilibrium crystal shape of c-ZrO_2 nano-particles, calculated according to the Wulff construction by minimizing the surface energy, can be described as a truncated octahedron showing a mixture of (001) and (111) surfaces (Fig. 6b).

The Wulff theorem states that for a given volume, the crystal exposes different surface orientations in order to

Table 2 Low index surface Gibbs free energy of ceria and cubic zirconia

Material	Surface index	Surface Gibbs free energy (eV)	Surface energy difference (eV)	References
CeO_2	(001)	0.230	0.193	[20]
	(111)	0.037		
c-ZrO_2	(001)	0.189	0.116	This work
	(111)	0.073		

minimize the surface energy. In the case of highly textured thin films on single crystal substrates, the epitaxial stress can modify the extension of the different facets present in the stress-free equilibrium shape, but it cannot create new facets [22]. In this sense, we can say that the analysis proposed here for nano-particles may give an indication of the local preferential surface orientation of a thin film at thermodynamic equilibrium.

In both cases (CeO_2 and c-ZrO_2), the (111) surface is stable if compared with other low-index orientations. Even though the growth is driven toward the (001) direction by crystalline constrains, the (111) facet is strongly favoured for CeO_2, which will arrange the lattice in order to expose the (111) surface. As a result, the favoured (001) surface will consist on a mosaic of (111)-facet octahedra (001)-oriented, thus enhancing the surface roughness. On the other hand, c-ZrO_2, having a smaller difference between the (001) and (111) surface energies will expose a mixture of these two orientations originating smoother surfaces.

The growth mechanism of ceria- and zirconia-based superlattices

Multilayered hetero-structures and superlattices made by coupling SDC or YSZ with an insulating phase [9, 11, 12] have been fabricated to study the conducting properties of the interfaces [23]. To the best of our knowledge, for these hetero-structures, high crystalline quality (biaxial texture, no grain boundary) has been achieved only for (001)-oriented YSZ–STO multilayers grown by PLD on STO substrates [24], and for (001)-oriented CeO_2–YSZ hetero-structures grown on MgO [25, 26].

Highly (111)-oriented hetero-structures of Gd-doped ceria and Gd-doped zirconia, with minor phases from other orientations, were obtained on C-cut sapphire by molecular beam epitaxy [10]. In this case, the authors reported the observation of a streaky RHEED pattern up to a film thickness of about 40 nm with additional weak features corresponding to polycrystalline structure with increasing the thickness of the deposition.

The literature does not report any detailed study of the growth mechanism of these hetero-structures, particularly in the case of the (001) orientation.

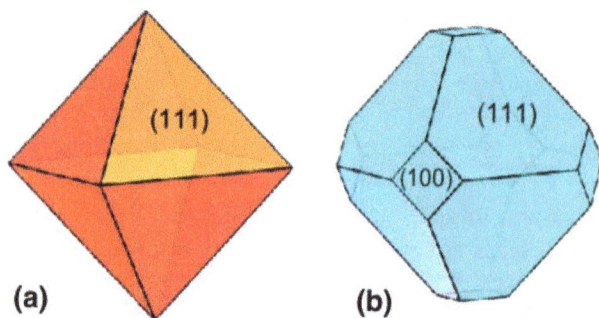

Fig. 6 Morphology of the equilibrium crystal shape nano-particles of CeO_2 (**a**) and c-ZrO_2 (**b**) obtained by the minimization of the Gibbs free energy, as indicated by the Gibbs-Wulff theorem

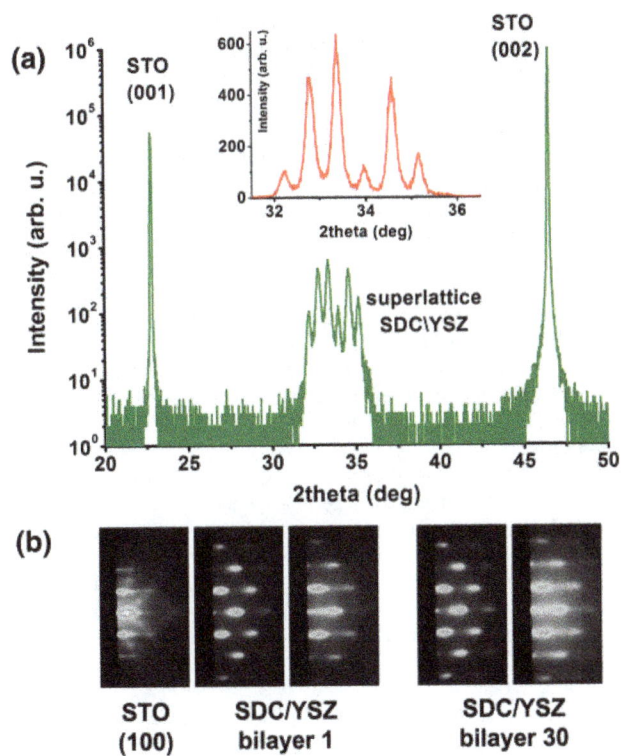

Fig. 7 a XRD analysis of an SDC/YSZ superlattice grown on (001)-oriented STO substrate. **b** RHEED patterns acquired during the growth of the superlattice along the (100) in-plane orientation of the STO substrate

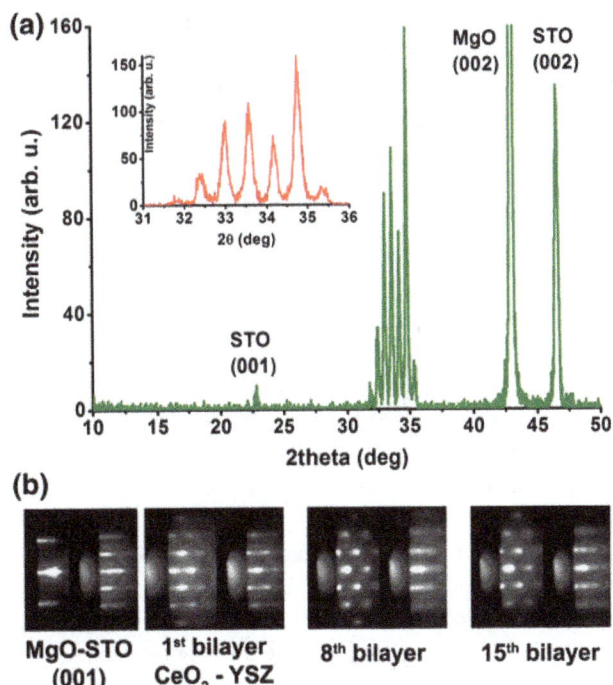

Fig. 8 a XRD analysis of a CeO$_2$/YSZ superlattice grown on (001)-oriented MgO substrate using a 5 nm thick STO buffer layer. **b** RHEED patterns acquired during the growth of the superlattice along the (100) in-plane orientation of the MgO substrate

For this work, we fabricated superlattices made of SDC and YSZ on (001)-oriented STO substrates and superlattices made of CeO$_2$ and YSZ on (001)-oriented MgO substrates. The growth of each layer was monitored by RHEED.

The SDC/YSZ superlattice consisted of 30 bilayers of the two materials with single layer thickness of about 30 unit cells, as estimated from the deposition rate measured by XRR. Figure 7a shows the XRD analysis of such a superlattice. The superlattice was epitaxially (001)-oriented with the STO substrate and the XRD plot showed the typical superlattice peaks, as shown in the inset in Fig. 7a.

The RHEED patterns of several SDC/YSZ bilayers were recorded during the growth of the superlattice. Figure 7b shows the electron reflection patterns relative to the first and the 30th final bilayers. RHEED diagnostic revealed that the two materials, SDC and YSZ, showed the two different growth mechanisms described above along this crystallographic orientation, i.e. a 3D growth for SDC and a quasi-2D growth for YSZ.

Figure 8 shows the XRD plot (Fig. 8a) and the RHEED patterns (Fig. 8b) of the CeO$_2$/YSZ superlattice that consisted of 15 CeO$_2$/YSZ bilayers grown on MgO using a thin buffer layer of STO, about 5 nm thick. According to the XRR measurement of the deposition rate, the expected

thickness of each individual layer was about 15 unit cells. The very well-defined superlattice features allowed identifying the diffraction satellite peaks up to the third order. For the fabrication of this sample, MgO was selected as deposition substrate due to its good insulating properties at high temperatures that allow the electrical characterization along the planar direction of the hetero-interfaces. The STO buffer layer provides the suitable lattice matching for the epitaxial growth [5].

An almost ideal 2D layer-by-layer growth with cube-on-cube symmetry was observed for the STO buffer layer on MgO. Also in this case, a regular array of spots characterized the RHEED patterns of all the CeO$_2$ layers, while streaky patterns were observed for all the YSZ layers.

Both Fig. 7 (SDC–YSZ on STO) and Fig. 8 (CeO$_2$–YSZ on STO-buffered MgO) show a mechanism of 2D surface reconstruction of the ceria layers determined by the YSZ layer growing above. A similar mechanism of roughness suppression induced by the deposition of a second layer was already observed during the growth of superlattice structures [27].

Conclusions

Thin films of CeO$_2$, 15 % Sm-doped CeO$_2$ (SDC), as well as 8 mol% yttria-doped zirconia (YSZ) were grown on different single crystalline substrates. RHEED diagnostic allowed identifying two distinct growth mechanisms for the ceria films along the (001) and (111) growth directions. A 3D growth mechanism, associated with a more pronounced surface roughness, clearly characterized the growth of the (001)-oriented surfaces, while a quasi-2D growth was observed for the (111)-oriented surfaces. The 3D (001)-oriented growth of CeO_2 and SDC was observed over a relatively wide range of deposition temperatures, laser energy densities, oxygen background pressures, and deposition rates, suggesting that this growth mechanism is intrinsically related with the physicochemical properties of this material. On the contrary, YSZ films showed an almost ideal 2D layer-by-layer growth along the (001) orientation, even in the presence of relatively unfavourable crystalline matching with the substrate, such as in the case of films grown on MgO.

The two distinct growth mechanisms along the (001) orientation were, to the best of our knowledge, for the first time correlated with the low index surface Gibbs free energies of the two materials, computed by first-principles DFT. Our calculation showed that both crystalline structures favour the (111) surface that strongly minimizes the Gibbs free energy, but the difference between the (001) and (111) surface energies of CeO_2 is twice that of cubic ZrO_2.

The enhanced roughness of the (001)-oriented doped or undoped CeO_2 surfaces should be taken into account in particular as far as the growth of highly textured multi-layered hetero-structures is concerned. In this work, we have shown that the ceria-based layers preserve their characteristic 3D growth mechanism also when comprised in super-structures with YSZ leading to potentially significant effects on the interface morphology. Therefore, especially for CeO_2 and SDC, much better surface (and interface) quality might be achieved by developing the expertise needed to obtain coherent film epitaxy and biaxial texture along less common crystallographic directions, for instance on pseudo-cubic (110)- or (111)-oriented surfaces.

Acknowledgments This work was partly supported by the World Premier International Research Centre Initiative of MEXT, Japan.

References

1. Esposito, V., Traversa, E.: Design of electroceramics for solid oxides fuel cell applications: playing with ceria. J. Am. Ceram. Soc. **91**, 1037–1051 (2008)
2. Wachsman, E.D., Lee, K.Y.: Lowering the temperature of solid oxide fuel cells. Science **334**, 935–939 (2011)
3. Chen, L.B.: Yttria-stabilized zirconia thermal barrier coatings—a review. Surf. Rev. Lett. **13**, 535–544 (2006)
4. Trovarelli, A.: Catalysis by ceria and related materials. In: Catalytic science series, vol. 2. World Scientific Publishing Co, Singapore (2002)
5. Sanna, S., Esposito, V., Pergolesi, D., Orsini, A., Tebano, A., Licoccia, S., Balestrino, G., Traversa, E.: Fabrication and electrochemical properties of epitaxial samarium-doped ceria films on $SrTiO_3$-buffered MgO substrates. Adv. Funct. Mater. **19**, 1713–1719 (2009)
6. Goebel, M.C., Gregori, G., Guo, X., Maier, J.: Boundary effects on the electrical conductivity of pure and doped cerium oxide thin films. Phys. Chem. Chem. Phys. **12**, 14351–14361 (2010)
7. Goebel, M.C., Gregori, G., Maier, J.: Mixed conductivity in nanocrystalline highly acceptor doped cerium oxide thin films under oxidizing conditions. Phys. Chem. Chem. Phys. **13**, 10940–10945 (2011)
8. Sillassen, M., Eklund, P., Pryds, N., Johnson, E., Helmersson, U., Bøttiger, J.: Low-temperature superionic conductivity in strained yttria-stabilized zirconia. Adv. Funct. Mater. **20**, 2071–2076 (2010)
9. Korte, C., Peters, A., Janek, J., Hesse, D., Zakharov, N.: Ionic conductivity and activation energy for oxygen ion transport in superlattices—the semicoherent multilayer system YSZ (ZrO_2 + 9.5 mol% Y_2O_3)/Y_2O_3. Phys. Chem. Chem. Phys. **10**, 4623–4635 (2008)
10. Azad, S., Marina, O.A., Wang, C.M., Saraf, L., Shutthanandan, V., McCready, D.E., El-Azab, A., Jaffe, J.E., Engelhard, M.H., Peden, C.H.F., Thevuthasan, S.: Nanoscale effects on ion conductance of layer-by-layer structures of gadolinia-doped ceria and zirconia. Appl. Phys. Lett. **86**, 1319061 (2005)
11. Korte, C., Schichtel, N., Hesse, D., Janek, J.: Influence of interface structure on mass transport in phase boundaries between different ionic materials. Monatsh. Chem. **140**, 1069–1080 (2009)
12. Schichtel, N., Korte, C., Hesse, D., Zakharov, N., Butz, B., Gerthsenc, D., Janek, J.: On the influence of strain on ion transport: microstructure and ionic conductivity of nanoscale YSZ/Sc2O3 multilayers. Phys. Chem. Chem. Phys. **12**, 14596–14608 (2010)
13. Wang, C.M., Thevuthasan, S., Peden, C.H.F.: Interface structure of an epitaxial cubic ceria film on cubic zirconia. J. Am. Ceram. Soc. **86**, 363–365 (2003)
14. Ikegawa, S., Motoi, Y.: Growth of CeO_2 thin films by metal-organic molecular beam epitaxy. Thin Solid Films **281**, 60–63 (1996)
15. Zaitsev, A.G., Ockenfuss, G., Guggi, D., Wördenweber, R., Krüger, U.: Structural perfection of (001) CeO_2 thin films on (1102) sapphire. J. Appl. Phys. **81**, 3069–3072 (1997)
16. Bera, D., Kuchibhatla, S.V.N.T., Azad, S., Saraf, L., Wang, C.M., Shutthanandan, V., Nachimuthu, P., McCready, D.E., Engelhard, M.H., Marina, O.A., Baer, D.R., Seal, S., Thevuthasan, S.: Growth and characterization of highly oriented gadolinia-doped ceria (111) thin films on zirconia (111)/sapphire (0001) substrates. Thin Solid Films **516**, 6088–6094 (2008)
17. Kurian, J., Naito, M.: Growth of epitaxial CeO_2 thin films on r-cut sapphire by molecular beam epitaxy. Physica C **492**, 31–37 (2004)
18. Nandasiri, M.I., Nachimuthu, P., Varga, T., Shutthanandan, V., Jiang, W., Kuchibhatla, S.V.N.T., Thevuthasan, S., Seal, S., Kayani, A.N.: Influence of growth rate on the epitaxial orientation and crystalline quality of CeO_2 thin films grown on Al2O3 (0001). J. Appl. Phys. **109**, 013525 (2011)
19. Manning, P.S., Sirman, J.D., De Souza, R.A., Kilner, J.A.: The kinetics of oxygen transport in 9.5 mol % yttria stabilized zirconia. Solid State Ionics **100**, 1–10 (1997)

20. Fronzi, M., Soon, A., Delley, B., Traversa, E., Stampfl, C.: Stability and morphology of cerium oxide surfaces in an oxidizing environment: a first-principles investigation. J. Chem. Phys. **131**, 104701 (2009)

21. Xia, X., Oldman, R., Catlow, R.: Computational modeling study of bulk and surface of yttria-stabilized cubic zirconia. Chem. Mater. **21**, 3576–3585 (2009)

22. Muller, P., Kern, R.: Equilibrium nano-shape change induced by epitaxial stress: effect of surface stress. Surf. Sci. **457**, 229–253 (2000)

23. Fabbri, E., Pergolesi, D., Traversa, E.: Ionic conductivity in oxide heterostructures: the role of interfaces. Sci. Technol. Adv. Mater. **11**, 054503–054512 (2010)

24. Garcia-Barriocanal, J., Rivera-Calzada, A., Varela, M., Sefrioui, Z., Iborra, E., Leon, C., Pennycook, S.J., Santamaria, J.: Colossal ionic conductivity at interfaces of epitaxial $ZrO_2:Y_2O_3/SrTiO_3$ heterostructures. Science **321**, 676–680 (2008)

25. Pergolesi, D., Tebano, A., Fabbri, E., Balestrino, G., Licoccia, S., Traversa, E.: Pulsed lased deposition of superlattices based on ceria and zirconia. ECS Trans. **35**, 1125–1130 (2010)

26. Pergolesi, D., Fabbri, E., Cook, S.N., Roddatis, V., Traversa, E., Kilner, J.A.: Tensile lattice distortion does not affect oxygen transport in yttria-stabilized zirconia-CeO_2 heterointerfaces. ACS Nano (2012).

27. Tebano, A., Balestrino, G., Lavanga, S., Martellucci, S., Medaglia, P.G., Paoletti, A., Pasquini, G., Petrocelli, G., Tucciarone, A.: Reflection high-energy electron diffraction oscillations during epitaxial growth of artificially layered films of $(BaCuO_x)m/(CaCuO_2)n$. Physica C **355**, 335–340 (2001)

Microstructure and thermoelectric properties of Y$_x$Al$_y$B$_{14}$ samples fabricated through the spark plasma sintering

Satofumi Maruyama · Toshiyuki Nishimura ·
Yuzuru Miyazaki · Kei Hayashi ·
Tsuyoshi Kajitani · Takao Mori

Abstract Excellent control in p- and n-type transport characteristics was previously obtained for the thermoelectric Y$_x$Al$_y$B$_{14}$ compounds through Al flux method. In this study, new attempts were made to reduce their grain sizes to obtain dense samples and to possibly lower the thermal conductivity. Introducing the reduction of grain sizes into Y$_x$Al$_y$B$_{14}$ samples was attempted by two methods; one was through mechanical grinding, and the other was by synthesizing Y$_x$Al$_y$B$_{14}$ via Y$_{0.56}$B$_{14}$ (denoted as "vYB-YAlB$_{14}$"). Mechanical grinding using ball milling with Si$_3$N$_4$ pots and balls was found not to be an efficient way to decrease the grain size because of contamination of Si$_3$N$_4$. In contrast, vYB-YAlB$_{14}$ samples were successfully synthesized. Through the synthesis of Y$_{0.56}$B$_{14}$, the boron network structure was first formed. Afterward, Y$_x$Al$_y$B$_{14}$ was obtained by adding Al in the boron network structure through a heat treatment. Due to shorter heating time at lower temperature, the grain sizes were discovered to be smaller than that of Al flux method. The decrease of grain size was found to be beneficial for the densification of Y$_x$Al$_y$B$_{14}$ and the decrease of its thermal conductivity.

S. Maruyama · T. Nishimura · T. Mori
International Center for Materials Nanoarchitectonics (MANA),
National Institute for Materials Science (NIMS),
1-1 Namiki, Tsukuba 305-0044, Japan

S. Maruyama · Y. Miyazaki · K. Hayashi · T. Kajitani
Department of Applied Physics, Graduate School
of Engineering, Tohoku University, 6-6-05 Aoba, Aramaki,
Aoba-ku, Sendai 980-8579, Japan

T. Mori (✉)
University of Tsukuba, 1-1-1 Tennodai, Tsukuba,
Ibaraki 305-8577, Japan
e-mail: MORI.Takao@nims.go.jp

Keywords Thermoelectric · Boride · n-type · Thermal
conductivity · Grain size

Introduction

The development of thermoelectric materials has recently been carried out with great intensity because of the possibility for useful energy conversion of waste heat [1]. Thermoelectric performance is evaluated by the dimensionless figure of merit $ZT = S^2T/\rho\kappa$, where S, ρ, κ and T are the Seebeck coefficient, the electrical resistivity, the thermal conductivity and the absolute temperature, respectively. Boron icosahedra cluster compounds are good candidates for high-temperature thermoelectric materials because they exhibit intrinsic low thermal conductivity and are stable at high temperature [2–5]. Boron carbide is one of the attractive p-type thermoelectric materials for the high-temperature region [6]. Metal doped β-boron compounds have been also investigated as possible thermoelectric materials [7, 8]. In addition, novel boron icosahedra cluster compounds like RB$_{44}$Si$_2$ (R = rare earth) [9–12], B$_6$S$_{1-x}$ [13] are being investigated. Rare earth borocarbonitrides, RB$_{22}$C$_2$N, RB$_{17}$CN and RB$_{28.5}$C$_4$, were discovered to be the first boron icosahedra cluster containing compounds that exhibit intrinsic n-type thermoelectric materials [14, 15]. Some of the recently discovered novel borides have been found not to be easy to obtain dense samples, and various studies to remedy this have been carried out; for example, mechanical grinding and sintering the sample through the spark plasma sintering (SPS) treatment with sintering aids such as metals, rare earth tetra brides and carbides [16–19].

Recently, thermoelectric properties of Y$_x$Al$_y$B$_{14}$ have been investigated [20] and Y$_x$Al$_y$B$_{14}$ was found to exhibit excellent p–n control with large absolute values of the Seebeck coefficient through the control of the Al occupancy y. To increase figure of merit, we attempt to reduce grain size by means of mechanical grinding and also by a change of synthesis method. The reduction of grain size is expected to help densification of $Y_xAl_yB_{14}$ and decrease thermal conductivity. Since $Y_xAl_yB_{14}$ is usually synthesized with high-temperature molten Al flux method [21], it is normally difficult to reduce the grain sizes through the synthesis conditions. As a result, $Y_xAl_yB_{14}$ is also difficult to densify. Therefore, we focused on the boron atomic network structure of $Y_xAl_yB_{14}$. $Y_{0.56}B_{14}$ (YB_{25}) has a similar atomic network structure and is synthesized by the borothermal reduction [22, 23]. As both compounds have similar boron atomic network structures, we aimed to sinter $Y_xAl_yB_{14}$ via $Y_{0.56}B_{14}$ which possibly can act like a precursor. In this work, we tried to decrease grain size of $Y_xAl_yB_{14}$ samples and investigate effects of the grain size on the thermoelectric properties.

Experiment

Polycrystalline samples of $Y_xAl_yB_{14}$ were synthesized using two methods: one is the high-temperature Al flux method [20, 21] with mechanical grinding using ball milling. Starting materials of YB_4, B and excess Al, serving as flux, were mixed and pressed using cold isostatic pressing. This mixture was heated around 1,633 K. After heating, sample was crushed in a Si_3N_4 mortar and excess Al was dissolved using NaOH. This sample was denoted as "$YAlB_{14}$." The $YAlB_{14}$ sample was crushed by ball mill at 200 rpm for 30 min putting balls and ethanol together into a Si_3N_4 pot. This mechanical grinded sample was denoted as "MG-$YAlB_{14}$." Another method was the synthesis of $Y_xAl_yB_{14}$ via $Y_{0.56}B_{14}$ (denoted as "vYB-$YAlB_{14}$"). To prevent grain growth through the Al flux method, we sintered $Y_{0.56}B_{14}$ that has a similar crystal structure with $Y_xAl_yB_{14}$, and then Al was added to $Y_{0.56}B_{14}$. Synthesis of $Y_{0.56}B_{14}$ has previously been reported by the borothermal reduction in vacuum as follows [22, 23];

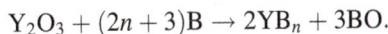

$$Y_2O_3 + (2n+3)B \rightarrow 2YB_n + 3BO.$$

After synthesis of $Y_{0.56}B_{14}$, Al was added to $Y_{0.56}B_{14}$ and the mixture was heated around 1,573 K for 4 h in vacuum. After heating, samples were crushed in a S_3N_4 mortar and excess Al was dissolved using NaOH. All samples were pressed and compacted using SPS treatment at several sintering temperature ranged from 1,673 to 1,823 K.

X-ray diffraction (XRD) measurements using Rigaku Ultima-3 with Cu Kα radiation were performed to characterize the samples and to determine the detailed crystal structure by means of Rietveld refinement using Rietan-FP

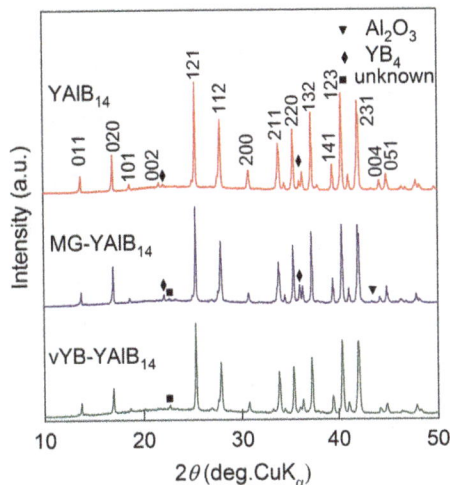

Fig. 1 XRD patterns of $YAlB_{14}$, MG-$YAlB_{14}$ and vYB-$YAlB_{14}$ after SPS treatment. Peak indexes are labeled to main peaks that are isolated or with strong intensities as some peaks are overlapped each other

software [24]. Grain sizes of samples before SPS treatment were checked using a laser diffraction particle size analyzer. The microstructures of the SPS sintered samples are observed using a scanning electron microscope (SEM). The electrical resistivity and the Seebeck coefficient were measured with an ULVAC ZEM-2 using the four-probe method and differential method, respectively. To determine the thermal conductivity, the thermal diffusivity coefficients were measured by the laser flash method and the specific heat was measured by Quantum Design, PPMS.

Result and discussion

XRD patterns of $YAlB_{14}$, MG-$YAlB_{14}$ and vYB-$YAlB_{14}$ samples after SPS treatment are shown in Fig. 1. In the XRD patterns of MG-$YAlB_{14}$ and vYB-$YAlB_{14}$ samples, a small unknown peak was observed around 22.6°. It is considered to be attributed to a very small amount of an unknown secondary phase. In the XRD pattern of Fig. 1, through the ball milling treatment, any secondary phases from the material of the pots or balls were not observed. Although a very small unknown peak was detected, the vYB-$YAlB_{14}$ samples were successfully synthesized. According to the Rietveld analysis, the composition of vYB-$YAlB_{14}$ was $Y_{0.54}Al_{0.59}B_{14}$ and it was almost the same as $YAlB_{14}$ and MG-$YAlB_{14}$ whose composition was estimated at $Y_{0.56}Al_{0.57}B_{14}$. In Fig. 1, the MG-$YAlB_{14}$ sample was observed to have a slight amount of a secondary phase of Al_2O_3. We surmise that during the mechanical grinding, the sample might have been oxidized and Al_2O_3 subsequently appears through the SPS treatment.

The dispersion of the grain sizes are shown in Fig. 2. Before mechanical grinding, the dispersion was broad and

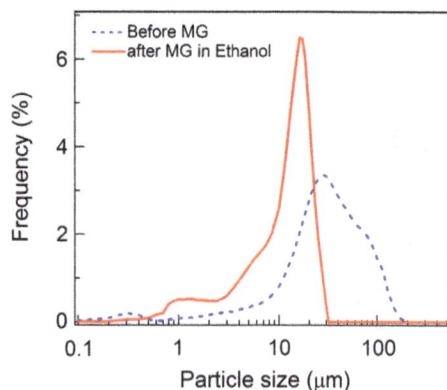

Fig. 2 Dispersion of grain sizes before and after the mechanical grinding

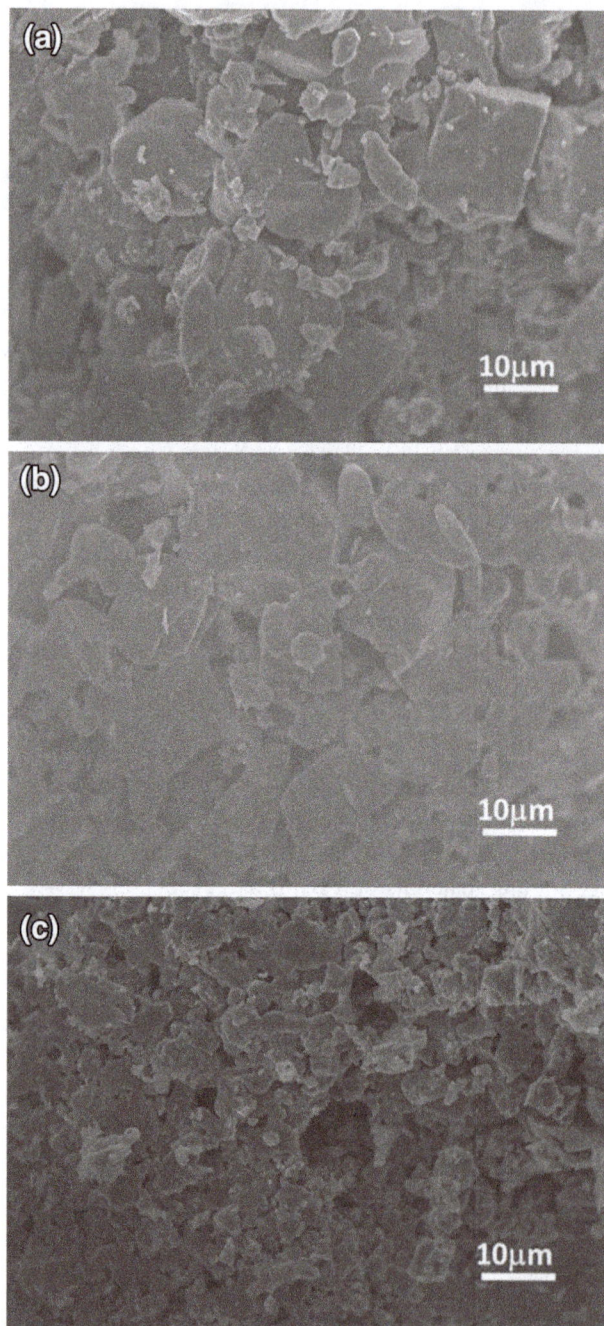

Fig. 3 SEM images of MG-YAlB$_{14}$ samples sintered at 1,773 K **a** and 1,823 K **b** and vYB-YAlB$_{14}$ sintered at 1,673 K **c** after SPS treatment

grain size of 50 % of cumulative diameter by the volume grain size distribution (d50) was 28 μm. Through the mechanical grinding treatment at 200 rpm for 30 min, the dispersion became sharper and the grain size of d50 decreased to 12 μm. To decrease grain size further, rotation speed and grinding time were increased. However, with the increase of these parameters, a secondary phase of Si$_3$N$_4$ appeared. During the SPS treatment, Si$_3$N$_4$ reacted with Y$_x$Al$_y$B$_{14}$ and the Al occupancy decreased. As the Al occupancy is strongly related to the thermoelectric properties, the Si$_3$N$_4$ contamination should be avoided. Thus, we could not increase the rotation speed and grinding time further.

SEM images of SPS sintered MG-YAlB$_{14}$ and vYB-YAlB$_{14}$ samples are shown in Fig. 3. MG-YAlB$_{14}$ samples in Fig. 3a, b are sintered at 1,773 and 1,823 K under pressure of 150 MPa, respectively. Although the observed grain sizes in Fig. 3a, b were almost same size around 10 μm, grain boundaries disappeared in Fig. 3b and the sintering was largely promoted by the increase of the SPS sintering temperature. This implies that higher sintering temperature leads to densify the samples.

The SEM image of vYB-YAlB$_{14}$ sample after SPS treatment is shown in Fig. 3c. There can be observed some agglomeration of grains and also pores. Through the SPS treatment, the agglomeration was sintered preferentially. To achieve further dense and homogeneous sample, agglomeration should be avoided. Interestingly, the observed average grain size was around 5 μm, which was smaller than those of MG-YAlB$_{14}$ samples. This result can be explained by considering the synthesis route. In the synthesis of the YAlB$_{14}$, we used the Al flux, which was assumed to promote grain growth. In fact, the grain size of d50 of the YAlB$_{14}$ before mechanical grinding was 28 μm as shown in Fig. 2. Even after the mechanical grinding, the grain size of d50 was 12 μm. However, in the synthesis method of the vYB-YAlB$_{14}$ sample, the boron network structure was first formed through the synthesis of Y$_{0.56}$B$_{14}$, and then afterward, Al was added to the boron network structure through a heat treatment. Since the heat treatment was of shorter time and lower temperature comparing to the Al flux method, we surmise that grain growth of vYB-YAlB$_{14}$ sample was suppressed. Although we could not measure dispersion of the grain sizes of vYB-YAlB$_{14}$ due to the agglomeration, the grain size of the

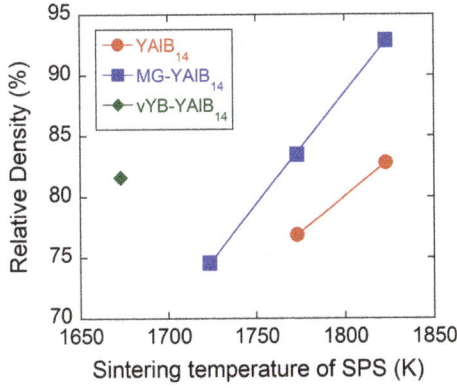

Fig. 4 SPS temperature dependence of the relative density of $YAlB_{14}$, MG-$YAlB_{14}$ and vYB-$YAlB_{14}$ samples

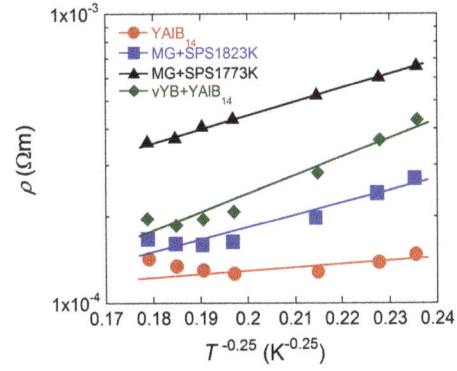

Fig. 5 Temperature dependence of the electrical resistivity of $YAlB_{14}$, MG-$YAlB_{14}$ and vYB-$YAlB_{14}$ samples. The logarithmic of ρ is plotted versus $T^{-0.25}$. The *lines* are guides to the eye

vYB-$YAlB_{14}$ seemed to be smaller than that of the $YAlB_{14}$ sample.

The relationship between relative densities of samples and SPS sintering temperature is plotted in Fig. 4. The relative density of the $YAlB_{14}$ samples was increased with increasing the SPS sintering temperature and reached 83 % at 1,823 K without special treatment for reducing grain sizes. In MG-$YAlB_{14}$ samples, relative densities were also largely increased with the increase of the SPS sintering temperature. In addition, compared to the $YAlB_{14}$ sample, sample density was largely increasing from 83 to 93 % at 1,823 K. For the vYB-$YAlB_{14}$ sample, relative density was 82 % even for the lower SPS sintering temperature at 1,673 K and almost as same relative density as $YAlB_{14}$ sample sintered at 1,823 K. With the decrease of the grain sizes, relative densities tend to increase and sintering temperature tend to become lower, due to the increase of the specific surface area of the samples. Therefore, from our results, we can conclude that the decrease of the grain size leads to lower the SPS temperature to obtain dense $Y_xAl_yB_{14}$ samples.

The temperature dependence of electrical resistivity of samples is shown in Fig. 5. The electrical resistivity of $Y_xAl_yB_{14}$ was previously observed [20] to follow Mott's variable range hopping mechanism, $\log\rho \propto T^{-0.25}$ [25, 26]. The temperature dependences of electrical resistivities observed to not follow this dependency and show curvatures due to the existence of grain boundaries, pores and secondary phases. The electrical resistivities of MG-$YAlB_{14}$ samples were higher than that of $YAlB_{14}$ sample in spite of the higher relative density. The MG-$YAlB_{14}$ samples seemed to be oxidized, and this can be considered as one of the reasons for the higher electrical resistivity. The electrical resistivity of the vYB-$YAlB_{14}$ sample was also increasing. According to the SEM image of vYB-$YAlB_{14}$ sample in Fig. 3c, grain boundaries and pores can be observed and they increase the electrical resistivity.

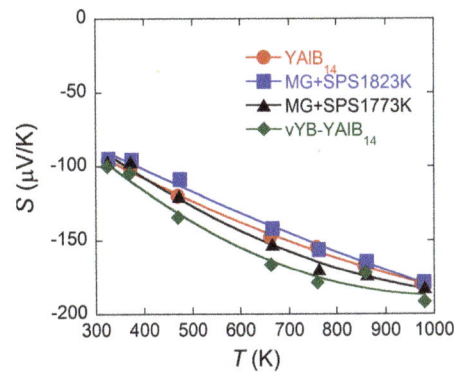

Fig. 6 Temperature dependence of the Seebeck coefficient of $YAlB_{14}$, MG-$YAlB_{14}$ and vYB-$YAlB_{14}$ samples

The temperature dependence of the Seebeck coefficient is shown in Fig. 6. The Seebeck coefficient in variable range hopping has been investigated, for example, by Zvyagin [27]. Assuming linear density of state, the Seebeck coefficient has the following dependency:

$$S \propto (T_0 T)^{1/2} d(\ln D(E))/dE|_{E_F}$$

The Seebeck coefficients of samples also tended to be proportional to $T^{1/2}$ and exhibited around -190 $\mu V/K$ at 1,000 K. Previously, we found the Al occupancies of samples relate with T_0 values [20]. Since the Al occupancies were almost the same value between samples here, little difference of the temperature dependence of the Seebeck coefficient was observed. Although the Seebeck coefficient can be increased through the reduction of the grain size [28, 29], we conclude the Seebeck coefficient is not changed by the reduction of grain sizes among samples obtained in this study.

Thermal conductivities of all samples are plotted in Fig. 7. The thermal conductivity of $YAlB_{14}$ sample was decreasing with the increase of the temperature and exhibited 3.6 W/mK at 1000 K. The thermal conductivity

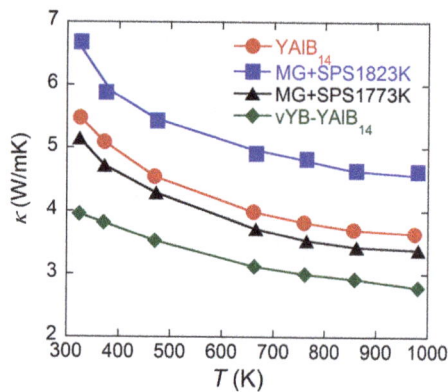

Fig. 7 Temperature dependence of the thermal conductivity of YAlB$_{14}$, MG-YAlB$_{14}$ and vYB-YAlB$_{14}$ samples

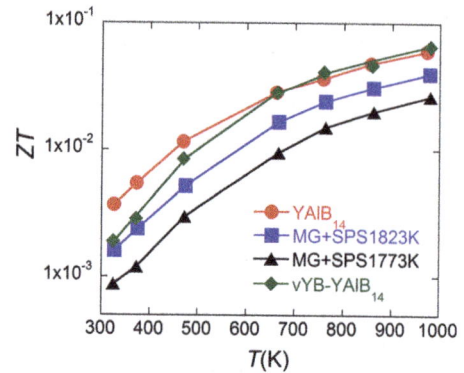

Fig. 8 Temperature dependence of the figure of merits of YAlB$_{14}$, MG-YAlB$_{14}$ and vYB-YAlB$_{14}$ samples

of the MG-YAlB$_{14}$ sample sintered at 1,773 K slightly decreased compared to YAlB$_{14}$ due to the decrease of grain sizes, i.e., the increase of the grain boundary. It leads to increase the phonon scattering and to decrease the thermal diffusivity. On the other hand, the thermal conductivity of the MG-YAlB$_{14}$ sample sintered at 1,823 K was higher than that of YAlB$_{14}$ due to the disappearance of grain boundaries. The thermal conductivity of vYB-YAlB$_{14}$ sample decreased with the increase of the temperature and exhibited 2.8 W/mK at 1,000 K. The thermal conductivity of vYB-YAlB$_{14}$ was the lowest among the samples although the relative density is the same as YAlB$_{14}$ and MG-YAlB$_{14}$ sintered at 1,773 K. The observed decrease of the thermal conductivity is caused by the large reduction in the lattice contribution. Considering both the SEM image and the result of thermal conductivity measurement, we can conclude that the synthesis method of the vYB-YAlB$_{14}$ sample is most effective to increase phonon scattering and to decrease thermal conductivity.

Temperature dependence of ZT value is plotted in Fig. 8. The ZT value of YAlB$_{14}$ sample increased with the increase of the temperature and exhibited 0.060 at 1000 K. ZT values of the MG-YAlB$_{14}$ samples decreased relative to YAlB$_{14}$ due to the decrease of the electrical resistivity, while the thermal conductivity did not decrease largely, because grain sizes were not decreased sufficiently through the mechanical grinding. The ZT value of vYB-YAlB$_{14}$ was slightly increasing in high-temperature range and exhibited 0.066 at 1000 K, because the thermal conductivity was decreased. However, as the electrical resistivity of vYB-YAlB$_{14}$ sample increased compared to YAlB$_{14}$ sample, ZT value of vYB-YAlB$_{14}$ was not enhanced so much. As we wrote in the parts of the results of the SEM and the electrical resistivity measurements, grain boundaries and pores can be observed in vYB-YAlB$_{14}$ sample. Due to these reasons, the electrical resistivity of vYB-YAlB$_{14}$ increased. Although the increase of ZT is within

the general error, since the values are those experimentally obtained, the difference can be considered to represent a tendency, at least indicating that the grain size control approach is promising. Although ZT values of vYB-YAlB$_{14}$ were still low due to the low density, further improvements are expected through the optimization of the synthesis condition.

Conclusion

In this study, the reduction of grain sizes was introduced to Y$_x$Al$_y$B$_{14}$ samples by two methods. One was the mechanical grinding using ball milling with Si$_3$N$_4$ pots and balls. However, it was found not to be an efficient way to reduce the grain size because of the contamination of Si$_3$N$_4$. Thermoelectric performance of mechanical grinded sample (MG-YAlB$_{14}$) was found not to be improved due to the secondary phase and the insufficiency of the decrease of the grain size. The other was the synthesis of Y$_x$Al$_y$B$_{14}$ sample via Y$_{0.56}$B$_{14}$ (vYB-YAlB$_{14}$). The vYB-YAlB$_{14}$ sample was successfully synthesized, and the grain sizes were discovered to be smaller than that of Al flux method. Considering both MG-YAlB$_{14}$ and vYB-YAlB$_{14}$, it was found that through the reduction of grain sizes, the relative densities after SPS treatment tend to increase at the same SPS sintering temperature. Thermal conductivity of vYB-YAlB$_{14}$ sample was discovered to be the lowest among the samples thanks to the small grain size, i.e., the increase of the grain boundaries. As a result, ZT value of vYB-YAlB$_{14}$ sample showed a slight enhancement in the high-temperature region. With further reduction of grain sizes through improvement of the synthesis process of the vYB-AlB$_{14}$ and optimization of SPS conditions to consolidate, further improvements are expected.

Acknowledgments TM was partially supported by a grant from AOARD.

References

1. Koumoto, K., Mori, T.: Thermoelectric Nanomaterials; Materials Design and Applications, Springer Series in Materials Science, vol. 182. Springer, Heidelberg (2013)

2. Mori, T.: Higher borides. In: Gschneidner Jr, K.A., Bunzli, J.-C., Pecharsky, V. (eds.) Handbook on the Physics and Chemistry of Rare-Earths, vol. 38, p. 105. North-Holland, Amsterdam (2008)

3. Slack, G.A., Oliver, D.W., Horn, F.H.: Thermal conductivity of boron and some boron compounds. Phys. Rev. B **4**, 1714–1720 (1971)

4. Cahill, D.G., Fischer, H.E., Watson, S.K., Pohl, R.O.: Thermal properties of boron and boride. Phys. Rev. B **40**, 3254–3260 (1989)

5. Mori, T.: Thermal conductivity of a rare-earth B12-icosahedral compound. Phys. B **383**, 120–121 (2006)

6. Wood, C., Emin, D.: Conduction mechanism in boron carbide. Phys. Rev. B **29**, 4582–4587 (1984)

7. Nakayama, T., Shimizu, J., Kimura, K.: Thermoelectric properties of metal-doped β-rhombohedral boron. J. Solid State Chem. **154**, 13–19 (2000)

8. Kim, H., Kimura, K.: Vanadium concentration dependence of thermoelectric properties of β-rhombohedral boron prepared by spark plasma sintering. Mater. Trans. **52**, 41–48 (2011)

9. Mori, T.: High temperature thermoelectric properties of B_{12} icosahedral cluster-containing rare earth boride crystal. J. Appl. Phys. **97**, 093703 (2005)

10. Mori, T., Martin, J., Nolas, G.: Thermal conductivity of $YbB_{44}Si_2$. J. Appl. Phys. **102**, 073510 (2007)

11. Mori, T., Berthebaud, D., Nishimura, T., Nomura, A., Shishido, T., Nakajima, K.: Effect of Zn doping on improving crystal quality and thermoelectric properties of borosilicides. Dalton Trans. **39**, 1027–1030 (2010)

12. Berthebaud, D., Sato, A., Michiue, Y., Mori, T., Nomura, A., Shishido, T., Nakajima, K.: Effect of transition element doping on crystal structure of rare earth borosilicides $REB_{44}Si_2$. J. Solid State Chem. **184**, 1682–1687 (2011)

13. Sologub, O., Matsusita, Y., Mori, T.: An α-rhombohedral boron related compound with sulfur: synthesis structure and thermoelectric properties. Scripta Mater. **68**, 289–292 (2013)

14. Mori, T., Nishimura, T.: Thermoelectric properties of homologous p- and n-type boron-rich borides. J. Solid State Chem. **179**, 2908–2915 (2006)

15. Mori, T., Nishimura, T., Yamaura, K., Takayama-Muromachi, E.: High temperature thermoelectric properties of a homologous series of n-type boron icosahedra compounds: a possible counterpart to p-type boron carbide. J. Appl. Phys. **101**, 093714 (2007)

16. Berthebaud, D., Nishimura, T., Mori, T.: Thermoelectric properties and spark plasma sintering of doped $YB_{22}C_2N$. J. Mater. Res. **25**, 665–669 (2010)

17. Berthebaud, D., Nishimura, T., Mori, T.: Microstructure and thermoelectric properties of dense $YB_{22}C_2N$ samples fabricated through spark plasma sintering. J. Electron. Mater. **40**, 682–686 (2011)

18. Prytuliak, A., Mori, T.: Effect of transition-metal additives on thermoelectric properties of $YB_{22}C_2N$. J. Electron. Mater. **40**, 920–925 (2011)

19. Prytuliak, A., Maruyama, S., Mori, T.: Anomalous effect of vanadium boride seeding on thermoelectric properties of $YB_{22}C_2N$. Mater. Res. Bull. **48**, 1972–1977 (2013)

20. Maruyama, S., Miyazaki, Y., Hayashi, K., Kajitani, T., Mori, T.: Excellent p-n control in a high temperature thermoelectric boride. Appl. Phys. Lett. **101**, 152101 (2012)

21. Korsukova, M.M., Lundstrom, T., Tergenius, L.-E., Gurin, V.N.: The crystal structure of defective $YAlB_{14}$ and $ErAlB_{14}$. J. Alloy Compd. **187**, 39–47 (1992)

22. Tanaka, T., Okada, S., Yu, Y., Ishizawa, Y.: A new yttrium boride: YB_{25}. J. Solid State Chem. **133**, 122–124 (1997)

23. Mori, T., Zhang, F.X., Tanaka, T.: Synthesis and magnetic properties of binary boride REB_{25} compounds. J. Phys.: Condens. Matter **13**, L423–L430 (2001)

24. Izumi, F., Momma, K.: Three-dimensional visualization in powder diffraction. Solid State Phenom. **130**, 15–20 (2007)

25. Efros, A.L., Pollak, M.: Electron–Electron Interactions in Disordered Systems. North-Holland, Amsterdam (1985)

26. Mott, N.F.: Conduction in glasses containing transition metal ions. J. Non-Cryst. Solids **1**, 1–17 (1968)

27. Zvyagin, I.P.: On the theory of hopping transport in disordered semiconductors. Phys. Status Solidi B **58**, 443–449 (1973)

28. Kishimoto, K., Yamamoto, K., Koyanagi, T.: Influences of potential barrier scattering on the thermoelectric properties of sintered n-type PbTe with a small grain size. Jpn. J. Appl. Phys. **42**, 501–508 (2003)

29. Heremans, J.P., Thrush, C.M., Morelli, D.T.: Thermopower enhancement in lead telluride nanostructures. Phys. Rev. B **70**, 115334 (2004)

6

Enhanced photocatalytic hydrogen production from water–methanol mixture using cerium and nonmetals (B/C/N/S) co-doped titanium dioxide

N. Vinothkumar · Mahuya De

Abstract In the present study, photocatalytic hydrogen production from water/methanol solution was investigated over cerium and nonmetal (B/C/N/S) co-doped titanium dioxide catalyst under visible light irradiation. The cerium and nonmetal co-doped titania photocatalysts were prepared by co-precipitation and characterized by surface area and pore size analysis, X-ray diffraction analysis, diffuse reflectance UV–Vis spectroscopy analysis, and photoluminescence analysis. The UV–Visible spectra showed that incorporation of cerium and nonmetals to TiO_2 resulted in narrow band gap and improved absorption of visible light. The band gap energy of co-doped samples depended on the properties of nonmetals. Photoluminescence studies showed that the radiative recombination rates of photo-generated electron–hole pairs were effectively suppressed by the addition of cerium and nonmetals and contributed to higher activity. The highest hydrogen production of 206 µmol/h was obtained for Ce–N–TiO_2 sample, which can be attributed to the higher surface area, higher absorption of visible light, and higher separation efficiency of electron–hole pairs in Ce–N–TiO_2.

Keywords Photocatalysts · Hydrogen · Visible light · Cerium · Nonmetals

Introduction

The energy economy of the world mainly depends on the fossil fuels such as coal, natural gas, and petroleum products. Gradual depletion of fossil fuels and fast-growing energy demand has necessitated to develop alternative fuels, which should also be pollution-free, storable, and economical [1]. Hydrogen is considered as a promising alternative fuel for the future. It has higher energy content and high heating value compared to other fuels such as methane, methanol, gasoline, and diesel [2]. At present, hydrogen is mainly produced by steam reforming of methane or naphtha. Various alternative processes such as electrolysis, thermolysis, thermochemical reactions, photolysis, photoelectro-chemical, photocatalytic, and biochemical processes are being studied for the production of hydrogen. But only few of these methods are efficient and economically viable [3]. The production of hydrogen from water using solar radiation is one of the potential routes to achieve clean, low-cost, and eco-friendly fuel. Several oxide materials, such as TiO_2, Fe_2O_3, ZnO, ZrO_2, $K_4Nb_6O_{17}$, $K_2La_2Ti_3O_{10}$, $BaTi_4O_9$, and Ta_2O_5, have been studied for the photodecomposition of water [4–8]. TiO_2 has been reported to be an excellent photocatalyst for water-splitting reaction. Advantages of TiO_2 include chemical inertness, photostability, and low cost. However, the activity of TiO_2 is limited in UV region because of its wide band gap (3.2 eV). The visible light ($\lambda > 400$ nm) constitutes the major fraction of solar spectrum reaching the earth surface. For better utilization of the sunlight, it is essential to develop stable photocatalytic systems with suitable band gap to absorb most of the solar spectrum. In order to develop an efficient titania-based visible light active photocatalyst, various methods have been reported such as doping with metals and nonmetals, dye

N. Vinothkumar · M. De (✉)
Department of Chemical Engineering, Indian Institute
of Technology, Guwahati 781 039, Assam, India
e-mail: mahuya@iitg.ernet.in

sensitization as well as coupling with other semiconductors [9]. The dopant impurities create additional energy levels within the band gap of titania acting either as donor or acceptor level. An extensive review has been done on visible light activity of titania by Pelaez et al. [10]. Photocatalytic activities of TiO_2 doped with rare earth metals and nonmetals are mostly reported for the degradation of pollutants. Xu et al. [11] investigated various rare earth-doped TiO_2 samples for the degradation of nitrite and observed an enhanced photoabsorption in the visible light region. Ce-doped TiO_2 materials were also reported to show high activity under visible light irradiation for photocatalytic degradation of dye and phenol derivatives [12–14]. The $N–TiO_2$ [15], and $S–TiO_2$ [16] were studied for the photodegradations of methylene blue under UV/Visible light irradiation, while $C–TiO_2$ [17] was reported to be active for the degradation of phenol. The titania co-doped with rare earth metals and nitrogen are also reported to show an enhanced activity for the decomposition of pollutants [18, 19]. Though doped TiO_2 catalysts were extensively investigated for the degradation of pollutants, comparatively fewer studies have been reported for photocatalytic hydrogen production [20–22]. In this study, the photocatalytic activities of TiO_2 co-doped with cerium and nonmetals were studied for hydrogen evolution from water–methanol mixture under visible light irradiation. The effect of various nonmetals such as B, C, N, and S on hydrogen production was studied. The physical and optical properties of the photocatalysts affecting the activity were investigated using various characterization techniques such as surface and pore analysis, XRD, UV–Vis, and photoluminescence spectroscopy.

Experimental

Preparation of photocatalyst

Cerium and nonmetal co-doped TiO_2 samples were prepared by co-precipitation method. Titanium tetra isopropoxide (TTIP) and cerium nitrate hexahydrate were used as precursors for titania and ceria, respectively, and were obtained from Sigma-Aldrich. The boric acid, glucose, urea, and thiourea were used as sources for boron, carbon, nitrogen, and sulfur, respectively, and were procured from Merck. The isopropanol and ammonia solution were also procured from Merck.

Requisite amount of TTIP was mixed with isopropanol under continuous stirring condition for 10 min to form solution A. Solution B was prepared by dissolving the required amount of cerium nitrate hexahydrate and respective precursors of nonmetals in deionized water. The loadings in co-doped samples for Ce and nonmetals were 2.5 and 1 mol% of titanium, respectively. The loadings of Ce (x) and nonmetals (y) are defined as

$$x = \frac{Ce(mol)}{Ti(mol)} \times 100 = 2.5 \text{ mol}\% \quad \text{and}$$

$$y = \frac{NM(mol)}{Ti(mol)} \times 100 = 1 \text{ mol}\%.$$

Aqueous ammonia (25 wt%) and solution B were simultaneously added dropwise to the solution A with continuous stirring. Addition of ammonia was continued until complete precipitation occurred. The resultant mixture was stirred for 2 h at room temperature and was placed in a water bath at 363 K for overnight. The precipitate was filtered and dried in an oven at 373 K for 24 h. The as-prepared samples were calcined at 723 K for 3 h. In text, nonmetal co-doped samples are referred as Ce–NM–TiO_2 (NM = B, C, N, S). For reference, cerium-doped titania were prepared by dissolving required amount of cerium nitrate hexahydrate in deionized water (solution B). The amount of Ce was 2.5 mol% of Ti, and sample is represented as Ce–TiO_2 in text. The solution B was added to solution A along with aqueous ammonia as described earlier to obtain cerium-doped titania samples. Similar aging condition and calcination temperature were maintained for the preparation of Ce–TiO_2 sample. Undoped titania was also prepared for reference by similar procedure without the addition of cerium or nonmetals and represented by TiO_2 in text.

Characterization of photocatalyst

The surface areas of the samples were determined from N_2 isotherms collected at 77 K using a Beckman Coulter SA^{TM} 3100 analyzer. Prior to the experiments, the samples were degassed at 423 K for a period of 2 h. The pore size distribution of the photocatalysts was determined by Barrett–Joyner–Halenda (BJH) method. The X-ray diffraction (XRD) patterns were recorded on a Bruker D2 phaser X-ray diffractometer using Ni-filtered Cu K_α as radiation source ($\lambda = 1.5406$ Å) in the 2θ range from 20 to 80° at a scanning rate of 1°/min. A beam voltage and beam current of 40 kV and 40 mA were used, respectively. The diffuse reflectance UV–vis spectra of the photocatalysts were measured on a PerkinElmer Lambda 750 instrument with a 60-mm labsphere specular reflectance accessory at room temperature. The baseline correction was done using a calibrated sample of barium sulfate as reference. The scan parameter was set with slit size 2 nm and scan in the spectral range of 250–650 nm. The photoluminescence (PL) measurements were taken in a Thermo Spectronic (Aminco Bowman Series 2) instrument with a Xe lamp as the excitation source at room temperature. The powders were dispersed in ethanol, and the emission spectra were collected at an excitation wavelength of 325 nm. The same quantity was used for recording the PL spectra of all the samples. The entrance and exit slit widths were fixed as same for all the measurements. Field emission scanning electron microscopic analysis was done using instrument of ZEISS-Sigma. For FESEM analysis, the samples were dispersed in a solvent and deposited in an aluminum foil that was mounted on a sample holder for gold coating. The composition of the samples was determined by EDS analysis (energy-dispersive X-ray spectroscopy) equipped with SEM instrument (LEO 1430 VP).

Photocatalytic activity studies

The activities of the photocatalysts were examined using the experimental setup shown in Fig. 1. The reaction was typically carried out by adding 0.2 g catalyst to solution of water (25 ml) and methanol (1 ml) under stirred condition. Prior to irradiation, the reaction mixture was de-aerated with N_2 gas (50 ml/min) for 30 min to completely remove the dissolved oxygen. Then, the reaction mixture was irradiated with a 500-W tungsten halogen lamp (Halonix, India), placed approximately 15 cm away from the reactor, as source of visible light. The emission spectrum of the lamp was measured in front of the reactor using Ocean optics USB4000 spectrometer as shown in Figure S1 in the

supplementary information. The temperature of the reaction mixture was maintained at 307 K using a water circulation bath, which also acted as an IR filter. The evolved gas was collected in an inverted burette by water displacement method. The gas mixture was analyzed using a gas chromatograph (Varian, CP 3800) equipped with carbosieve SII column and thermal conductivity detector. After completion of the photocatalytic reaction, the liquid solution was analyzed for the detection of intermediates (formaldehyde and formic acid) by HPLC (SHIMADZU, C18 column). To confirm the photocatalytic activity of the catalysts, experiments were conducted first in the absence of light with catalysts and next in the presence of light without catalyst. In both cases, no hydrogen evolution was observed. The hydrogen evolution was observed only when the reaction was carried out in the presence of both light and catalyst. In all cases, same feed mixture was used. This confirmed the photocatalytic activity of the catalysts.

Results and discussion

Characterization of prepared catalysts

The actual composition of the prepared photocatalysts was determined by EDS, and the results are shown in Table S1

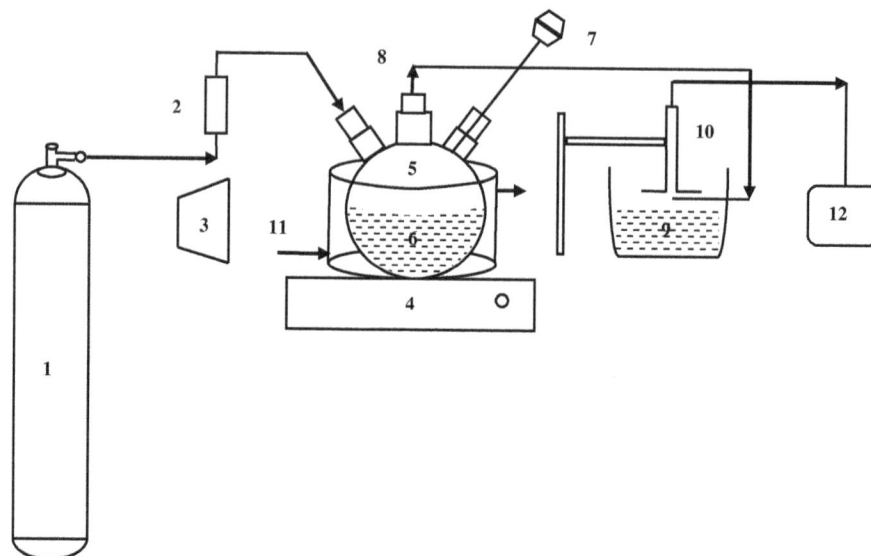

Fig. 1 Schematic representation of experimental setup for photocatalytic water splitting

1. N_2 gas
2. Flowmeter
3. Halogen lamp
4. Magnetic stirrer
5. Reactor
6. Reactant mixture
7. Thermocouple
8. Product gas outlet
9. Water bath
10. Gas collector
11. Cooling water inlet
12. Gas Chromatograph

Table 1 Physicochemical and crystalline properties of undoped TiO_2, Ce–TiO_2, and Ce–NM–TiO_2 (NM = B, C, N, S) samples

Samples	BET surface area (m^2/g)	Pore volume (cc/g)	Lattice parameters Anatase		Lattice distortion	d spacing (Å)
			a (Å) (101)	c (Å) (004)		
TiO_2	84	0.45	3.7435	9.4217	0.378	3.4790
Ce–TiO_2	70	0.36	3.7451	9.4599	0.268	3.4822
Ce–B–TiO_2	87	0.57	3.7907	9.5197	0.271	3.5128
Ce–C–TiO_2	53	0.67	3.7659	9.5080	0.387	3.5013
Ce–N–TiO_2	158	0.06	3.7936	9.7181	0.301	3.2240
Ce–S–TiO_2	39	0.18	3.7727	9.7014	1.391	3.5174

of supplementary information. The actual cerium content in samples varied in the range of 2.39–2.49 mol%, and nonmetal content varied in the range of 0.8–0.97 mol% (Table S2 of supplementary information). Thus, the actual composition agreed well with the intended composition of the photocatalysts within experimental error.

The surface area and pore volumes of undoped TiO_2, Ce–TiO_2, Ce–NM–TiO_2 (NM = B, C, N, S) samples are shown in Table 1. Surface area and pore volume of titania decreased with the addition of ceria. Structure modification and partial pore blockage in the presence of cerium may be responsible for decrease in surface area and pore volume in cerium-doped samples. The isotherm of Ce–TiO_2 and corresponding pore size distribution are shown in Fig. 2a. The Ce–TiO_2 sample exhibited a type IV isotherm with a H1-type hysteresis loop associated with open-ended cylindrical pores. The pore size distribution of Ce–TiO_2 showed the presence of both mesopores and macropores with pore diameter in the range of 3–148 nm. Co-doping of titania with nonmetals along with cerium resulted in increase in surface area as shown in Table 1. The co-doped Ce–N–TiO_2, Ce–C–TiO_2, and Ce–B–TiO_2 exhibited Type II isotherm with H3 hysteresis loop corresponding to slit-shaped pores (Fig. 2b). For sulfur co-doped sample, H1 hysteresis loop was observed, which indicated the presence of open cylindrical pores. Figure 2b shows that the pore size distribution varied significantly with nonmetals. For Ce–N–TiO_2 and Ce–B–TiO_2, narrow pore distribution was obtained, with 39 and 47 % pores being below 10 nm, respectively. The pore size distributions of Ce–C–TiO_2 and Ce–S–TiO_2 co-doped samples were much broader. The presence of significant amount of pores in the range of 20–50 nm may have contributed to their lower surface area. The surface area and pore size distribution results showed that the presence of nonmetal had significant effect on the stabilization of porous network. The maximum effect was observed for Ce–N–TiO_2 with highest surface area.

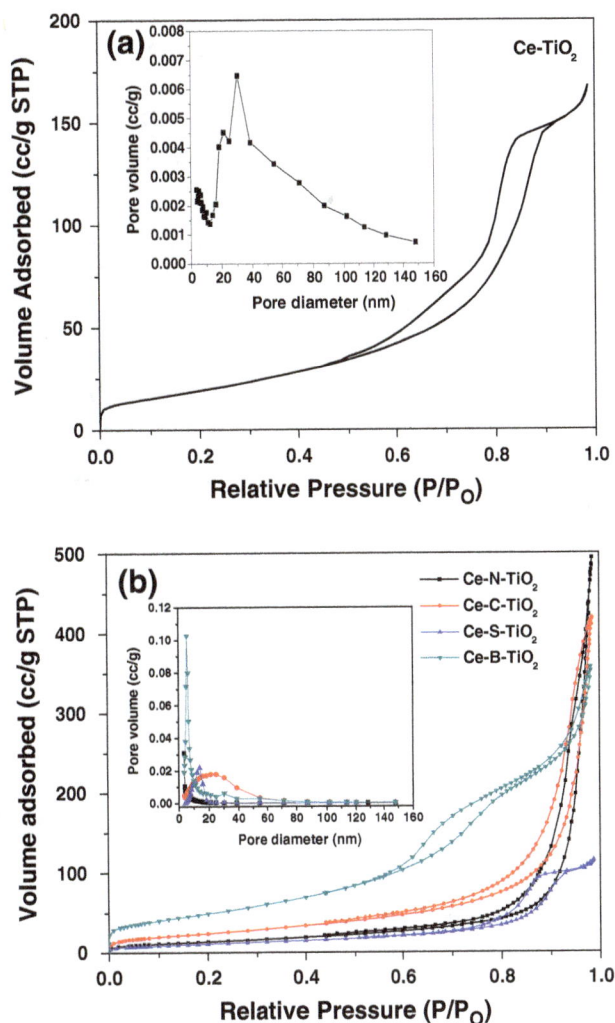

Fig. 2 N_2 adsorption–desorption isotherms of Ce–TiO_2 and Ce–NM–TiO_2 (NM = B, C, N, S) samples (*inset* pore size distribution)

The X-ray diffraction patterns of Ce–TiO_2 and Ce–NM–TiO_2 are shown in Fig. 3. All the samples, except Ce–N–TiO_2, were dominated by anatase phase ($2\theta = 25.5°$, 37°, 48°, 53.8°, 55.1°, 64°, JCPDS 21-1272). For Ce–N–TiO_2, significant amount of rutile phase ($2\theta = 28°$, 36.1°, 41.5°, JCPDS 21-1276) was observed. In general, the phase transformation of anatase to rutile occurs at temperature above 873 K and anatase phase is expected to be dominant in samples calcined at lower temperature of 723 K, used in this study. Accordingly, anatase phase was dominant in XRD profiles of all samples except Ce–N–TiO_2. For Ce–N–TiO_2 sample, the presence of significant amount of rutile phase suggests that nitrogen can catalyzes the anatase–rutile phase transformation significantly at lower temperature [23]. In addition, Ce–S–TiO_2 co-doped sample was more amorphous in nature as observed from broad XRD peaks of low intensity compared to that of other nonmetal-doped samples. No characteristic peaks of CeO_2

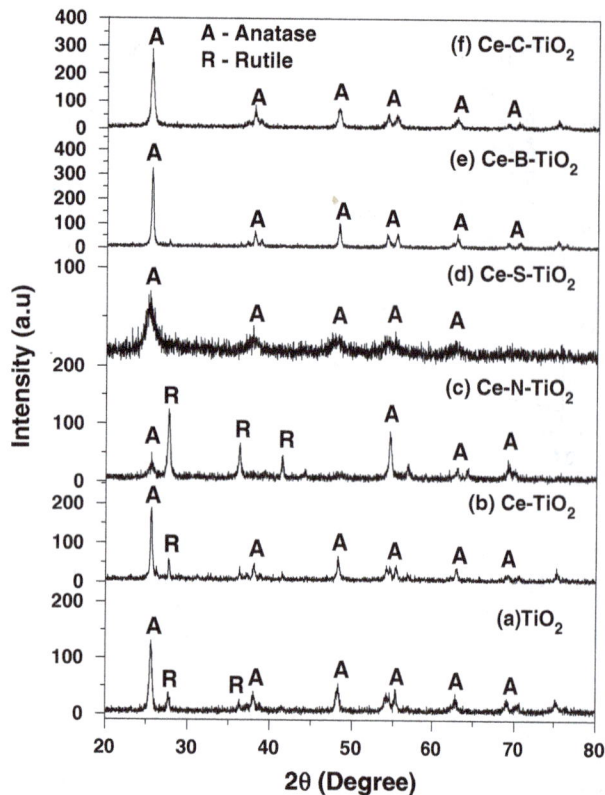

Fig. 3 XRD patterns of TiO$_2$, Ce–TiO$_2$, and Ce–NM–TiO$_2$ (NM = B, C, N, S) samples

phases were observed in the XRD profiles, which suggest that the ceria was well dispersed in TiO$_2$ matrix or were below detection limit. In addition, no peaks due to any Ce–Ti mixed oxides were observed in any of the samples.

On addition of Ce to TiO$_2$, cerium ions can substitute Ti^{4+} sites or can occupy the interstitial sites with Ce–O–Ti linkages [24]. But this is expected to be associated with distortion in titania lattice due to larger size of Ce^{4+} (IR = 0.101 nm) and Ce^{3+} (IR = 0.111 nm) compared to Ti^{4+} (IR = 0.068 nm) ions. The lattice distortion values are shown in Table 1 and correspond to maximum strain observed in crystal lattice due to distortion by incorporation of dopant metals. The lattice distortion was calculated using the formula defined by Stokes and Wilson as $\varepsilon = \beta/(4 \tan \theta)$, where β is the full width at half maximum of diffracted peaks and θ is the Bragg angle of the [h k l] reflection [25]. Maximum value of lattice distortion was obtained, as expected for Ce–S–TiO$_2$ sample, since sulfur has largest ionic radii among the nonmetals used. The corresponding change in lattice parameters for tetragonal titania lattice is also included in Table 1. With the addition of Ce in TiO$_2$, due to the lattice distortion, the lattice parameters increased compared to that of undoped titania. Co-addition of nonmetals resulted in further increase in lattice parameter values.

The FESEM images of undoped TiO$_2$, Ce–TiO$_2$, Ce–NM–TiO$_2$ are shown in Fig. 4. The images showed a definite change in morphology of the samples when cerium and nitrogen were added to titania suggesting that the dopants played a prominent role in the development of material structure during co-precipitation. This resulted in different physical properties as was observed in Table 1. Particles of irregular shape and size were observed for undoped titania, whereas ceria-doped titania, prepared by co-precipitation, consisted of particles of regular size and shape. The particles were spherical in the range of 100–200 nm. The morphology again changed when titania was co-doped with cerium and nonmetals. All the co-doped Ce–NM–TiO$_2$ samples were agglomerated in nature.

The UV–Visible absorption spectra of TiO$_2$, Ce–TiO$_2$, and Ce–NM–TiO$_2$ are shown in Fig. 5. The Ce–TiO$_2$ sample showed intense absorption bands in the visible light region ranging from 400 to 500 nm. The band gap energies of the samples were determined by extrapolating the rising part of the onset of the absorption edge to the x-axis (λ, nm). The values of λ was then used in the Planck's Einstein equation to calculate the band gap, $E_g = hc/\lambda$, where E_g is band gap energy in eV, h is Planck's constant in eV. s, c is the speed of light in m/s, and λ is the absorption wavelength in nm [26]. The calculated band gap energy values for undoped TiO$_2$ and Ce–TiO$_2$ were 3.11, and 2.81, respectively. The decreased band gap energy values of cerium-doped sample indicated the formation of new energy levels within the TiO$_2$ band gap and resulted in redshift of the absorption edge. This shift can be attributed to the incorporation of Ce 4f levels into the TiO$_2$ crystal structure just below the conduction band of TiO$_2$ and thereby reducing the effective band gap [27]. Addition of nonmetal to cerium-doped TiO$_2$ samples resulted in further shift in the band gap energy to longer wavelength. Significant reductions in the optical band gap energies from 3.11 eV for undoped TiO$_2$ to about 2.88 eV for Ce–B–TiO$_2$, 2.66 eV for Ce–C–TiO$_2$, 2.29 eV for Ce–N–TiO$_2$, and 2.24 eV for Ce–S–TiO$_2$ were observed for co-doped samples. For the co-doped catalysts, the band gap energy was affected by the formation of additional impurity states within titania matrix by the interaction of p-orbitals of the nonmetal dopants (B, C, N, S) and 2p-orbitals of oxygen [28–31]. The relative position of the new states with respect to valence band of titania depends on the properties of dopants and its position within the matrix. Whether the nonmetal dopant would occupy anionic or cationic substitutional sites or interstitial site depends upon its size, chemical valence and electronegativity compared that of the host TiO$_2$. The probability that a dopant will substitute oxygen depends on its electronegativity [32]. As difference in electronegativity values of dopant and oxygen becomes smaller, the probability that dopant will substitute oxygen becomes higher. Substitutional dopants form the new band states at higher energy level compared to interstitial dopant

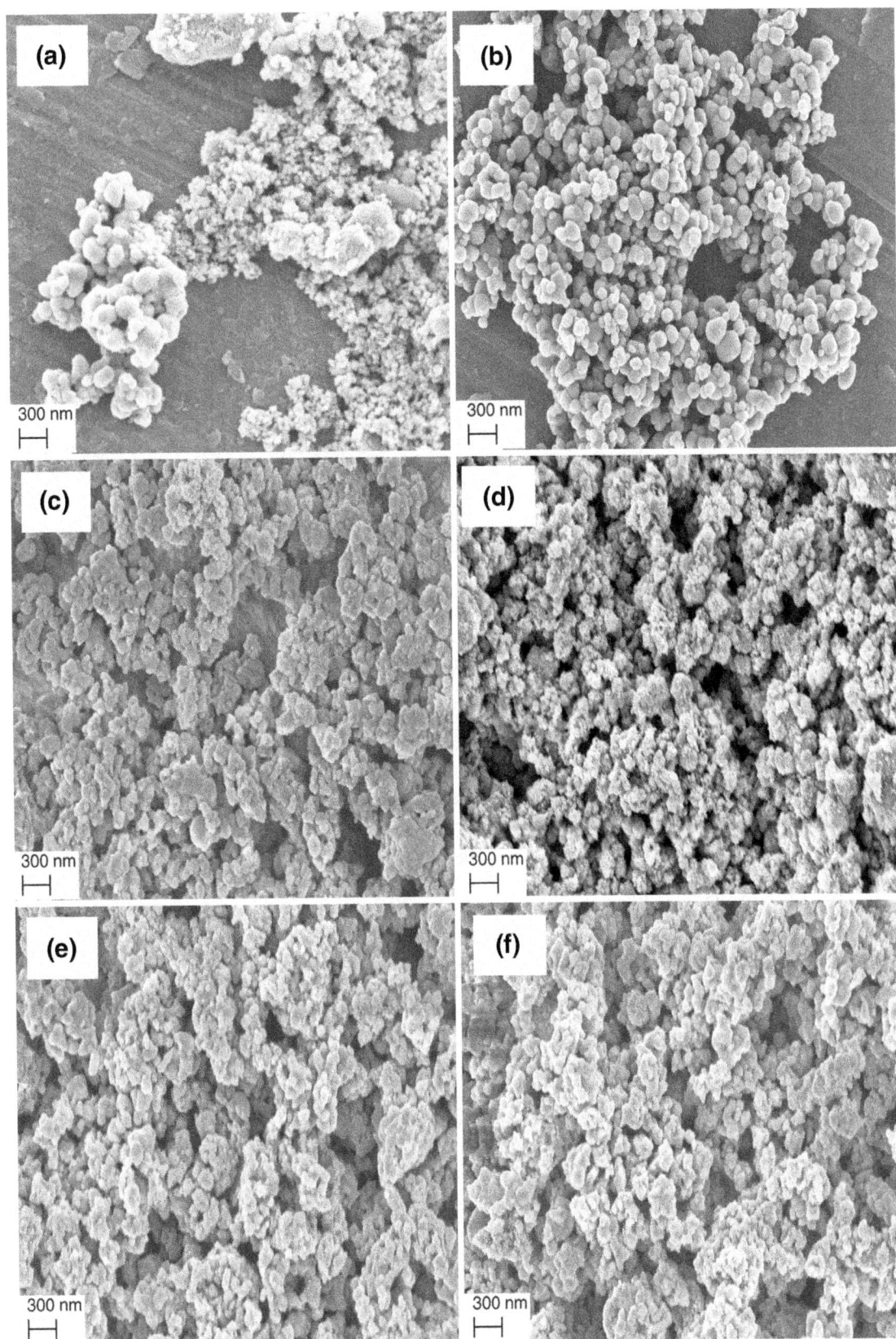

Fig. 4 FESEM images of samples **a** TiO$_2$, **b** Ce–TiO$_2$, **c** Ce–N–TiO$_2$, **d** Ce–S–TiO$_2$, **e** Ce–B–TiO$_2$, and **f** Ce–C–TiO$_2$

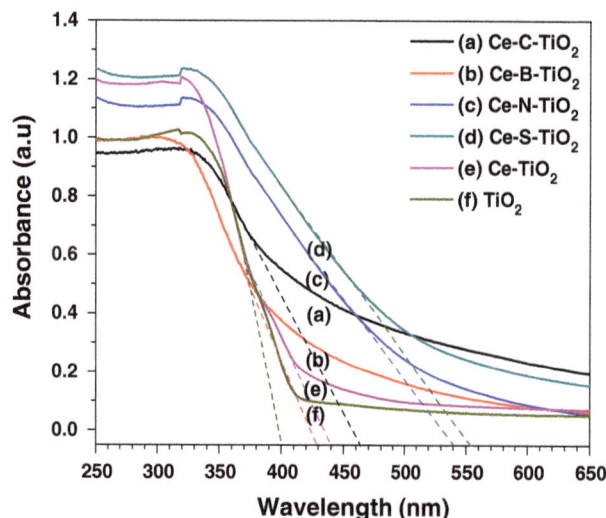

Fig. 5 UV–Visible diffuse reflectance spectra of TiO_2, Ce–TiO_2, and Ce–NM–TiO_2 (NM = B, C, N, S) samples

Fig. 6 Photoluminescence spectra of TiO_2, Ce–TiO_2, and Ce–NM–TiO_2 (NM = B, C, N, S) samples

[33]. For boron co-doped catalyst, it is difficult to replace Ti^{4+} sites by B^{3+} due to the smaller ionic radius of B^{3+} (IR = 0.023 nm) compared to Ti^{4+} (IR = 0.068 nm). Therefore, boron may replace either an oxygen atom or incorporate in the interstitial position in TiO_2 matrix. But due to the lower electronegativity of boron, its probability of occupying the interstitial sites is higher. Similarly, carbon can also substitute oxygen or occupy an interstitial site. In nitrogen co-doped sample, the substitution of oxygen sites is most probable due to their closer electronegativity values [34]. Substitutions of S in both anionic and cationic sites of TiO_2 have been reported in the literature [35, 36]. However, anionic sulfur substitution by S^{2-} (IR = 0.17 nm) might be difficult because of its larger ionic radius than O^{2-} (IR = 0.122 nm). But the cationic substitution of Ti^{4+} (IR = 0.068 nm) by S^{6+} (IR = 0.029 nm) would be more favorable due to smaller size of latter. When TiO_2 is doped with sulfur, either the mixing of S $3p$ states with valence band O $2p$ states or the formation of isolated p-orbitals of S above the valence band maximum of TiO_2 contributes toward narrowing the band gap [37].

The radiative recombination of electron–hole pairs in photocatalysts can be studied by the photoluminescence (PL) emission spectra. Figure 6 shows the room-temperature photoluminescence spectra of the TiO_2, Ce–TiO_2, and Ce–NM–TiO_2 samples in the range of 350–600 nm. For all the samples, six emission peaks appeared at 360, 396, 450, 468, 481, and 492 nm wavelengths and the corresponding transition energy are 3.44, 3.13, 2.76, 2.65, 2.8, and 2.52 eV, respectively. In the literature, the peak in the range of 360–370 nm has been assigned to either anatase particle <10 nm [38] or to self-trapped excitons localized in TiO_6 octahedral [39]. In this study, the emission peak at

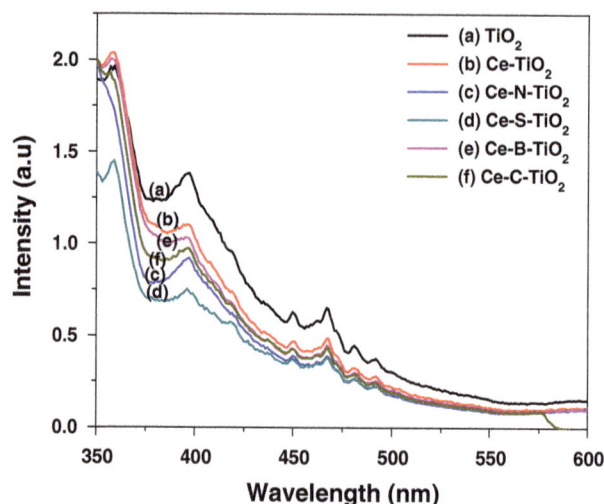

360 nm can be ascribed to the latter as particle is much larger than 10 nm. The emission signal at 396 nm (3.13 eV) corresponds to the band gap transition of the anatase [38]. This agrees well with the band gap energy of titania calculated from UV–Vis spectra. The excitonic emissions in titania result due to the presence of defects and oxygen vacancies. The bands at 450 and 468 nm can be attributed to free excitons, and emission peaks at longer wavelength of 481 and 492 nm can be attributed to the bound excitons [40, 41]. The bound excitons result due to the presence of oxygen vacancies [40, 42, 43]. It can be observed from the Fig. 6 that Ce–TiO_2 had lower PL intensity compared to TiO_2. In Ce–TiO_2, cerium dopant-induced energy level in TiO_2 band structure, as discussed earlier, serves as an electron trap as represented by Eq. (1) resulting in non-radiative recombinations.

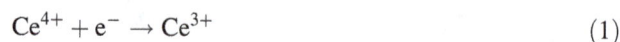

$$Ce^{4+} + e^- \rightarrow Ce^{3+} \tag{1}$$

This reduces the rate of radiative recombination of photogenerated electrons with holes in valence band of titania as shown in Fig. 6. The PL emission intensities of the cerium and nonmetal co-doped samples decreased in the order of TiO_2 > Ce–TiO_2 > Ce–B–TiO_2 > Ce–C–TiO_2 > Ce–N–TiO_2 > Ce–S–TiO_2. In the cerium and nonmetal co-doped samples, the oxygen vacancies generated by substituting nonmetal dopants can act as additional effective traps for photo-induced electrons, thereby reducing the radiative recombination rate of charge carriers further. The photogenerated electrons are trapped by oxygen vacancies by non-radiative combinations [42]. The decreasing emission intensity order can be attributed to increased oxygen vacancies resulted by increasing substitution effects from B to N as discussed in earlier section. In sulfur co-doped samples, the

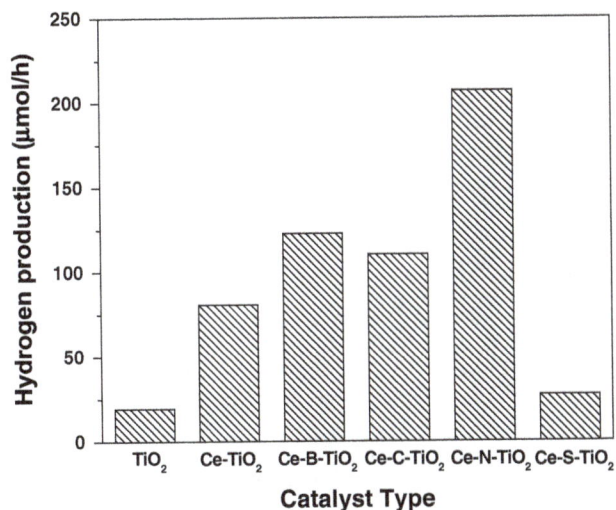

Fig. 7 Hydrogen evolution over TiO_2, $Ce–TiO_2$, and $Ce–NM–TiO_2$ (NM = B, C, N, S) samples

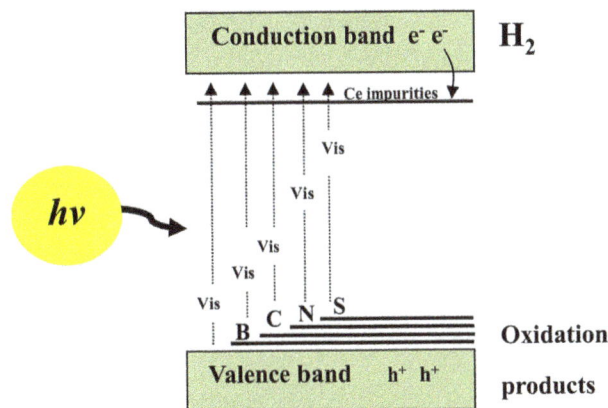

Fig. 8 Schematic representation of hydrogen production over cerium and nonmetal co-doped TiO_2 photocatalysts

presence of isolated cationic S^{6+} dopant state may further act as surface trap states for photogenerated electrons, thereby resulting in a significant decrease in the recombination rate and lowering the PL emission.

Photocatalytic activity studies

The photocatalytic hydrogen production from water–methanol mixture for TiO_2, $Ce–TiO_2$, and $Ce–NM–TiO_2$ (NM = B, C, N, S) catalysts is shown in Fig. 7. The activity of undoped TiO_2 was very low but $Ce–TiO_2$ showed significant photocatalytic activity and high hydrogen evolution. The high activity of cerium-doped TiO_2 can be attributed to their ability of absorption of light in visible light region as shown in Fig. 5. Addition of nonmetals such as B, C, and N to $Ce–TiO_2$ samples further enhanced hydrogen evolution activity. This may be attributed to higher surface area, enhanced visible light absorption as well as lower recombination rate of electron–hole pairs in co-doped samples. The highest activity of $Ce–N–TiO_2$ among co-doped samples may be attributed to its higher surface area, higher visible light absorption, and lower recombination rate of electron–hole pairs compared to other nonmetals, B and C, doped samples. The activity of $Ce–C–TiO_2$ should be higher than $Ce–B–TiO_2$ due to higher absorption of visible light and lower recombination of photogenerated electron–hole pairs for former as shown by absorption spectra (Fig. 5) and PL spectra (Fig. 6), respectively. But slightly higher hydrogen evolution observed for $Ce–B–TiO_2$ compared to that of $Ce–C–TiO_2$ may be attributed to higher surface area of former providing more active sites. Much lower hydrogen production was observed for $Ce–S–TiO_2$ sample, although it has enhanced absorption in the visible light region and efficient

charge separation. The lower activity of $Ce–S–TiO_2$ can be attributed to its very low surface area. Moreover, the S may be in isolated state and that may lead to lower mobility of the holes and limits the number of charge carriers reaching the catalyst surface. This may have also resulted in lower photocatalytic activity in S-doped samples.

The apparent quantum efficiency of the photocatalysts was calculated using the method followed by Sasikala et al. [44]. The apparent quantum efficiency values of TiO_2, $Ce–TiO_2$, $Ce–B–TiO_2$, $Ce–C–TiO_2$, $Ce–N–TiO_2$, and $Ce–S–TiO_2$ were 1.6, 6.7, 10.2, 9.2, 17.2, and 2.2 %, respectively. The calculated values agreed with the reported apparent quantum efficiency range of 2–22 % for different photocatalysts [45]. For comparison, the photocatalytic activity of $N–TiO_2$ was also measured. The apparent quantum efficiency of $N–TiO_2$ (8.6 %) was slightly higher than that $Ce–TiO_2$ (6.7 %) but the efficiency increased significantly for co-doped $Ce–N–TiO_2$ to 17.2, suggesting that both doping components were contributory to photocatalytic activity, as discussed earlier, giving higher hydrogen production. The results are shown in Supplementary Figure S2.

The durability test of $Ce–N–TiO_2$ for photocatalytic H_2 production from water–methanol mixture was performed for 10 h irradiation period with hydrogen evolution being recorded at every 1 h as shown in Figure S3 of supplementary information. The total hydrogen evolution of 737 μmol was obtained after 10 h of irradiation time. The observed results showed that the $Ce–N–TiO_2$ had high H_2 evolution rate as well as desired durability.

Based on the observed band gap energies of co-doped samples, the position of the impurity states and probable transitions of electrons are schematically shown in Fig. 8. In $Ce–TiO_2$, the photogenerated electrons are excited from the valence band to extended conduction band where electrons are trapped effectively by Ce^{4+}/Ce^{3+} couple. This results in the accumulation of electrons in extended conduction band, which can serve as hydrogen formation site. On addition of nonmetals to cerium-doped titania, the excitation of electrons can also takes place from nonmetal-induced impurity levels resulting in the absorption of visible light. The reduction and oxidation reactions occurring in conduction band and valance band, respectively, can be represented as follows [46, 47].

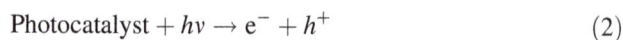

$$Photocatalyst + hv \rightarrow e^- + h^+ \qquad (2)$$

Reduction reaction

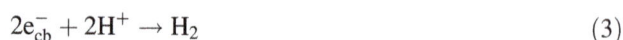

$$2e_{cb}^- + 2H^+ \rightarrow H_2 \qquad (3)$$

Oxidation reactions:

$$CH_3OH + 2h_{vb}^+ \rightarrow 2H^+ + HCHO \qquad (4)$$

$$HCHO + H_2O \rightarrow HCO_2H + H_2 \qquad (5)$$

$$HCO_2H + h_{vb}^+ \rightarrow CO_2 + H^+ \qquad (6)$$

In addition to proposed hole-mediated path as shown in the above reaction scheme, methanol can also enhances hydrogen generation by reacting with hydroxyl radicals generated from photocatalytic water splitting [48]. The oxidation of methanol results in the formation of HCHO, HCOOH, and CO_2. For oxidation of formaldehyde to formic acid, the extra oxygen atom is provided by water through hydroxyl radical. Intermediate products were only detected in liquid phase, and no intermediate products were detected in gaseous phase in the present study.

Conclusions

Cerium and nonmetal co-doped samples synthesized by co-precipitation method were found to be potential photocatalysts for hydrogen generation under visible light radiation. Doping of titania with cerium induced distinct redshift of the absorption edge in visible light region. Nonmetal modification further enhanced the visible light absorption and separation efficiency of photogenerated electron–hole pairs. The effective band gap energies of co-doped samples depended on the properties of nonmetals. The surface area and phase distribution were also modified in the presence of nonmetal. Highest hydrogen evolution rate of 206 μmol/h was observed for cerium–nitrogen co-doped sample. The higher photoactivty of Ce–N–TiO_2 can be attributed to higher surface area, higher light absorption, and efficient charge separation.

References

1. Ashokkumar, M.: An overview on semiconductor particulate systems for photoproduction of hydrogen. Int. J. Hydrog. Energy **23**, 427–438 (1998)
2. Balat, H., Kirtay, E.: Hydrogen from biomass—present scenario and future prospects. Int. J. Hydrog. Energy **35**, 7416–7426 (2010)
3. Holladay, J.D., Hu, J., King, D.L., Wang, Y.: An overview of hydrogen production technologies. Catal. Today **139**, 244–260 (2009)
4. Kudo, A.: Photocatalysis and solar hydrogen production. Pure Appl. Chem. **79**, 1917–1927 (2007)
5. Takata, T., Tanaka, A., Hara, M., Kondo, J.N., Domen, K.: Recent progress of photocatalysts for overall water splitting. Catal. Today **44**, 17–26 (1998)
6. Gurunathan, K., Maruthamuthu, P.: Photogeneration of hydrogen using visible light with undoped/doped α-Fe_2O_3 in the presence of methyl viologen. Int. J. Hydrog. Energy **20**, 287–295 (1995)
7. Wang, X., Liu, G., Lu, G.Q., Cheng, H.-M.: Stable photocatalytic hydrogen evolution from water over ZnO–CdS core–shell nanorods. Int. J. Hydrog. Energy **35**, 8199–8205 (2010)
8. Kudo, A., Miseki, Y.: Heterogeneous photocatalyst materials for water splitting. Chem. Soc. Rev. **38**, 253–278 (2009)
9. Ni, M., Leung, M.K.H., Leung, D.Y.C., Sumathy, K.: A review and recent developments in photocatalytic water-splitting using TiO_2 for hydrogen production. Renew. Sustain. Energy Rev. **11**, 401–425 (2007)
10. Pelaez, M., Nolan, N.T., Pillai, S.C., Seery, M.K., Falaras, P., Kontos, A.G., Dunlop, P.S.M., Hamilton, J.W.J., Byrne, J.A., O'Shea, K., Entezari, M.H., Dionysiou, D.D.: A review on the visible light active titanium dioxide photocatalysts for environmental applications. Appl. Catal. B **125**, 331–349 (2012)
11. Xu, A., Gao, Y., Liu, H.: The preparation, characterization, and their photocatalytic activities of rare-earth-doped TiO_2 nanoparticles. J. Catal. **207**, 151–157 (2002)
12. Tong, T., Zhang, J., Tian, B., Chen, F., He, D., Anpo, M.: Preparation of Ce–TiO_2 catalysts by controlled hydrolysis of titanium alkoxide based on esterification reaction and study on its photocatalytic activity. J. Colloid Interface Sci. **315**, 382–388 (2007)
13. Silva, A.M.T., Silva, C.G., Drazic, G., Faria, J.L.: Ce-doped TiO_2 for photocatalytic degradation of chlorophenol. Catal. Today **144**, 13–18 (2009)
14. Fan, C., Xue, P., Sun, Y.: Preparation of nano-TiO_2 doped with cerium and its photocatalytic activity. J. Rare Earths **24**, 309–313 (2006)
15. Asahi, R., Morikawa, T., Ohwaki, T., Aoki, K., Taga, Y.: Visible-light photocatalysis in nitrogen-doped titanium oxides. Science **293**, 269–271 (2001)
16. Ohno, T., Akiyoshi, M., Umebayashi, T., Asai, K., Mitsui, T., Matsumura, M.: Preparation of S-doped TiO_2 photocatalysts and their photocatalytic activities under visible light. Appl. Catal. A **265**, 115–121 (2004)
17. Yun, H.J., Lee, H., Joo, J.B., Kim, N.D., Kang, M.Y., Yi, J.: Facile preparation of high performance visible light sensitive photo-catalysts. Appl. Catal. B **94**, 241–247 (2010)
18. Shen, X.Z., Liu, Z.C., Xie, S.M., Guo, J.: Degradation of nitrobenzene using titania photocatalyst co-doped with nitrogen and cerium under visible light illumination. J. Hazard. Mater. **162**, 1193–1198 (2009)
19. Huang, D.G., Liao, S.J., Zhou, W.B., Quan, S.Q., Liu, L., He, Z.J., Wan, J.B.: Synthesis of samarium- and nitrogen-co-doped TiO_2 by modified hydrothermal method and its photocatalytic performance for the degradation of 4-chlorophenol. J. Phys. Chem. Solids **70**, 853–859 (2009)

20. Moon, S.-C., Mametsuka, H., Tabata, S., Suzuki, E.: Photocatalytic production of hydrogen from water using TiO_2 and B/TiO_2. Catal. Today **58**, 125–132 (2000)

21. Khan, S.U.M., Al-Shahry, M., Ingler Jr, W.B.: Efficient photochemical water splitting by a chemically modified n-TiO_2. Science **297**, 2243–2245 (2002)

22. Sreethawong, T., Laehsalee, S., Chavadej, S.: Comparative investigation of mesoporous- and non-mesoporous-assembled TiO_2 nanocrystals for photocatalytic H_2 production over N-doped TiO_2 under visible light irradiation. Int. J. Hydrog. Energy **33**, 5947–5957 (2008)

23. Sidheswaran, M., Tavlarides, L.L.: Characterization and visible light photocatalytic activity of cerium- and iron-doped titanium dioxide sol–gel materials. Ind. Eng. Chem. Res. **48**, 10292–10306 (2009)

24. Yu, T., Tan, X., Zhao, L., Yin, Y., Chen, P., Wei, J.: Characterization, activity and kinetics of a visible light driven photocatalyst: cerium and nitrogen co-doped TiO_2 nanoparticles. Chem. Eng. J. **157**, 86–92 (2010)

25. Llordés, A., Palau, A., Gázquez, J., Coll, M., Vlad, R., Pomar, A., Arbiol, J., Guzmán, R., Ye, S., Rouco, V., et al.: Nanoscale strain-induced pair suppression as a vortex-pinning mechanism in high-temperature superconductors. Nat. Mater. **11**, 329–336 (2012)

26. Sreethawong, T., Suzuki, Y., Yoshikawa, S.: Synthesis, characterization, and photocatalytic activity for hydrogen evolution of nanocrystalline mesoporous titania prepared by surfactant-assisted templating sol–gel process. J. Solid State Chem. **178**, 329–338 (2005)

27. Magesh, G., Viswanathan, B., Viswanath, R.P., Varadarajan, T.K.: Photocatalytic behavior of CeO_2–TiO_2 system for the degradation of methylene blue. Indian J. Chem. **48A**, 480–488 (2009)

28. Zhao, W., Ma, W., Chen, C., Zhao, J., Shuai, Z.: Efficient degradation of toxic organic pollutants with $Ni_2O_3/TiO_{2-x}B_x$ under visible irradiation. J. Am. Chem. Soc. **126**, 4782–4783 (2004)

29. Sathish, M., Viswanathan, B., Viswanath, R.P.: Characterization and photocatalytic activity of N-doped TiO_2 prepared by thermal decomposition of Ti–melamine complex. Appl. Catal. B **74**, 307–312 (2007)

30. Yu, C., Yu, J.C.: A simple way to prepare C–N-codoped TiO2 photocatalyst with visible-light activity. Catal. Lett. **129**, 462–470 (2009)

31. Tian, F., Liu, C., Zhao, W., Wang, X., Wang, Z., Yu, J.C.: Cationic S-doped anatase TiO_2: a DFT study. J. Comput. Sci. Eng. **1**, 33–41 (2011)

32. Lu, J., Jin, H., Dai, Y., Yang, K., Huang, B.: Effect of electronegativity and charge balance on the visible-light-responsive photocatalytic activity of nonmetal doped anatase TiO_2. Int. J. Photoenergy. Article ID 928503, 8 pp (2012)

33. Di Valentin, C., Pacchioni, G.: Trends in non-metal doping of anatase TiO_2: B, C, N and F. Catal. Today **206**, 12–18 (2013)

34. Kitano, M., Matsuoka, M., Ueshima, M., Anpo, M.: Recent developments in titanium oxide-based photocatalysts. Appl. Catal. A **325**, 1–14 (2007)

35. Umebayashi, T., Yamaki, T., Itoh, H., Asai, K.: Band gap narrowing of titanium dioxide by sulfur doping. Appl. Phys. Lett. **81**, 454–456 (2002)

36. Yu, J.C., Ho, W., Yu, J., Yip, H., Wong, P.K., Zhao, J.: Efficient visible-light-induced photocatalytic disinfection on sulfur-doped nanocrystalline titania. Environ. Sci. Technol. **39**, 1175–1179 (2005)

37. Xiang, Q., Yu, J., Jaroniec, M.: Nitrogen and sulfur co-doped TiO2 nanosheets with exposed 001 facets: synthesis, characterization and visible-light photocatalytic activity. Phys. Chem. Chem. Phys. **13**, 4853–4861 (2011)

38. Li, Y., Lee, N.-H., Hwang, D.-S., Song, J.S., Lee, E.G., Kim, S.-J.: Synthesis and characterization of nano titania powder with high photoactivity for gas-phase photo-oxidation of benzene from $TiOCl_2$ aqueous solution at low temperatures. Langmuir **20**, 10838–10844 (2004)

39. Hu, C., Zhang, R., Xiang, J., Liu, T., Li, W., Li, M., Duo, S., Wei, F.: Synthesis of carbon nanotube/anatase titania composites by a combination of sol–gel and self-assembly at low temperature. J. Solid State Chem. **184**, 1286–1292 (2011)

40. Yu, J., Hai, Y., Jaroniec, M.: Photocatalytic hydrogen production over CuO-modified titania. J. Colloid Interface Sci. **357**, 223–228 (2011)

41. Zhang, Y.X., Li, G.H., Jin, Y.X., Zhang, Y., Zhang, J., Zhang, L.D.: Hydrothermal synthesis and photoluminescence of TiO_2 nanowires. Chem. Phys. Lett. **365**, 300–304 (2002)

42. Wang, E., Yang, W., Cao, Y.: Unique surface chemical species on indium doped TiO_2 and their effect on the visible light photocatalytic activity. J. Phys. Chem. C **113**, 20912–20917 (2009)

43. Selvam, K., Balachandran, S., Velmurugan, R., Swaminathan, M.: Mesoporous nitrogen doped nano titania—a green photocatalyst for the effective reductive cleavage of azoxybenzenes to amines or 2-phenyl indazoles in methanol. Appl. Catal. A **413–414**, 213–222 (2012)

44. Sasikala, R., Sudarsan, V., Sudakar, C., Naik, R., Sakuntala, T., Bharadwaj, S.R.: Enhanced photocatalytic hydrogen evolution over nanometer sized Sn and Eu doped titanium oxide. Int. J. Hydrog. Energy **33**, 4966–4973 (2008)

45. Jin, Z., Zhang, X., Li, Y., Li, S., Lu, G.: 5.1% Apparent quantum efficiency for stable hydrogen generation over eosin-sensitized CuO/TiO_2 photocatalyst under visible light irradiation. Catal. Commun. **8**, 1267–1273 (2007)

46. Wu, N.-L., Lee, M.-S.: Enhanced TiO_2 photocatalysis by Cu in hydrogen production from aqueous methanol solution. Int. J. Hydrog. Energy **29**, 1601–1605 (2004)

47. Choi, H.-J., Kang, M.: Hydrogen production from methanol/water decomposition in a liquid photosystem using the anatase structure of Cu loaded TiO_2. Int. J. Hydrog. Energy **32**, 3841–3848 (2007)

48. Chiarello, G.L., Ferri, D., Selli, E.: Effect of the CH_3OH/H_2O ratio on the mechanism of the gas-phase photocatalytic reforming of methanol on noble metal-modified TiO_2. J. Catal. **280**, 168–177 (2011)

7

SPS-sintered NaTaO$_3$–Fe$_2$O$_3$ composite exhibits enhanced Seebeck coefficient and electric current

Wilfried Wunderlich · Takao Mori ·
Oksana Sologub

Abstract NaTaO$_3$—50 wt% Fe$_2$O$_3$ composite ceramics showed a large Seebeck voltage of −300 mV at a temperature gradient of 650 K yielding a constant Seebeck coefficient of more than −500 μV/K over a wide temperature range. We report for the first time that spark plasma sintering (SPS) at low temperature (870 K) could maintain the short-circuit current of −80 μA, which makes this thermoelectric material a possible candidate for high-temperature applications up to 1,623 K. The reason for the good performance is the interface between Fe$_2$O$_3$ and surrounding NaTaO$_3$ perovskite. When SPS is used, constitutional vacancies disappeared and the electric conductivity increases remarkably yielding ZT of 0.016.

Keywords Thermoelectricity · Seebeck coefficient · Ceramics · Spark−plasma sintering · Interface

Introduction

The pressing problem of CO$_2$ increase and climate change requires the search for new energy sources such as the thermoelectric power generators (TEG), which can turn waste heat into usable electricity when operating at high temperatures. The research for new thermoelectric ceramic materials began in the last decade and Nb-doped SrTiO$_3$

[1, 2] NaCoO$_3$ [3], and CaCoO$_3$ [4] were found to have a remarkable figure-of-merit ZT. They are already successfully established in devices for high-temperature electric generators. A detailed band structure study of the perovskite-based Nb-doped SrTiO$_3$ material has emphasized the combination of large and small effective masses as the reason for the large Seebeck coefficient [5, 6]. While Co-based perovskites [7–9] have been investigated also as potential thermoelectric materials, our search for new materials yielded to the NaTaO$_3$ perovskite material, which is known as efficient photo catalyst for splitting water [10, 11] and its large effective electron mass [5].

The composite material NaTaO$_3$–Fe$_2$O$_3$ shows a large Seebeck voltage of −300 mV at a temperature gradient of $\Delta T = 650$ K with linear temperature dependence [10–15] and is stable up to 1,623 K [12, 15]. Yet its large resistivity has to be lowered for increasing the power factor and figure-of-merit. We have reported previous results on spark plasma sintering (SPS) experiments [15], where a remarkable increase in electric conductivity was achieved, but the Seebeck voltage has dropped. The conclusion was that either the interface structure or the microstructure has changed due to the high-temperature plasma, the vacuum, diffusion from the carbon crucible, or during the subsequent fast cooling and are responsible for the decrease of the Seebeck coefficient. Further findings were that composites processed from Fe$_2$O$_3$ and NaTaO$_3$, or additions of NaFeO$_3$ deteriorate the electric conductivity and yield to an insulator. The reason for the good performance of this composite material is the interface between Fe$_2$O$_3$ and NaTaO$_3$ with perovskite structure. The largest Seebeck voltage was measured when the second phase Fe$_2$O$_3$ reaches an amount of 50 mol% [10], which is just the percolation limit when the second phase starts to surround the perovskite phase NaTaO$_3$. Metallic behavior with high

W. Wunderlich (✉)
Department of Material Science, Tokai University, Hiratsuka-shi 259-1292, Japan
e-mail: wi-wunder@rocketmail.com

T. Mori · O. Sologub
Nat. Inst. Mat. Sci. (NIMS), Int. Center Mat. Nanoarchitec. (MANA), Tsukuba 305-0044, Japan

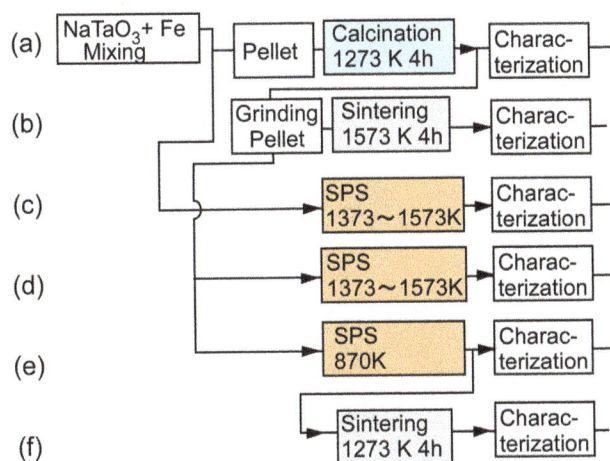

Fig. 1 The processing procedure yielded to different routes as indicated **a** calcination, **b** conventional sintering, **c** SPS at 1,373–1,573 K, **d** calcination 1,273 K in air, then SPS at 1,373–1,573 K, **e** calcination 1,273 K in air, SPS at 870 K, and **f** same as **e** with additional sintering at 1,273 K for 4 h

carrier concentration was recently found at similar NaTaO$_3$/SrTiO$_3$ perovskite interfaces [16].

Hence, the goal of this paper is to gain deeper insight in the materials behavior with the goal to improve both, Seebeck voltage and electric conductivity. For optimum densification, a second sintering step is required after calcination and grinding. Therefore, this paper describes processing of these composite ceramics on different routes and compare both, thermoelectric and microstructural properties of the resolved specimens.

Experimental procedure

Powders in μm size of NaTaO$_3$ and Fe (Fine Chemicals Ltd., Japan) were weighed according to the desired weight ratio of 50 mol-% Fe$_2$O$_3$ and mixed in a mortar for at least 10 min. Then the mixture was put in a steel cylinder with 15-mm diameter and cold-pressed with a stress of 50 MPa. These pellets were used in the following different synthesis methods as sketched in Fig. 1. The conventional calcination and sintering route are shown as paths (a) and (b) in Fig. 1 and details have been described in [12–14]. A sliced specimen with 10 × 2 × 2 mm dimensions was measured using the thermoelectric multi-measuring device ZEM3 (Ulvac Ltd., Japan) as described in the following section.

The next straight-forward step is to try SPS on the cold-pressed pellet [route (c) in Fig. 1] [15]. The report on such specimens showed improved resistivity, but poor Seebeck voltage [15]. The present paper focusses on SPS sintering. After calcination, the specimen were crushed, then the powder mixed again and put into a 15 mm graphite cylinder and finally attached in the Doctor Sinter 1080 SPS

device (Syntex Sumitomo Heavy Industries, Ltd). Two regimes were tested and are marked as high and low temperature routes (d) and (e) in Fig. 1, namely 1,373–1,573 K and 870 K, respectively. The maximum pressure of 80 MPa was applied and kept constant, while temperature, spark plasma voltage and current were increased as described in detail in [15]. The plasma chamber was first evacuated, and then the sintering was performed at 1 atm Ar pressure. The duration of sintering was kept constant as 600 s.

The obtained specimens were characterized concerning their microstructure and composition using a Hitachi 3200-N scanning electron microscope (SEM) operated at 20 kV and equipped with an electron dispersive spectrometer (EDS, Noran Ltd.). The thermoelectric voltage was measured against nickel wires with a distance of 10 mm in a home-made device when applying a temperature difference between the micro-ceramic heater (Sakaguchi Ltd. MS 1000) up to 1,273 K, and maintaining the cold end at around 473 K, as reported elsewhere [12–14]. The electric multi-meter measurement devices (Sanwa PC510) recorded the data directly on a computer. The densities of the specimens were calculated from their mass-to-volume ratio, where the specimen dimensions were measured by a caliper.

Results and discussion

At first, a conventionally calcined and sintered specimen as described in [12] was measured for the first time by using a ZEM3. The thermoelectric measurements as displayed in Fig. 2 confirmed the relatively high resistivity, which decreases as a function of temperature. At $T = 1,000$ K the resistivity reached as less as 0.05 Ω m, as shown in Fig. 2a, about one order of magnitude better than previously reported values [12]. The Seebeck coefficient reaches −0.6 mV/K (Fig. 2b) and is almost constant over a wide temperature range from 600 to 1,000 K, allowing such specimen to be used as reference specimens for calibrating a measuring device. These measurements are in good agreement with previously published ones [12–14], and hence justify the validity of the Seebeck coefficient measurements under large temperature gradient at least for this materials with its constant Seebeck coefficient over a large temperature range. The power factor $S^2 \sigma$ as displayed in Fig. 2c reaches a value of 6 × 10^{-6} W/m K, which is yet one or two orders of magnitude smaller than comparable oxides [1–4]. The reason is that the resistivity remains still unacceptable large as concluded from this measurement and motivates the SPS processing trial reported in the following section. When closing the open circuit with different load resistors (1–100 kΩ), the electric current was

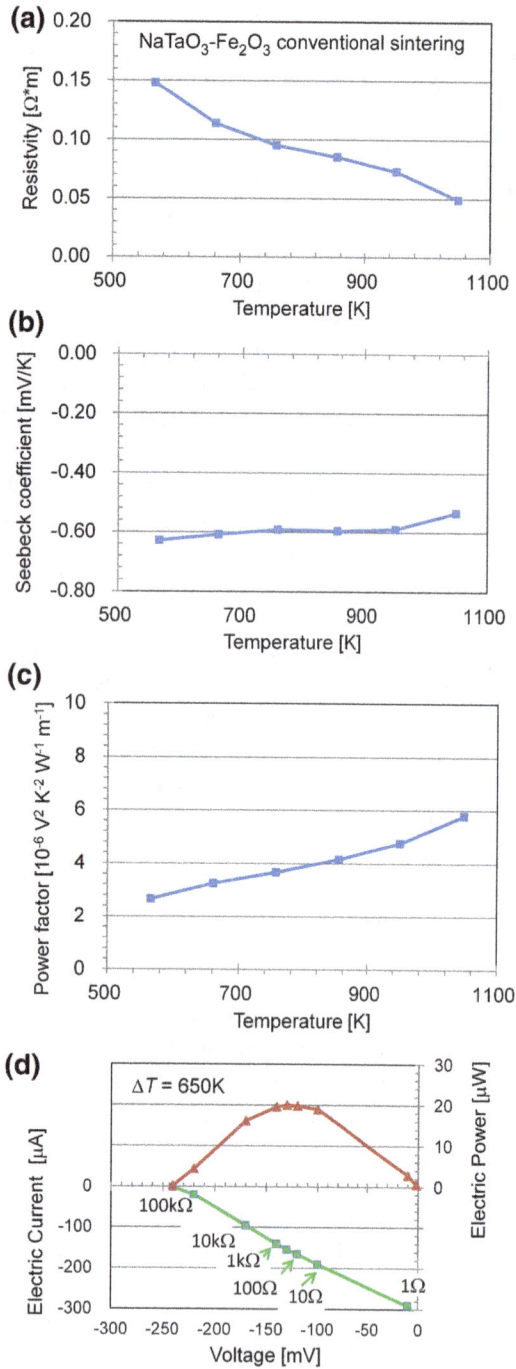

Fig. 2 Thermoelectric properties of NaTaO$_3$-50 mol% Fe$_2$O$_3$ produced by calcination and conventional sintering **a** resistivity, **b** Seebeck coefficient, **c** power factor $S^2 \sigma$, **d** closed circuit current as a function of the Seebeck voltage and generated electric power at $\Delta T = 650$ K

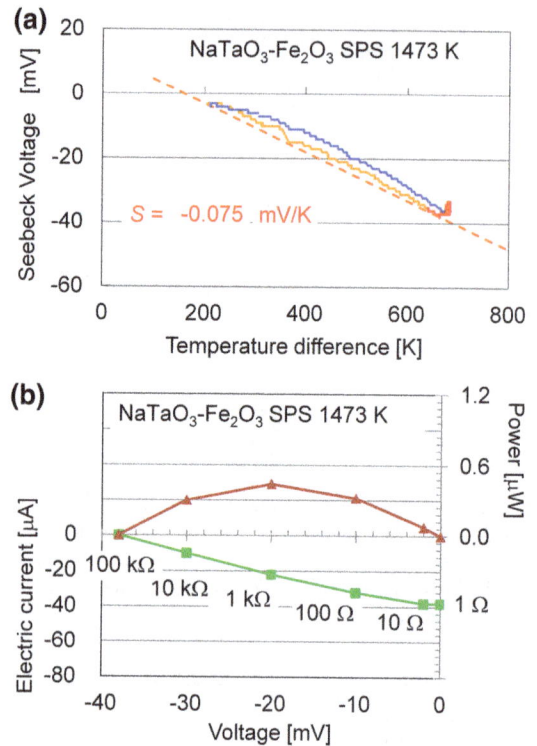

Fig. 3 a Seebeck voltage of a NaTaO$_3$-50 mol% Fe$_2$O$_3$ specimen, which was calcined and then SPS sintered at 1,473 K. **b** Electric current as a function of Seebeck voltage at $\Delta T = 650$ K when load resistors are applied

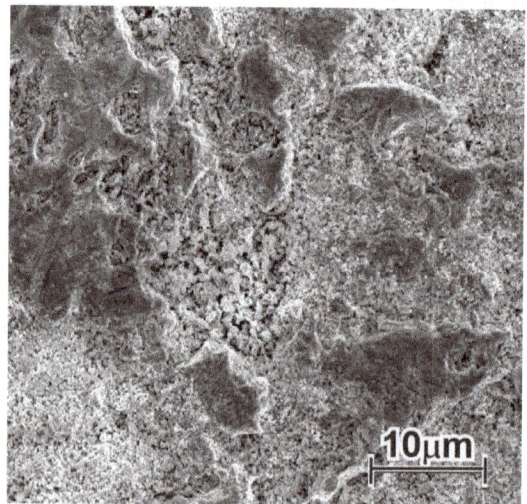

Fig. 4 Microstructure of the SPS specimen sintered at 1,473 K

measured. The maximum in this closed circuit current was -0.32 mA, and the maximum Seebeck voltage was -240 mV with a linear dependence (Fig. 2d). This relation yields at an intermediate operation point at $U = -130$ mV, $I = -154$ µA to a maximum power of $P = U \times I = 20$ µW for a temperature difference $\Delta T = 650$ K, which is

still a factor of 10 below state-of-the-art oxides [3, 4]. By assuming a thermal conductivity of $\kappa = 1.7$ W/m K such as typical for comparable perovskites, we end up with a figure-of-merit in the order of ZT = 0.016, still low, but promising.

In order to improve the electric conductivity, we tried the SPS sintering at high temperatures 1,373, 1,473 and

Table 1 Density and thermoelectric properties of the NaTaO$_3$-50 mol % Fe$_2$O$_3$ composites as a function of processing method and sintering temperature

Processing method	Sinter temperature (K)	Density (kg/m^3)	Seebeck coefficient (mV/K)	Electric current (μA)	Maximal power (μW)
(a) [12]	1,273	2.55	−0.50	−50	2.5
(b) [12, 13], Fig. 1	1,473	2.82	−0.62	−320	20
(c) [15]	1,473	4.36	−0.001	−10	0.01
(d) Fig. 3	1,473	4.43	−0.075	−40	0.5
(e) Fig. 5a	870	2.88	−0.50	−10	0.1
(f) Fig. 5b	1,273	3.56	−0.50	−80	4

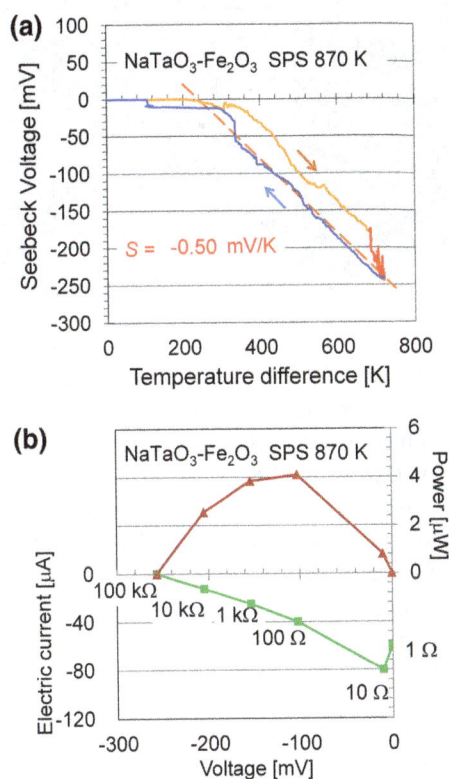

Fig. 5 Thermoelectric characterization of the NaTaO$_3$-50 mol% Fe$_2$O$_3$ calcined and **a** SPS sintered at 870 K only, and **b** sintered two times, with SPS at 870 K and 1,273 K 4 h. **a** Seebeck voltage as function of a large temperature gradient, where the *arrows* mark the heating and cooling, and the *slope* marks the Seebeck coefficient. **b** Electric current and power as a function of Seebeck voltage at $\Delta T = 650$ K, when load resistors are applied as marked

route (c) in Fig. 1. One or more of the factors, vacuum, Ar pressure, spark plasma, the surrounding carbon container, or the fast cooling rate, are apparently bad conditions, which are deteriorating both, the Seebeck voltage and electric conductivity, probably by reducing the oxygen content or changing the interface structure.

Microstructural characterization by SEM, as shown in Fig. 4, confirmed that large pores due to insufficient compaction disappeared, but smaller pores on a sub-micrometer scale are still present in some areas. Their shape and location between Fe$_2$O$_3$ (black) and NaTaO$_3$ (white) suggest that two mechanisms are occurring; at first, the solving Fe into NaTaO$_3$ perovskite lattice and then the oxidation of Fe to Fe$_2$O$_3$, both are responsible for these pores due to unbalanced stoichiometry. Nevertheless, the density has remarkably increased compared to the conventional sintering as shown in Table 1, which compares all experiments.

In order to improve the thermoelectric properties, we have processed another sample by calcination and subsequent SPS at low temperature of 870 K according to route (e) in Fig. 1, see also Table 1. This sample maintained the large Seebeck coefficient of −0.5 mV/K (Fig. 5a). By subsequent sintering at 1,273 K 4 h [step (f)] simultaneously, a maximum current of −80 μA was measured when closing the circuit with the load of a 10 Ω resistor. This is the first time that such specimens revealed good thermoelectric performance after SPS sintering, in spite of the fact that the output power is still one order of magnitude less than that of conventional sintered specimens, as shown in Fig. 5b. The specific resistivity decreased one order of magnitude and reached 0.015 Ω m at a temperature of 1,100 K, which means the figure-of-merit is ZT = 0.016. This fact confirms the benefits of using SPS sintering for closing pores and increasing density. In SEM micrographs, like that shown in Fig. 6, areas with stoichiometric frustration have almost vanished. The EDX mapping confirmed the homogenous distribution of chemical elements in the dark phase, Fe$_2$O$_3$, as well as in the bright one, NaTaO$_3$.

The results of this study are summarized in Table 1, which has the same notation of processing steps as Fig. 1. While solving Fe into NaTaO$_3$ and calcination requires

1,573 K on previously calcined specimens as sketched in Fig. 1c. The absolute value of the Seebeck voltage was reduced to −38 mV as shown in Fig. 3a, and also a lower absolute value of the closed circuit current of −40 μA (Fig. 3b). Compared to a maximum of the Seebeck coefficient of $S = -0.62$ mV/K for the conventional sintered NaTaO$_3$–Fe$_2$O$_3$ samples, the Seebeck coefficient decreased to $S = -0.075$ mV/K. The vanishing close circuit current can be explained by similar experience from a SPS sintering experiment without pre-calcining this material [15],

Fig. 6 Microstructure and element mapping by SEM-EDX of the specimen calcined and SPS sintered at 870 K the SPS specimen sintered at 1,473 K according to Fig.1e

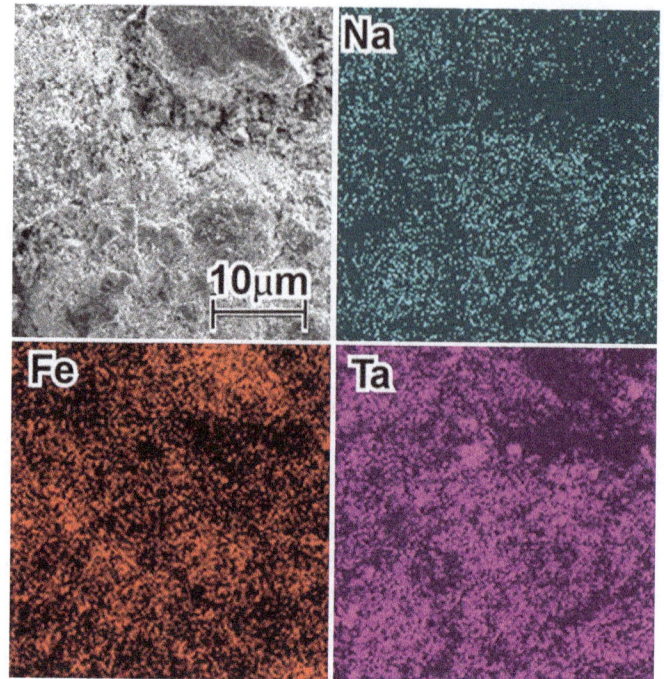

slow annealing [steps (a) and (b)], densification by maintaining a large negative Seebeck coefficient was confirmed here for the first time. The large electric conductivity is explained by the enlarged densification (Table 1) due to applied pressure during SPS sintering. The shrinkage factor c defined as $c = t_f/t_0$, where t_0 and t_f are the thicknesses before and after SPS, were estimated as 0.9, constant for all SPS experiments (c–e), and no anisotropies in S or σ were found.

On the other hand, the large thermo power can only be maintained, when the specimens are sintered by SPS at relatively low temperature below 900 K with subsequent sintering in air [steps (e) and (f)]. This fact also confirmed the self-repairing ability of this composite besides the long-term stability due to the fact that thermoelectric properties improved after each sintering cycle [12, 13]. Further investigations should confirm the detailed mechanism why a large negative Seebeck coefficient and a fairly large electric conductivity are achieved at the same time. The fact that the optimum composition fits well to the percolation composition [12], where the volume fraction of both phases is equal, as well as results from recent reports [2, 16] indicate that a confined two-dimensional electron gas (2DEG) at the heterogeneous interface is the reason for the good performance of this composite. Further improvement of ZT is expected from nano-scaled composites or super lattices or by finding suitable co-dopants.

Conclusions

The results of this SPS study on the $NaTaO_3$–Fe_2O_3 composite material confirmed the following facts.

(1) For the first time, we could achieve a large negative Seebeck voltage of this n-type thermoelectric and closed circuit current with remarkable power output by substantially decreasing the SPS sintering temperature from 1,273 to 870 K.

(2) The densities of the SPS-processed specimens were much higher than the cold-pressed specimens with fewer pores and cracks, leading to a significantly lower electrical resistivity.

(3) Sub-micrometer sized pores due to constitutional frustration almost disappeared confirming the homogeneous microstructure after calcination and subsequent SPS sintering at low temperature. While densification improved the electric conductivity significantly, the large thermo power can only be maintained by slow heating in air, low SPS sintering temperature and subsequent sintering at high temperature.

References

1. Ohta, S., Nomura, T., Ohta, H., Hosono, H., Koumoto, K.: Large thermoelectric performance of heavily Nb-doped $SrTiO_3$ epitaxial film at high temperature. Appl. Phys. Lett. **87**, 092108 (2005).

2. Ohta, H., Kim, S., Mune, Y., Mizoguchi, T., Nomura, K., Ohta, S., Nomura, T., Nakanishi, Y., Ikuhara, Y., Hirano, M., Hosono,

H., Koumoto, K.: Giant thermoelectric Seebeck coefficient of a two-dimensional electron gas in $SrTiO_3$. Nat. Mater. **6**, 129 (2007).

3. Terasaki, I., Sasago, Y., Uchinokura, K.: Large thermoelectric power in $NaCo_2O_4$ single crystals. Phys. Rev. B **56**, R12685 (1997).

4. Funahashi, R., Matsubara, I., Ikuta, H., Takeuchi, T., Mizutani, U., Sodeoka, S.: An Oxide single crystal with high thermoelectric performance in air. Jpn. J. Appl. Phys. **39**, L1127 (2000).

5. Wunderlich, W., Ohta, H., Koumoto, K.: Enhanced effective mass in doped $SrTiO_3$ and related perovskites. Phys. B **404**, 2202 (2009).

6. Shirai, K., Yamanaka, K.: Mechanism behind the high thermoelectric power factor of $SrTiO_3$ by calculating the transport coefficients. J. Appl. Phys. **113**, 053705 (2013).

7. Tomes, P., Robert, R., Trottmann, M., Bocher, L., Aguirre, M.H., Bitschi, A., Hejtmanek, J., Weidenkaff, A.: Synthesis and characterization of new ceramic thermoelectrics implemented in a thermoelectric oxide module. J. Elec. Mat. **39**(9), 1696 (2010).

8. Wunderlich, W., Fujiwara, H.: The difference between thermo- and pyroelectric Co-based RE-(=Nd, Y, Gd, Ce)-oxide composites measured by high-temperature gradient. J. Electron. Mater. **40**(2), 127–133 (2011).

9. Wunderlich, W.: Large Seebeck voltage of Co, Mn, Ni, Fe-ceramics. Adv. Ceram. Sci. Eng. (ACSE) 2(1), 9–15 (2013).

10. Kato, H., Kudo, A.: New tantalate photocatalysts for water decomposition into H and O_2. Chem. Phys. Lett. **295**, 487–492 (1998).

11. Kudo, A., Kato, H.: Effect of lanthanide-doping into $NaTaO_3$ photocatalysts for efficient water splitting. Chem. Phys. Lett. **331**, 373–377 (2000)

12. Wunderlich, W.: $NaTaO_3$ composite ceramics—a new thermoelectric material for energy generation. J Nucl. Mat. **389**(1), 57–61 (2009).

13. Wunderlich, W., Soga, S.: Microstructure and Seebeck voltage of $NaTaO_3$ composite ceramics with additions of Mn, Cr, Fe or Ti. J. Ceram. Proc. Res. **11**, 233–236 (2010)

14. Wunderlich, W., Baufeld, B.: Development of thermoelectric materials based on NaTaO3—composite ceramics. In: Wunderlich, W. (ed.) Ceramic materials, pp. 1–27. Intech Publisher, London (2010).

15. Wunderlich, W., Mori, T., Sologub, O., Baufeld, B.: SPS-sintering of $NaTaO_3$–Fe_2O_3 composites. J. Aust. Ceram. Soc. **47**(2), 57–60 (2011)

16. Nazir, S., Schwingenschlögl, U.: High charge carrier density at the NaTaO3/SrTiO3 hetero-interface. Appl. Phys. Lett. **99**, 073102 (2011).

8

Correlativity of the nitrogen oxide adsorption mechanism and crystal structure in hollandite-type compounds

Kenjiro Fujimoto · Chihiro Yamakawa ·
Yuki Yamaguchi · Shigeru Ito

Abstract Hollandite-type $K_{1.88}Ga_{1.88}Sn_{6.12}O_{16}$ and $K_{1.58}Ga_{1.58}Ti_{6.42}O_{16}$ powders obtained by the solid-state reaction method were examined by DRIFT spectroscopy to evaluate their NO adsorption properties. Two key findings were deduced from the refined structure information and evaluation of NO adsorption on a hollandite surface: first, the presence of oxygen in carrier gas contributed significantly to NO adsorption; second, the occupation probability and atomic coordinate of alkaline-metal at the end of the one-dimensional tunnel structure influenced the adsorption volume per unit cell.

Keywords Hollandite · One-dimensional ionic conductor · Catalyst · Nitrogen oxide · Adsorption

Introduction

Hollandite-type compounds have hitherto studied as the one-dimensional fast ionic conductor [1–3], as the nuclear waste immobilizers [4, 5] and as NO_x reduction catalysts [6, 7]. It is thought that these properties contribute significantly to the crystal structure. Figure 1 shows the crystal structure of hollandite-type compound. The general chemical formula for hollandite compounds can be described as $A_xM_yN_{8-y}O_{16}$ ($x \leq 2$, $y \leq 2$), where "A" usually consists of alkali or alkaline earth ions, "M", of di or trivalent cations, and "N", of tetravalent cations. And, the hollandite-type structure has a tetragonal symmetry and contains one-dimensional tunnels extended along a unique

axis with a lattice period of about 0.3 nm. The framework of a hollandite structure consists of double chains of metal–oxygen octahedra edge-shared with adjacent ones.

The potassium ions in the tunnels are well known to contribute to the one-dimensional fast ionic conductivity of Ti-type hollandites such as $K_{1.6}Al_{1.6}Ti_{6.4}O_{16}$ and $K_{1.6}Mg_{0.8}Ti_{7.2}O_{16}$ [1–3]. The correlations between physical properties and the crystal structure can be studied in hollandite structures by substituting the "N" element in $A_xM_yN_{8-y}O_{16}$ by Sn or by Ti.

The NO_x selective catalytic reduction properties of $K_2Ga_2Sn_6O_{16}$ and $K_{1.6}Ga_{1.6}Ti_{6.4}O_{16}$ have been evaluated using hydrocarbons as a reductant source in the presence of NO, C_3H_6 and O_2 [6, 7]. The NO_x conversion rates of $K_2Ga_2Sn_6O_{16}$ and $K_{1.6}Ga_{1.6}Ti_{6.4}O_{16}$ are estimated to be 40 and 10 %, respectively, at 350 °C. The active site of nitrogen oxide adsorption can be determined by studying the crystal structure in detail by single crystal X-ray diffraction and evaluating the adsorption form by infrared spectroscopy. Fujimoto et al. [8, 9] previously studied the single crystal growth and structure refinement of $K_{1.88}Ga_{1.88}Sn_{6.12}O_{16}$, $K_{1.98}Fe_{1.98}Sn_{6.02}O_{16}$ [10], and $K_{1.59}Ga_{1.59}Ti_{6.41}O_{16}$ [11] using the flux slow-cooling method. In the present study we prepared $K_xGa_xSn_{8-x}O_{16}$ (KGSO, $x \sim 2$) and $K_yGa_yTi_{8-y}O_{16}$ (KGTO, $y \sim 1.6$) powders and observed their NO adsorption properties to elucidate the NO adsorption mechanism on hollandite surfaces.

Experimental

Preparation of KGSO and KGTO powders

KGSO and KGTO powders were prepared by the conventional solid-state reaction method. The starting

K. Fujimoto (✉) · C. Yamakawa · Y. Yamaguchi · S. Ito
Department of Pure and Applied Chemistry, Faculty of Science and Technology, Tokyo University of Science, Yamazaki 2641, Noda, Chiba 278-8510, Japan
e-mail: fujimoto_kenjiro@rs.tus.ac.jp

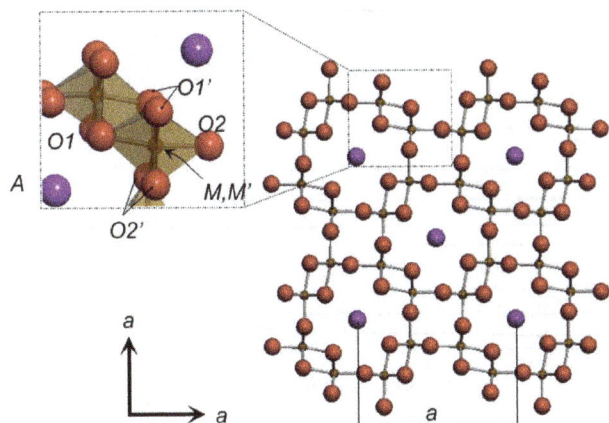

Fig. 1 Crystal structure of hollandite-type compound

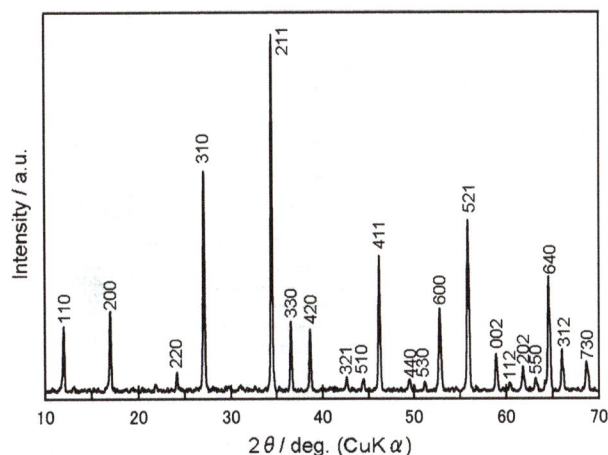

Fig. 2 Powder X-ray diffraction pattern of $K_{1.88}Ga_{1.88}Sn_{6.12}O_{16}$ prepared by the solid-state reaction method (@1,350 °C, 24 h)

materials were K_2CO_3, Ga_2O_3, SnO_2, and TiO_2 powders of 99.99 % purity. The KGSO powder was prepared by weighing out K_2CO_3, Ga_2O_3, and SnO_2 powders in a molar ratio of 1.05:1:3. The KGTO powder was prepared by weighing out K_2CO_3, Ga_2O_3, and TiO_2 powders in a molar ratio of 1.68:1.6:6.4. The mixtures were calcined at 950 °C for 1 h, ground, and sintered at either 1,350 °C for 24 h (KGSO) or 1,250 °C for 15 h (KGTO). The powders thus obtained were identified by the powder X-ray diffraction method using a Rigaku miniflex system with CuKα radiation ($\lambda = 0.15405$ nm). The chemical compositions were analyzed by ICP emission spectrometry (SHIMADZU ICPS-7510). In the chemical analysis, the single crystal aggregate was dissolved in concentrated HCl in a pressure vessel heated to about 100 °C.

Study of NO adsorption on hollandite surfaces

The NO adsorption behavior on the KGSO and KGTO powders was evaluated by diffuse reflectance infrared Fourier transform spectroscopy (DRIFTS). The combination of a selector and environmental chamber (Specac Ltd.) served as the diffuse reflectance apparatus. An environmental chamber consisting of a mounted furnace and ZnSe window was placed in the FT-IR spectrometer (FTS-45RD, BioRad Jpn. Co. Ltd.). KGSO and KGTO powders were preheated at 700 °C for 30 min with a gas mixture of N_2 (>99.9 %) and O_2 (>99.9 %) flowed through at a rate of 50 mL min^{-1}, to remove the adsorbate on the hollandite surfaces. After preheating at 700 °C, the background spectra were taken at 700, 600, 500, 400, 300, 200, 100 °C, and room temperature. The mixing ratio of N_2/O_2 was also switched from 100/0 to 80/20 to examine the correlation between the extent of NO adsorption and the amount of oxygen in the gas mixture. The pretreated powder was exposed to an NO (8,087 ppm) stream balanced with Ar (flow rate of 50 mL min^{-1}) at ambient temperature for

10 min. After the NO gas supply was stopped, the inside of the furnace was purged of the N_2/O_2 gas mixture for 10 min until the NO species in the gas phase became undetectable in the infrared spectrum. The powder exposed to the gas mixture was then heated stepwise to 100, 200, 300, 400, 500, 600, and 700 °C, to calculate the NO adsorption condition at each temperature by subtracting the background spectrum. Infrared spectra were recorded by accumulating 256 scans at a spectral resolution of 2 cm^{-1}.

Results and discussion

KGSO powder and its NO adsorption property

Figure 2 shows the powder X-ray diffraction pattern of KGSO powder after sintering at 1,350 °C for 24 h. The powder obtained showed a single phase of hollandite. The chemical composition of the KGSO powder was calculated to be $K_{1.88}Ga_{1.88}Sn_{6.12}O_{16}$ by ICP emission spectrometry and corresponded to the result for a single crystal.

Figures 3 and 4 show the DRIFT spectra in gas flows of N_2/O_2 at ratios of 100/0 and 80/20, respectively. The absorption bands of 1,800–1,950 cm^{-1} are attributable to the NO species in the gaseous phase. The NO adsorption on the KGSO surface was unobservable from the infrared spectra in Fig. 2, as all of the absorption bands were close to negligible for the $N_2/O_2 = 100/0$ gas flow, except for the NO species in the gaseous phase. When the carrier gas used included 20 % oxygen, absorption bands were observed around 1,400 and 1,750 cm^{-1} up to a temperature of 600 °C, as shown in Fig. 4. These results indicated that the NO adsorption on the hollandite compound surface required oxygen gas and manifested a pattern consistent with chemisorption.

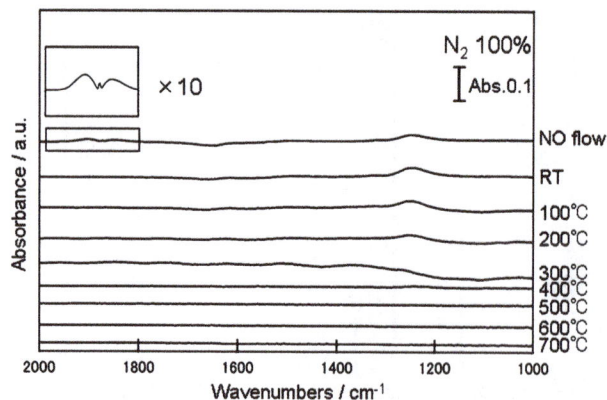

Fig. 3 DRIFT spectra of NO species on the $K_{1.88}Ga_{1.88}Sn_{6.12}O_{16}$ surface observed between room temperature and 700 °C ($N_2/O_2 = 100/0$)

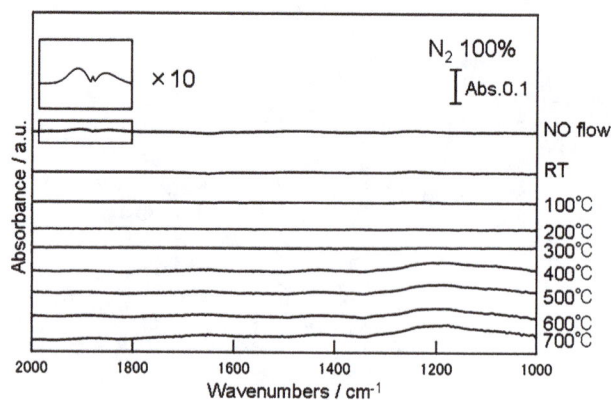

Fig. 6 DRIFT spectra of NO species on the $K_{1.58}Ga_{1.58}Ti_{6.42}O_{16}$ surface observed between room temperature and 700 °C ($N_2/O_2 = 100/0$)

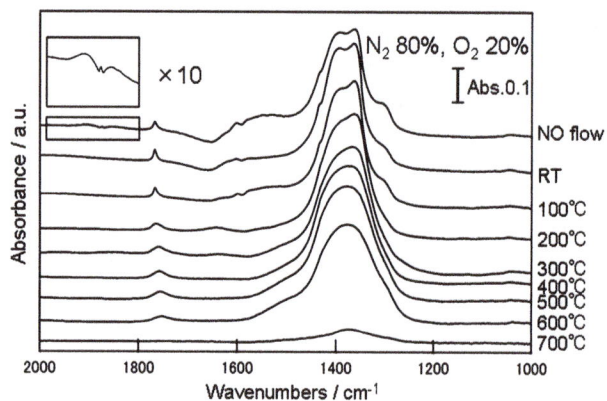

Fig. 4 DRIFT spectra of NO species on the $K_{1.88}Ga_{1.88}Sn_{6.12}O_{16}$ surface observed between room temperature and 700 °C ($N_2/O_2 = 80/20$)

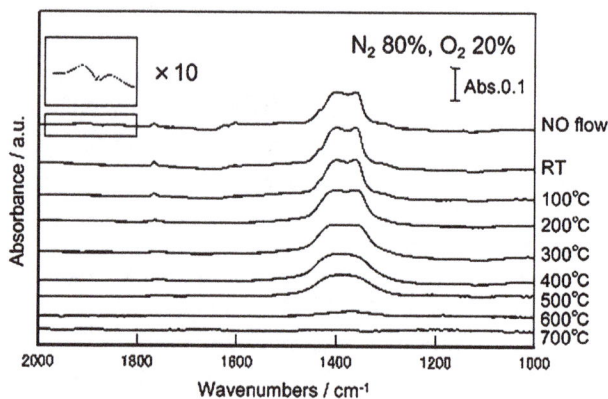

Fig. 7 DRIFT spectra of NO species on the $K_{1.58}Ga_{1.58}Ti_{6.42}O_{16}$ surface observed between room temperature and 700 °C ($N_2/O_2 = 80/20$)

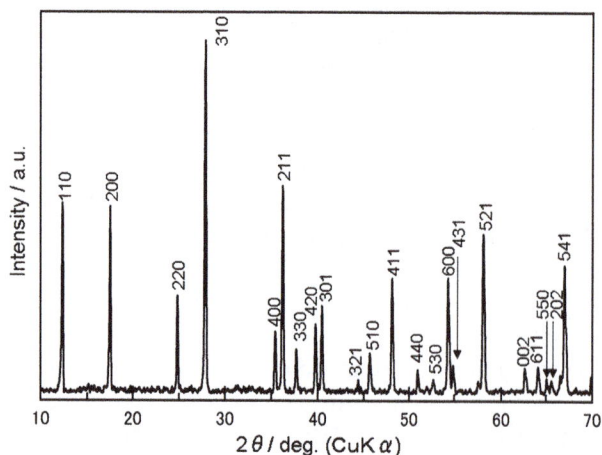

Fig. 5 Powder X-ray diffraction pattern of $K_{1.58}Ga_{1.58}Ti_{6.42}O_{16}$ prepared by the solid-state reaction method (@1,250 °C, 15 h)

KGTO powder and its NO adsorption property

Figure 5 shows the powder X-ray diffraction pattern of the KGTO powder after sintering at 1,250 °C for 15 h. The crystal structure of the sintered compound showed a single phase of hollandite. The chemical composition of the KGTO powder was calculated to be $K_{1.58}Ga_{1.58}Ti_{6.42}O_{16}$ by ICP emission spectrometry and corresponded to the result for a single crystal.

Figures 6 and 7 show DRIFT spectra in gas flow of N_2/O_2 at ratios of 100/0 and 80/20, respectively. In the former gas flow, at the gas composition ratio $N_2/O_2 = 100/0$, all of the absorption bands were close to negligible except for the NO species in the gaseous phase (see Fig. 6). When carrier gas included 20 % oxygen, absorption bands were observed around 1,400 and 1,750 cm^{-1} up to the temperature of 500 °C, as shown in Fig. 7.

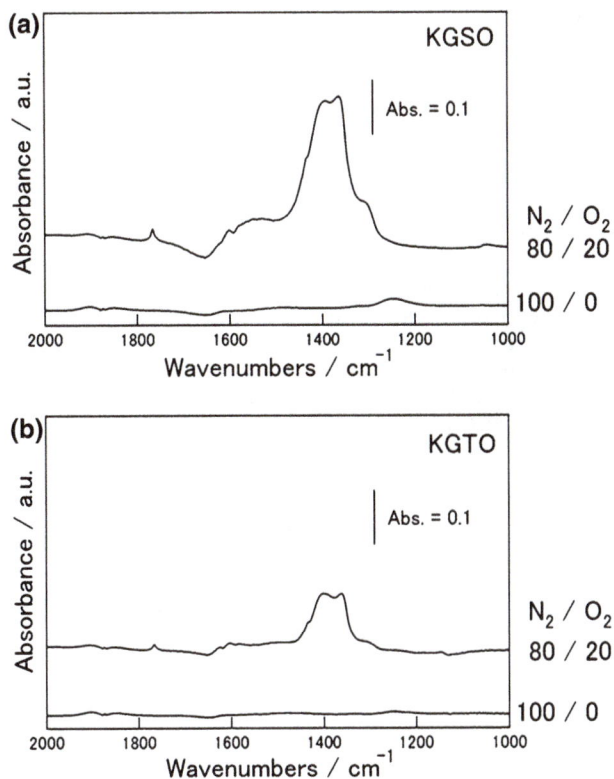

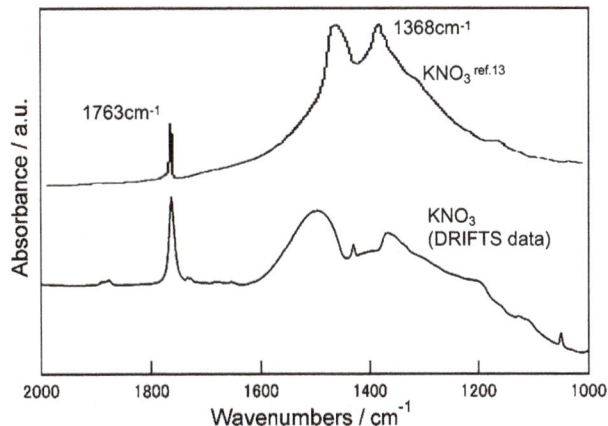

Fig. 9 DRIFT spectrum of KNO_3 powder and the IR spectra of KNO_3 by the Nujol mull method [13]

Fig. 8 DRIFT spectra of NO species adsorbed on (a) $K_{1.88}Ga_{1.88}Sn_{6.12}O_{16}$ powder and (b) $K_{1.58}Ga_{1.58}Ti_{6.42}O_{16}$ powder under different oxygen concentrations at room temperature

NO adsorption mechanism on the hollandite compound surfaces

Figure 8 shows DRIFT spectra observed just after the NO gas supply was stopped and the inside of the furnace was purged of the N_2/O_2 gas mixture. The absorbance intensity of NO adsorption clearly differed between the KGSO and KGTO surfaces when the N_2/O_2 gas mixture was 80/20.

Figure 9 shows the IR spectra of KNO_3 determined by the Nujol mull method [13] and the DRIFT spectrum of KNO_3 powder at room temperature. Nujol mull method is measured by grinding up the solid material, mixing it with liquid paraffin and nipping between NaCl or KBr plates. From these absorption bands defined as 1,763 and 1,368 cm^{-1}, it is thought that the NO adsorption form on hollandite surface is attributable to KNO_3.

Our group previously reported the structure refinement results for KGSO and KGTO single crystals [8, 9, 11]. From those results, we can conclude the following here:

1. The alkaline-metal site in hollandite-type KGSO and KGTO has the two atomic coordinates reported by Michiue [12], that is, a K1-site set in a special position located at the center of the bottleneck and a K2-site slightly shifted along the c-axis from the bottleneck center (one vacancy in the hollandite-tunnel makes two

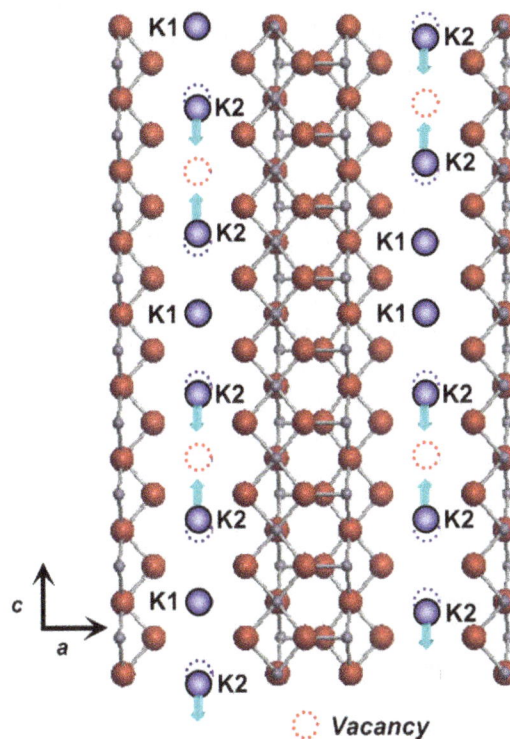

Fig. 10 Schematic image of K1-site and K2-site in one-dimensional tunnels of hollandite-type compound

K2-sites, to stabilize the crystal structure) as shown in Fig. 10.

2. The atomic coordinate of the potassium ion located nearest the surface in hollandite-type compound was assumed to be the K2-site, for the above reason. Under this scheme, the distance between the center of K2-site and surface was calculated to be 0.785 nm for KGTO and 0.766 nm for KGSO.

3. From the chemical formulas for KGSO and KGTO, the occupation probabilities for potassium ions at the K2-

sites were assumed to be 75 % (KGTO) and 90 % (KGSO).

Furthermore, from the crystallite size and the NO adsorption amount using temperature programmed desorption measurement of KGSO powder obtained by the solid-state reaction, the number of adsorbed NO molecules on hollandite surface was almost the same as the number of alkali ions at the end of the one-dimensional tunnel [14].

From a combination of previously reported data and these results, it is thought that NO gases are adsorbed as KNO_3 at the end of the one-dimensional tunnel by mediating oxygen gases in carrier gas. And, it is thought that the difference in adsorption intensity as shown in Fig. 9 depends on the occupation probability of the K2-site and distance and the distance between the K2-site and surface. In order to confirm these speculations, it is necessary to observe DRIFT spectrum using various crystal planes of single crystal and simulate adsorption behavior by molecular dynamics method and so on.

Conclusion

Hollandite-type $K_{1.88}Ga_{1.88}Sn_{6.12}O_{16}$ and $K_{1.58}Ga_{1.58}Ti_{6.42}O_{16}$ powders were prepared by the conventional solid-state reaction method to study the correlativity of the nitrogen oxide adsorption mechanism on the hollandite surface and the crystal structure information. The NO adsorption was observed by diffuse reflectance infrared Fourier transformed spectroscopy (DRIFTS). Absorption bands attributed to KNO_3 were observed as 1,763 and 1,368 cm^{-1}. NO adsorption on the KGSO and KGTO surfaces was observed up to temperatures of 600 and 500 °C, respectively. From structure refinement data and the NO adsorption property on the hollandite surface, we deduced that the presence of oxygen in the carrier gas contributed significantly to the NO adsorption, and that the occupation probability and atomic coordinate of the alkaline-metal at the end of the one-dimensional tunnel influenced the adsorption volume per unit cell.

References

1. Yoshikado, S., Ohachi, T., Taniguchi, I., Onoda, Y., Watanabe, M., Fujiki, Y.: Frequency-independent ionic conductivity of hollandite type compounds. Solid State Ion. 9(10), 1305–1309 (1983)
2. Yoshikado, S., Ohachi, T., Taniguchi, I., Onoda, Y., Watanabe, M., Fujiki, Y.: AC ionic conductivity of hollandite type compounds from 100 Hz to 37.0 GHz. Solid State Ion. 7(4), 335–344 (1982)
3. Yoshikado, S., Taniguchi, I., Watanabe, M., Onoda, Y., Fujiki, Y.: Frequency dependence of ionic conductivity in one-dimensional ionic conductors $K_{1.6}Mg_{0.8}Ti_{7.2}O_{16}$ and $K_{1.6}Al_{1.6}Ti_{6.4}O_{16}$. Solid State Ion. 79, 34–39 (1995)
4. Ringwood, A.E., Kesson, S.E., Ware, N.G., Hibberson, W., Major, A.: Immobilisation of high level nuclear reactor wastes in SYNROC. Nature 278, 219–223 (1979)
5. Mitamura, H., Matsumoto, S., Stewart, M.W.A., Tsuboi, T., Hashimoto, M., Vance, E.R., Hart, K.P., Togashi, Y., Kanazawa, H., Ball, C.J., White, T.J.: α-decay damage effects in curium-doped titanate ceramic containing sodium-free high-level nuclear waste. J. Am. Ceram. Soc. 77(9), 2255–2264 (1994)
6. Watanabe, M., Mori, T., Yamauchi, S., Yamamura, H.: Catalytic property of the hollandite-type 1-D ion-conductors: selective reduction of NO_x. Solid State Ion. 79, 376–381 (1995)
7. Mori, T., Yamauchi, S., Yamamura, H., Watanabe, M.: New hollandite catalysts for selective reduction of nitrogen monoxide with propene. Appl. Cat. A Gen. 129, L1–L7 (1995)
8. Fujimoto, K., Ito, S., Watanabe, M.: Crystal growth and refinement of $K_{1.88}Ga_{1.88}Sn_{6.12}O_{16}$ hollandite-type compound. Solid State Ion. 177, 1901–1904 (2006)
9. Fujimoto, K.: Private communication (in preparation)
10. Fujimoto, K., Takamori, K., Yamaguchi, K., Ito, S.: Single crystal growth and structure refinement of hollandite-type $K_{1.98}Fe_{1.98}Sn_{6.02}O_{16}$ (To be submitted)
11. Fujimoto, K., Yamakawa, C., Ito, S.: Crystal growth and structure refinement of hollandite-type $K_{1.59}Ga_{1.59}Ti_{6.41}O_{16}$. Solid State Ion. 184, 74–77 (2011)
12. Michiue, Y.: Superstructure of hollandite $K_xMg_{(8+x)/3}Sb_{(16-x)/3}O_{16}$ ($x \approx 1.76$). J. Solid State Chem. 180(6), 1840–1845 (2007)
13. Spectral Database for Organic Compounds.
14. Fujimoto, K., Suzuki, J., Harada, M., Awatsu, S., Mori, T., Watanabe, M.: Preparation of hollandite-type $K_xGa_xSn_{8-x}O_{16}$ thin film and NO adsorption behavior. Solid State Ion. 152–153, 769–775 (2002)

9

Grape pigment (malvidin-3-fructoside) as natural sensitizer for dye-sensitized solar cells

N. Gokilamani · N. Muthukumarasamy ·
M. Thambidurai · A. Ranjitha ·
Dhayalan Velauthapillai

Abstract TiO$_2$ nanocrystalline thin films have various applications, among them dye-sensitized solar cells are more promising and low-cost alternative to conventional inorganic photovoltaic devices. TiO$_2$ nanocrystalline thin films have been sensitized with the natural dye (malvidin-3-fructoside) extracted from grape fruits. The interaction between the semiconductor and the dye has been studied. The maximum absorption band of grape extract at 525 nm has been shifted to 545 nm after incorporation of the dye indicating interaction of the dye molecules with TiO$_2$. The grape extract has polyphenolic anthocyanin pigment (malvidin-3-fructoside), which has carboxylic and hydroxyl groups that can attach effectively to the surface of TiO$_2$ film. The dye-sensitized TiO$_2$-based solar cell sensitized using grape dye exhibited a J_{sc} of 4.06 mA/cm^2, V_{oc} of 0.43 V, FF of 0.33 and efficiency of 0.55 %. Natural dye-sensitized TiO$_2$ photo electrodes present the prospect to be used as an environment-friendly, low-cost alternative system.

Keywords Nanocrystalline · Sol–gel · TiO$_2$ thin films · Dye-sensitized solar cell

N. Gokilamani (✉) · N. Muthukumarasamy · A. Ranjitha
Department of Physics, Coimbatore Institute of Technology,
Coimbatore, India
e-mail: gokilarajasekar@gmail.com

M. Thambidurai
Department of Electrical and Computer Engineering, Global
Frontier Center for Multiscale Energy Systems, Seoul National
University, Seoul 151-744, Republic of Korea

D. Velauthapillai
Department of Engineering, University College of Bergen,
Bergen, Norway

Introduction

Nanocrystalline TiO$_2$ films have attracted scientific and technological interests because of their potential applications in the fields of photo electronic optical devices, solar cells, gas sensors, photocatalysts and biomaterials [1–3]. TiO$_2$ is the best photo-catalyst so far, that absorbs the light at shorter wavelength below 400 nm, due to its wide band gap. Dye-sensitized solar cell, a device converting light energy to electrical energy by imitating photosynthesis of plants [4], was firstly developed by Gratzel's group [5–7]. It is widely known as a low-cost and easy-assembly solar cell, in which both synthetic and natural dyes can be used as a sensitizer. Electrical energy is generated when the cell is exposed to sunlight. Electrons in dye molecules are excited and then injected to the conduction band of a wide band gap (TiO$_2$) n-type semiconductor on which the dye molecules adsorb. These electrons migrate through the host semiconductor particles until they reach the collector; also the holes simultaneously generated are reduced by a redox electrolyte or hole carrier at the back electrode (Pt). Because of low material cost, simple preparation, and lack of heavy metals, natural dyes are more favorable to be applied in such device [8–12].

Solar cells are the basic block of solar array where the absorption of light quanta of specific energy results in generation of charge carriers. Due to absorption of sun light, dye molecules get excited from the highest occupied molecular orbital's (HOMO) to the lowest unoccupied molecular orbital's (LUMO).

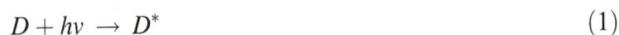

$$D + hv \rightarrow D^* \tag{1}$$

Once an electron injected into the conduction band of the wide band gap semiconductor nano structured TiO$_2$ film, the dye molecule (photosensitizer) become oxidized.

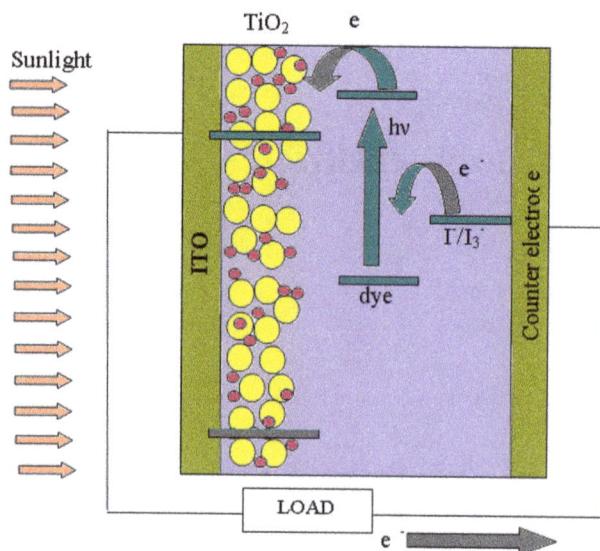

Fig. 1 Schematic diagram of dye-sensitized solar cell

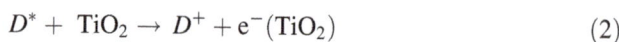

$$D^* + \ TiO_2 \rightarrow D^+ + e^- (TiO_2) \qquad (2)$$

The injected electron is transported between the TiO_2 nanoparticles and then gets extracted to a load where the work done is delivered as an electric energy.

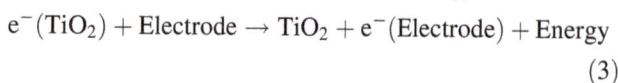

$$e^-(TiO_2) + Electrode \rightarrow TiO_2 + e^-(Electrode) + Energy \qquad (3)$$

To mediate electron between the TiO_2 photoelectrode and the platinum counter electrode, electrolyte containing I^-/I_3^- redox ions is used to fill the cell. Therefore, the oxidized dye molecules (photosensitizer) are regenerated by receiving electrons from the I^- ion redox mediator that will oxidized to I_3^- (Tri-iodide ions).

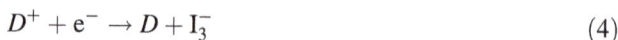

$$D^+ + e^- \rightarrow D + I_3^- \qquad (4)$$

The I_3^- substitutes the donated electron internally with that from the external load and gets reduced back to I^- ion.

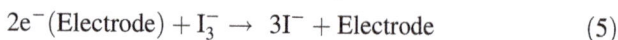

$$2e^-(Electrode) + I_3^- \rightarrow \ 3I^- + Electrode \qquad (5)$$

Therefore, generation of electric power in dye-sensitized solar cell causes no permanent chemical change or transformation. The conducting electrodes are prepared such that they posses low-sheet resistance and very high transparency in order to facilitate high solar cell performance [13]. Figure 1 shows the structure of a dye-sensitized solar cell.

Anthocyanins consists a large family of widespread flavonoids in plants and they are responsible for many fruit and floral colors that are observed in nature. Chemically, these flavonoids are most commonly based on six anthocyanidins: pelargonidin, cyanidin, peonidin, delphinidin, petunidin and malvidin [14]. Interest in the anthocyanins has increased because of their potential as natural colorants

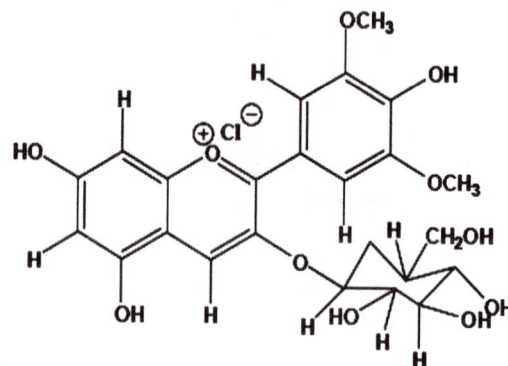

Fig. 2 Structure of malvidin-3-fructoside

and non-toxicity. They have bright attractive colors, and they are water-soluble. Acylated anthocyanin pigments show greater stability during processing and storage. Because of the presence of acylated groups in the structure of anthocyanins, it is believed that they can protect the oxonium ion from hydration, thereby prevent the formation of hemiketal (pseudobase) or chalcone forms. Anthocyanin pigments were found in 1835 for the first time. In 1916, Wehldale proposed anthocyanin flavanons formation pathway, since they were products of shikimic acid cycle. Willstatter and Zollinger [15, 16] reported malvidin 3-fructoside as the major pigment in grapes, and malvidin 3,5-diglucoside and malvidin as the minor ones. Anthocyanins from grapes include mono and diglucosides of five different aglycones with the addition of monoacylation [14]. The structure of malvidin 3-fructoside is shown in Fig. 2.

Sol–gel dip coating method has been used because of the low-cost equipments involved, lower growth temperatures and reproducibility. Many sensitizers have been used by different workers to sensitize TiO_2 nanoparticles for solar cell applications. In this paper, we report a simple route to synthesize TiO_2 nanoparticles. TiO_2 nanoparticles have been used as photoelectrode material to fabricate the dye-sensitized solar cells. The use of non-toxic natural pigments as sensitizers would definitely enhance the environmental and economic benefits of this alternative form of solar energy conversion. Natural dye-sensitized solar cells are a promising class of photovoltaic cells with the capability of generating green energy at low production cost since no vacuum systems or expensive equipment are required in their fabrication. In addition, natural dyes are abundant, easily extracted and safe materials.

Experimental

To prepare the photo-anode of dye-sensitized solar cells, the indium-doped tin oxide conducting (ITO) glass sheet

(Asahi Glass; Indium-doped SnO_2, sheet resistance: 15 U/ square) was first cleaned in a detergent solution using an ultrasonic bath for 15 min, rinsed with double distilled water and then dried. The matrix sol was prepared by mixing 0.695 ml of titanium (IV) isopropoxide with 15 ml of isopropanal at room temperature and stirred for half an hour. 0.135 ml of glacial acetic acid is added drop wise and stirred vigorously for 2 h to obtain a homogeneous mixture of TiO_2 sol. The prepared sol was deposited on the Indium-doped tin oxide (ITO) glass plate by sol–gel dip coating method. The film was dried at 80 °C for 30 min in air and then annealed at 500 °C in a muffle furnace. For grapes a dye extract preparation, well cleaned grape fruits were mixed with 250 ml ethanol and were kept for 12 h at room temperature. Then residual parts were removed by filtration. The ethanol fraction was separated and few drops of concentrated HCl was added so that the solution became deep red color (pH < 1). This was directly used as dye solution for sensitizing TiO_2 electrodes.

Crystallinity and phase analysis of the films were carried out using X-ray diffraction method (Rigaku Rint 2000 series). Energy dispersive X-ray analysis (EDAX) and structural studies have been carried out using transmission electron microscope (JEOL, JEM-2100). Transmission spectra have been recorded using UV–VIS–NIR spectro-photometer (Jasco V-570). Lithium iodide, iodine and acetonitrile purchased from Sigma-Aldrich have been used as received for the preparation of electrolyte. The redox electrolyte with $[I_3^-]/[I^-]$ 1:9 was prepared by dissolving 0.5 M LiI and 0.05 M I_2 in acetonitrile solvent. Since LiI is extremely hygroscopic, electrolytes were prepared in a dry room maintained at dew point of 60 °C. The counter electrode was prepared by using platinum chloride as follows: the H_2PtCl_6 solution in isopropanol (2 mg/ml) was deposited onto the ITO glass by spin coating method. TiO_2 electrode annealed at 500 °C was immersed in the extracted dye solution at room temperature for 24 h in the dark, since the grape dye extract is not stable under prolonged illumination of light. The electrode was then rinsed with ethanol to remove the excess dye present in the electrode and then the electrode was dried. The counter electrode was placed on the top of the TiO_2 electrode, such that the conductive side of the counter electrode facing the TiO_2 film with a spacer separating the two electrodes. The two electrodes were clamped firmly together using a binder clip. Now the prepared liquid electrolyte solution was injected into the space between the clamped electrodes. The electrolyte enters into the cell by capillary action. This resulted in the formation of sandwich type cell. Natural dye-sensitized TiO_2-based solar cells have been fabricated with area of 0.25 cm^2, and it was found that the cell efficiency was independent of cell area in this range as

reported by Yamazaki et al. [17]. The J–V characteristics of the cell were recorded using a Keithley 4200-SCS meter. A xenon lamp source (Oriel, USA) with an irradiance of 100 mW/cm^2 was used to illuminate the solar cell (equivalent to AM1.5 irradiation).

Results and discussion

Figure 3 shows the X-ray diffraction pattern of the sol–gel prepared TiO_2 films, annealed at 500 °C. A narrow peak at 25.35° corresponding to (101) reflection of the anatase phase of TiO_2 has been observed in the diffraction pattern. The observed peaks corresponds to (1 0 1), (0 0 4), (2 0 0), (1 0 5), (2 1 1) and (2 0 4) which represent only anatase phase of TiO_2. No peaks corresponding to the rutile or brookite phase has been observed. The grain size has been calculated using Scherer's formula [18].

$$D = K\lambda/\beta\cos\theta \qquad (6)$$

where D is the grain size, K is a constant taken to be 0.94, λ is the wavelength of the X-ray radiation ($\lambda = 1.51$ Å), β is the full width at half maximum and θ is the angle of diffraction. The grain size was found to be 33 nm for the film annealed at 500 °C. The annealing temperature facilitates the subsequent crystal growth process, accompanied by the diffusion of titania species forming big size anatase crystals formed by merging of some adjacent mesopores. At the same time, the spatial confinement by mesopore arrays controls the formation and growth of anatase phase, leading to a more or less uniform distribution of titania nanocrystals.

Energy dispersive X-ray (EDAX) spectra of nanocrystalline TiO_2 thin film is shown in Fig. 4. The compositional analysis of TiO_2 thin films shows the presence of 32.7 at%

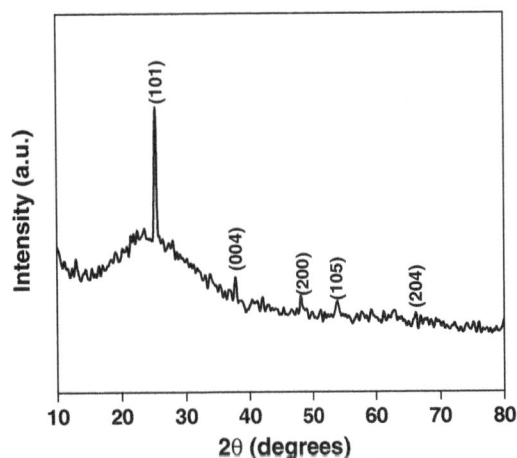

Fig. 3 X-ray diffraction pattern of annealed at 500 °C TiO_2 thin films

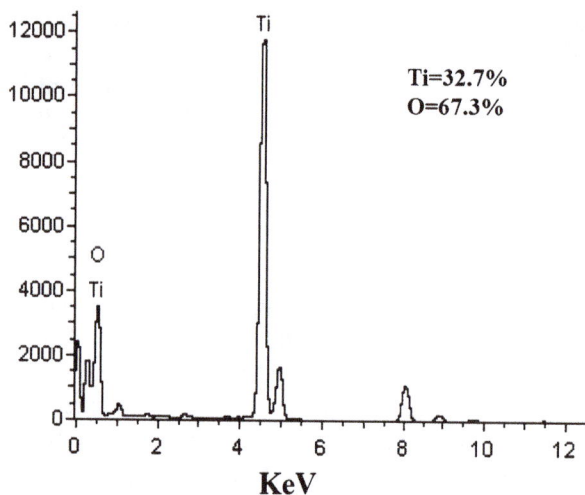

Fig. 4 The EDAX spectra of TiO$_2$ thin film

of Ti and 67.3 at% of O. Figure 5 shows the transmission electron microscope images of the TiO$_2$ thin film annealed at 500° C. Figure 5a shows the presence of close-packed agglomerated nanoparticles of uniform size which causes the mesoporous structure. This accumulation of nanoparticles creates narrow channels that may serve as electronic injection membranes [19]. The mesoporous structure, which provides a large surface area for adsorbing the dye, has been achieved in the present study, although no polymer was added to make pores. The prepared TiO$_2$ films exhibited good mechanical strength, had good adherence to the substrate and could not be easily erased by hand. Figure 5b shows the high-resolution transmission electron microscope (HRTEM) image of TiO$_2$ thin films. The interplanar distance has been calculated using the lattice fringes and is found to be 0.33 nm which corresponds to (101) lattice plane of anatase phase. Selective area electron diffraction pattern is used to learn about the crystal properties of a particular region. Figure 5c shows the selective area electron diffraction pattern of the nanocrystalline TiO$_2$ thin film and the presence of rings with discrete spots suggest that the TiO$_2$ nanocrystalline film is made of small particles of uniform size with anatase phase.

Figure 6a–c shows atomic force microscope images of the TiO$_2$ thin films annealed 400, 450 and 500 °C. The image shows well-defined particle-like features with granular topography and indicates the presence of small crystalline grains. Because of the heat treatment, the nano crystalline phase has been formed and this has led to the appearance of grains making the films to have higher surface roughness. The root mean square surface roughness of the film was found to be 19, 27 and 32 nm for the TiO$_2$ films annealed at 400, 450 and 500 °C.

To discriminate the local order characteristics of the TiO$_2$ films, we carried out non-resonant Raman

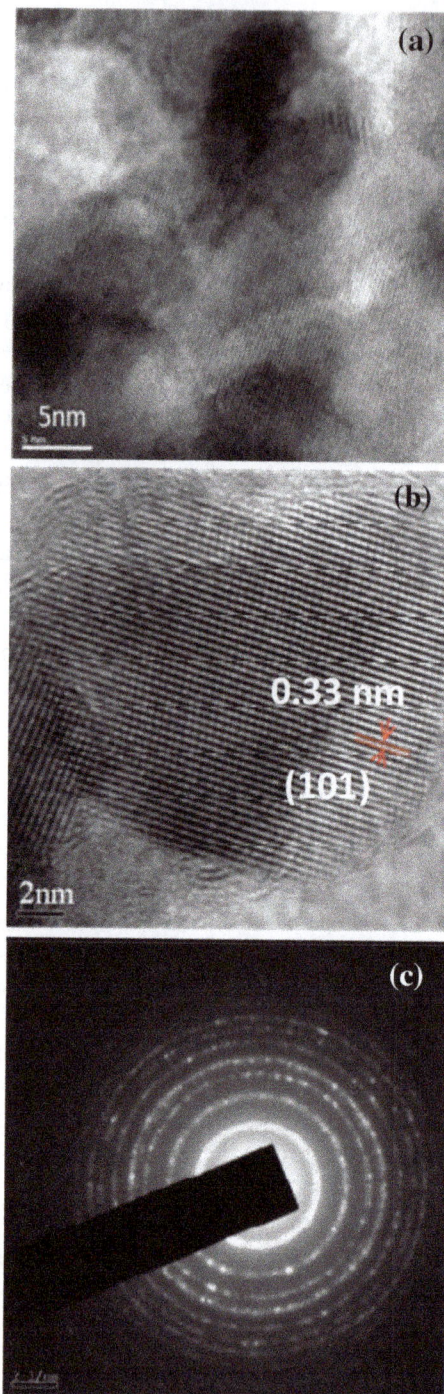

Fig. 5 a TEM image **b** HRTEM image and **c** SAED pattern of TiO$_2$ nanocrytalline thin film annealed at 500 °C

spectroscopy studies. The technique is non-destructive, capable to elucidate the titania structural complexity as peaks from each crystalline phase is clearly separated in frequency, and therefore the anatase and rutile phases are easily distinguishable. Figure 7 shows the Raman spectra of the nanocrystalline TiO$_2$ sample annealed at 500 °C. Raman peaks originate from the vibration of molecular

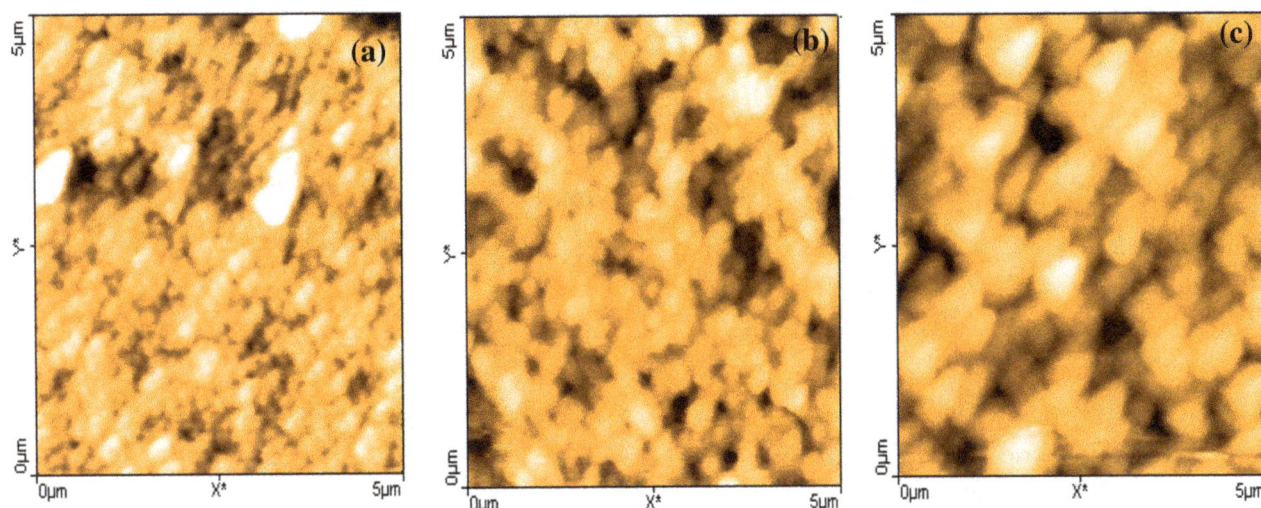

Fig. 6 AFM images of TiO$_2$ thin films annealed at **a** 400 °C, **b** 450 °C and **c** 500 °C

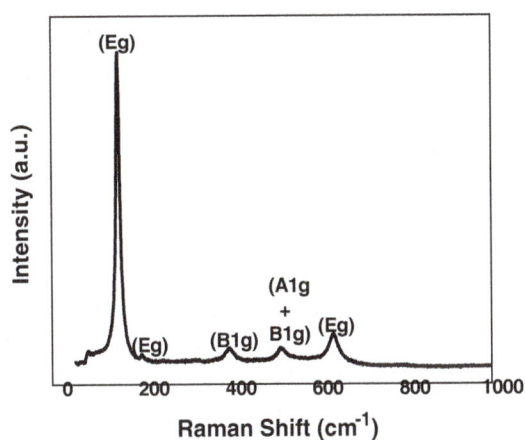

Fig. 7 The Raman spectra of TiO$_2$ thin film annealed at 500 °C

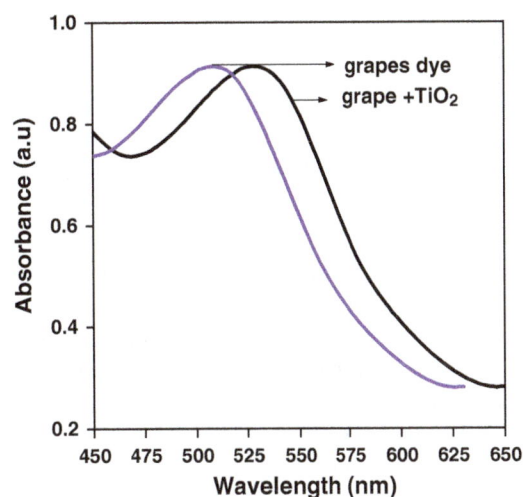

Fig. 8 Absorption spectra of grapes sensitized TiO$_2$ thin films

bonds, that is, vibrational mode Eg and A1g peaks, which are related to different crystal planes. According to factor group analysis, anatase has six Raman active modes (A1g + 2B1g + 3Eg). Ohsaka et al. [20] have reported the Raman spectra of an anatase TiO$_2$ and have stated that six allowed modes appear at 144 cm^{-1} (Eg), 197 cm^{-1} (Eg), 399 cm^{-1} (B1g), 513 cm^{-1} (A1g), 519 cm^{-1} (B1g), and 639 cm^{-1} (Eg). The vibrational peaks of the nanocrystalline TiO$_2$ samples at 143, 197, 396, 519, 638 cm^{-1} and the absence of overlapped broad peaks show that the material is well crystallized, with low number of imperfect sites. These peaks are unambiguously attributed to the anatase modification. A special attention must be given to the Raman peaks observed at 143, 197, 396, 519, 638 cm^{-1} which are all slightly shifted, probably due to a smaller size of TiO$_2$ nanocrystalline particles. The peak at 513 and 517 cm^{-1} are very close to each other. The main spectral features of samples annealed at different temperatures are closely similar which mean that the prepared samples

posses a certain degree of long range order of the anatase phase. The spectra vary symmetrically with grain size. It deteriorates with spectroscopic lines broadening and merging, line intensity decreases and position shifting with decrease in annealing temperature. However, the spectrum of sample showing the band broadening with decrease in intensity [21].

Optical absorption spectra of grape fruits and grape fruits absorbed TiO$_2$ nanocrystalline thin films are shown in Fig. 8. It is seen that the absorption peak of grape fruits extract is maximum at 510 nm. The presence of the absorbance peak at 530 nm in grape fruits sensitized TiO$_2$ nanocrystalline thin films confirms the incorporation of dye into the TiO$_2$ film. Malvidin-3-fructoside flavanoid is responsible for the absorption of light. This chemical attachment affects the energy levels of the highest occupied molecular level (HOMO) and the lowest unoccupied

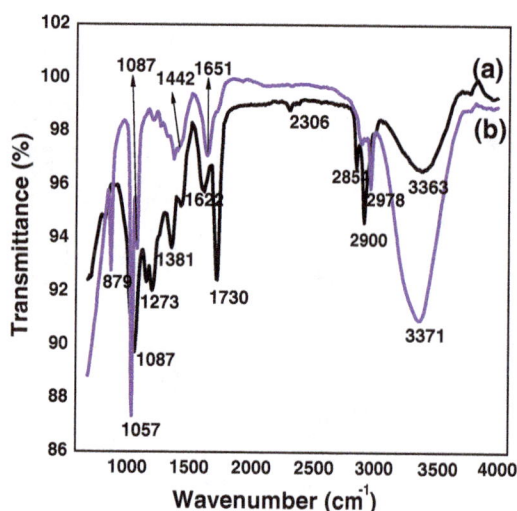

Fig. 9 FTIR spectra of *a* grape extract, *b* grape extract-sensitized TiO$_2$

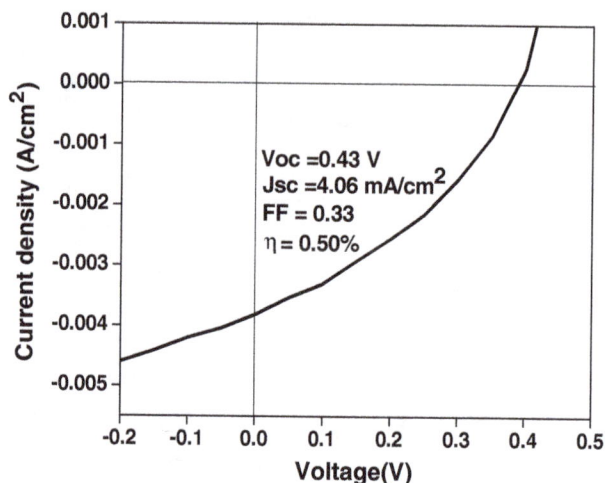

Fig. 10 J–V characteristics of dye-sensitized TiO$_2$-based solar cells

molecular level (LUMO) of the anthocyanidin molecule [22], which eventually affects the band gap of these materials and this results in a shift in the absorption peak of the absorption spectra. It is seen that absorption coefficients of grape fruits extract is about 14 times higher than that of the N-719 dye used in high efficiency dye-sensitized solar cells [23]. The absorption bands of the dye adsorbed TiO$_2$ semiconductor films are shifted to longer wavelength when compared to the absorption spectra of the dye solution as shown in Fig. 8. The intensity of light absorption has been enhanced due to the interfacial Ti–O coupling which exists between the C=O, C–H, C–O, O–H bonding of dye molecules and TiO$_2$ molecules [24].

Figure 9a, b shows the Fourier transform infrared spectra (FTIR) of grape extract, grape extract-sensitized TiO$_2$ nanocrystalline thin films. The spectra have been recorded using FTIR 1600 infrared spectrophotometer, operating in the wave number range 800–4,400 cm^{-1}. For the grape extract, the bands in the region of 1,651 cm^{-1} represents the amide group and the bands in the regions 2,900, 2,978 cm^{-1} represents the C–H bonding of the grape extract and a broad sharp peak is obtained in the region 3,371 cm^{-1} which corresponds to the hydrogen bonding. The bands in the region 1,381 cm^{-1} correspond to OH bonding. The bands in the region 1,273 and 1,087 cm^{-1} correspond to O–C bonding and the band at 879 cm^{-1} represents the C–H bonding in the grape extract. The grape extract-sensitized TiO$_2$ nanocrystalline thin films shows a peak at 1,442 cm^{-1} which corresponds to the absorption band of TiO$_2$ molecules [25]. The bands ranging from 2,854 to 3,363 cm^{-1} correspond to the C–H stretching due to methyl and methylene groups.

The J–V characteristics of TiO$_2$ nanocrystalline thin films sensitized with natural dyes is shown in Fig. 10. The

overall efficiency η of the grape dye cells are evaluated in terms of open circuit voltage (V_{oc}), short circuit current (J_{sc}) and fill factor (FF) using the formula

$$FF = \frac{J_{max} \times V_{max}}{J_{sc} \times V_{oc}} \tag{7}$$

$$\eta = \frac{J_{sc} \times V_{oc} \times FF}{P_{in}} \tag{8}$$

where J_{sc} is the short circuit photocurrent density (mA cm^{-2}), V_{oc} the open circuit voltage (volts), P_{in} is the intensity of the incident light ($P_{in} = 1$ Wcm^{-2}) and J_{max} (mA cm^{-2}) and V_{max} (volts) are the maximum current density and voltage in the J–V curve, respectively, at point of maximum power output. The solar cell sensitized with grape extract dye exhibited a power conversion efficiency of 0.55 % with a short circuit current density (J_{sc}) of 4.06 mA/cm^2, open circuit voltage (V_{oc}) of 0.43 V and fill factor (FF) of 0.33.

Conclusion

TiO$_2$ nanocrystalline thin films have been prepared by a simple sol–gel method. X-ray diffraction analysis reveals that the TiO$_2$ nanocrystalline thin films exhibit anatase structure. The dyes extracted from grapes strongly absorb visible light at 510 nm and the absorption bands of the dye adsorbed TiO$_2$ semiconductor films are shifted to longer wavelength and have been found to be suitable for the use as sensitizer in solar cells. The efficiency of the fabricated dye-sensitized solar cell using grapes extract is 0.55 %. Unlike artificial dyes, the natural ones are available, easy to prepare, low in cost, non-toxic, environmentally friendly and fully biodegradable.

References

1. Zhou, W., Du, G., Hu, P., Li, G., Wang, D., Liu, H., Wang, J., Robert, I., Lio, B.D., Jiang, H.: Nanoheterostructures on TiO_2 nanobelts achieved by acid hydrothermal method with enhanced photocatalytic and gas sensitive performance. J. Mater. Chem. **21**, 7937–7945 (2011)

2. Galstyan, V., Comini, E., Faglia, G., Sberveglieri, G.: TiO_2 nanotubes: recent advances in synthesis and gas sensing properties. Sensors (Basel). **13**(11), 14813–14839 (2014)

3. Gokilamani, N., Muthukumarasamy, N., Thambidurai, M., Ranjitha, A., Velauthapillai, D.: Utilization of natural anthocyanin pigments as photosensitizers for dye-sensitized solar cells. J. Sol–Gel. Sci. Technol. **66**, 212–219 (2013)

4. Hagfeldt, A., Didriksson, B., Palmqvist, T., Lindstrom, H., Sodergren, S., Rensmo, H., Lindqist, S.E.: Verification of high efficiencies for the Gratzel cell. A 7 % efficient solar cell based on dye-sensitized colloidal TiO_2 films. Sol. Energ. Mater. Sol. Cells **31**, 481–488 (1994)

5. Nazeeruddin, M.K., Kay, A., Rodicio, I., Humphry-Baker, R., Muller, E., Liska, P., Vlachopoulos, N., Gratzel, M.: Conversion of light to electricity by cis-X2bis(2,20-bipyridyl-4,40-dicarboxylate)ruthenium(II) charge-transfer sensitizers (X = Cl^-, Br^-, I^-, CN^-, and SCN^-) on nanocrystalline TiO_2 electrodes. J. Am. Chem. Soc. **115**, 6382–6390 (1993)

6. Barbe, C.J., Arendse, F., Comte, P., Jirousek, M., Lenzmanne, F., Shklover, V., Gratzel, M.: Nanocrystalline titanium oxide electrodes for photovoltaic applications. J. Am. Ceram. Soc. **80**, 3157–3171 (1997)

7. Kalyanasundaram, K., Gratzel, M.: Applications of functionalized transition metal complexes in photonic and optoelectronic devices. Coord. Chem. Rev. **177**, 347–414 (1998)

8. Gokilamani, N., Muthukumarasamy, N., Thambidurai, M., Ranjitha, A., Velauthapillai, D., Senthil, T.S., Balasundaraprabhu, R.: Dye-sensitized solar cells with natural dyes extracted from rose petals. J. Mater. Sci. Mater. Electron **24**, 3394–3402 (2013)

9. Thambidurai, M., Muthukumarasamy, N., Velauthapillai, D., Lee, C.: Synthesis and characterization of flower like ZnO nanorods for dye-sensitized solar cells. J. Mater. Sci. Mater. Electron **24**, 2367–2371 (2013)

10. Thambidurai, M., Muthukumarasamy, N., Velauthapillai, D., Lee, C.: Chemical bath deposition of ZnO nanorods for dye sensitized solar cell applications. J. Mater. Sci. Mater. Electron **24**, 1921–1926 (2013)

11. Thambidurai, M., Muthukumarasamy, N., Velauthapillai, D., Lee, C.: Synthesis of garland like ZnO nanorods and their application in dye sensitized solar cells. Mater. Lett. **92**, 104–107 (2013)

12. Gokilamani, N., Muthukumarasamy, N., Thambidurai, M., Ranjitha, A., Velauthapillai, D.: *Solanum nigrum* and *Eclipta alba* leaf pigments for dye sensitized solar cell applications. J. Sol–Gel. Sci. Technol. **69**(1), 17–20 (2014)

13. EbrahimJasim, Khalil: Natural dye-sensitized solar cell based on nanocrystalline TiO_2. Sains Malaysiana **41**(8), 1011–1016 (2012)

14. Heidari, R. , Khalafi, R. , Dolatabadzadeh, N.: Anthocyanin pigments of Siahe Sardasht grapes. J. Sci. Islamic Repub. Iran. **15**(2), 113–117 (2004)

15. Willstatter, R., Zollinger, E.H.: VI. Uber die Farbstoffe der Weintraube und der Heidelbeere. Ann. Chem. Liebigs. **408**, 83–109 (1915)

16. Willstatter, R., Zollinger, E.H.: XI. Uber die Farbstoffe der Weintraube und der Heidelbeere, II. Ann. Chem. Liebigs. **412**, 195–216 (1916)

17. Eiji, Y., Masaki, M., Naomi, N., Noritsugu, H., Masashi, S., Osamu, K.: Utilization of natural carotenoids as photosensitizers for dye sensitized solar cells. Sol. Energ. **81**, 512–516 (2007)

18. Sapanhel, L., Anderson, M.A.: Synthesis of porous quantum-size cadmium sulfide membranes: photoluminescence phase shift and demodulation measurements. J. Am. Chem. Soc. **112**, 2278–2284 (1990)

19. Cao, Y.M., Bai, Y., Yu, Q.J., Cheng, Y.M., Liu, S., Shi, D., Gao, F., Wang, P.: Dye-sensitized solar cells with a high absorptivity ruthenium sensitizer featuring a 2-(hexylthio) thiophene conjugated bipyridine. J. Phys. Chem. C **113**(15), 6290–6297 (2009)

20. Ohsaka, Toshiaki: Temperature dependence of the Raman spectrum in anatase TiO_2. J. Phys. Soc. Jpn. **48**, 1661–1668 (1980)

21. Šćepanović, M.J., Grujić-Brojčin, M., Dohčević-Mitrović, Z.D., Popović, Z.V.: Characterization of anatase TiO_2 nanopowder by variable–temperature Raman spectroscopy. Sci. Sinter. **41**, 67–73 (2009)

22. Sheng, Meng, Jun, Ren, Efthimios, Kaxiras: Natural dyes adsorbed on TiO_2 nanowire for photovoltaic applications: enhanced light absorption and ultrafast electron injection. Nano. Lett. **8**(10), 3266–3272 (2008)

23. Senthil, T.S., Muthukumarasamy, N., Velauthapillai, D., Agilan, S., Thambidurai, M., Balasundaraprabhu, R.: Natural dye (cyanidin 3-O-glucoside) sensitized nanocrystalline TiO_2 solar cell fabricated using liquid electrolyte/quasi-solid-state polymer electrolyte. Renew. Energ. **36**, 2484–2488 (2011)

24. Fernando, J.M.R.C., Senadeera, G.K.R.: Natural anthocyanins as photosensitizers for dye sensitized solar devices. Curr. Sci. **95**, 663–666 (2008)

25. Gratzel, M.: Perspectives for dye-sensitized nanocrystalline solar cells. Prog. Photovolt. Res. Appl. **8**, 171–185 (2000)

10

A comparative study of CO$_2$ sorption properties for different oxides

Sushant Kumar · Surendra K. Saxena

Abstract It is essential to capture CO$_2$ from flue gas stream, which is considered as one of the prime reasons of global warming. Although various CO$_2$ capture technologies already exist, most of these techniques are still unfit to be employed at a large scale. In the past one decade, oxides have emerged as a strong candidate to capture CO$_2$ for post-, pre- and oxy-combustion conditions. Oxides combine with CO$_2$ present in the flue gas and form carbonate, which when heated regenerates the oxides and thus liberates almost pure stream of CO$_2$. The selection criteria for potential oxides entail their CO$_2$ capture capacity, absorption rate, thermal stability, regeneration heat, cost and structural properties. Thus, here, we review the technical merit of all the suitable oxides which can be used for CO$_2$ capture at any fossil fuel burning plants.

Keywords CO$_2$ capture technology · Carbonate · Capture capacity · Thermal stability · Regeneration heat · Structural properties

Introduction

The problem of carbon dioxide (CO$_2$) emission from various fossil fuel burning plants (such as power, steel, cement and gasifiers) has become an important issue that may tremendously affect our future survival. Carbon dioxide, one of the greenhouse gasses, contributes to the global warming effect once released into the atmosphere [1, 2].

S. Kumar (✉) · S. K. Saxena
Center for the Study of Matter at Extreme Conditions, College of Engineering and Computing, Florida International University, Miami, FL 33199, USA
e-mail: skuma002@fiu.edu

CO$_2$ is thus far regarded as the most important contributor to global warming and accounts for 64 % of the increased greenhouse effect [3]. At present, there is ample scientific evidence to support the claim that global warming is human induced [4]. Many researchers believe that if such extensive use of fossil fuels continues for another 50 years, the CO$_2$ concentration will rise to 580 ppm, which would trigger a severe climate change.

The atmospheric concentration of CO$_2$ rose from 280 ppm in 1,800 to 397.8 in early 2014. Figure 1a presents the world energy-related CO$_2$ emissions. According to International Energy Outlook (IEO) 2013, the emission is projected to grow from 31.2 billion metric tons in 2010 to 45.5 billion metric tons in 2040. As expected, much of the growth is attributed to the developing countries (non-OECD). Non-OECD countries heavily rely on the supply of fossil fuels to meet their energy demand. The CO$_2$ emissions from the use of different fuel types such as coal, liquid fuels and natural gas are shown in Fig. 1b. Effluent emissions from coal use account for 44 % in 2010 and are projected to increase to 47 % in 2020–2030, before dropping marginally to 45 % in 2040. Liquid fuels have the slowest growth, resulting in an increment of only 3.5 billion metric tons of CO$_2$ from 2010 to 2040. The world consumption of natural gas is growing more rapidly than that of coal or liquid fuel. As natural gas has relatively low carbon intensity, its projected contribution to world energy-related CO$_2$ emissions is only 22 % in 2040.

Currently, most of the world energy requirement for transportation and heating (two-thirds of the primary energy demand) is satisfied through petroleum and natural gas. These two fuels are generally chosen due to the relative ease of transportation of liquid or gaseous forms. It is noteworthy that the combustion of hydrocarbon fuels for transportation and heating contributes over half of all

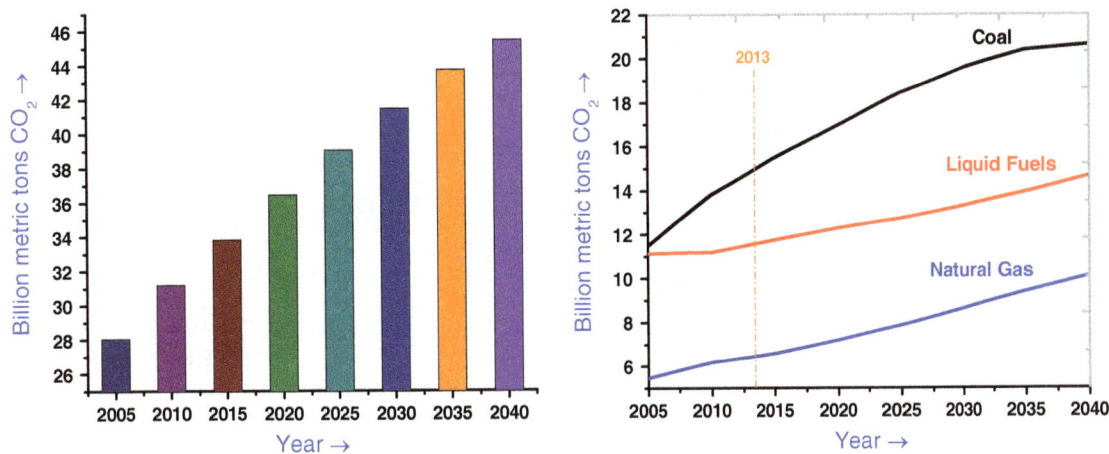

Fig. 1 World energy-related carbon dioxide emissions by **a** year and **b** fuel type

greenhouse gas emissions and a large fraction of air pollutant emissions. The current carbon emissions have to be reduced by at least a factor of three to maintain the CO_2 level in the atmosphere while meeting the ever-increasing global energy demand.

Therefore, it is essential to capture CO_2 from the industrial flue gas stream, one of the leading contributors of the anthropogenic gasses [5]. The development of carbon capture and sequestration (CCS) technologies can be seen as a viable solution to mitigate the vast carbon emission. The separation of a pure CO_2 stream, combined with a well-managed geological storage site, is being considered as a mitigation option for climate change. The CCS technologies could be applied using available technologies as many of the components in these systems are already in use. However, it is widely accepted that there is a large scope for cost reduction and energy efficiency improvements in CO_2 capture systems.

Currently, various CO_2 capture technologies exist including physical absorption [6, 7], chemical absorption [8, 9], adsorption [10] and membranes [11]. However, they are still far from being considered as a technologically viable solution. The obvious bottlenecks are the huge amount of flue gas which needs to be treated as well as low mass transfer rate during the processes. Among all the proposed techniques, chemical absorption using aqueous alkanolamine solutions is the most conventional way for CO_2 capture. But these amine-based solutions suffer severe drawbacks such as high energy consumption in regeneration process, high corrosive rate of equipment, essential pretreatment steps, large absorber volume and high sorbent cost. Because of these limitations of the chemical absorption techniques, an alternative way for post-, oxy- or pre-combustion CO_2 capture at fossil fuel burning power plants, mainly a solid adsorption/absorption method, can be proposed.

A schematic diagram on Fig. 2 is depicted to demonstrate the concept of cyclic CO_2 capture process by metal oxides. The metal oxides (MO), such as CaO and MgO, combine with CO_2 to form thermodynamically stable carbonates. Metal carbonates, when heated, liberate pure stream of CO_2 gas and regenerate the oxides. The combined use of exothermic carbonation reaction and endothermic regeneration reaction forms a cyclic process. Eventually, the generated pure CO_2 gas can either be sequestered underground or used for enhanced oil recovery [12]. In general, any metal oxides that can qualify for CO_2 capture application must be abundant in earth's crust, react with CO_2 at low temperature, require low regeneration energy, should have suitable reaction kinetics and must form a carbonate that is stable in the environment at ambient conditions. Porous oxides, namely alkali and alkaline-earth metals, have been reported as promising candidates for CO_2 capture. These metal oxides possess properties such as long durability, good mechanical strength, wide availability, and low cost since they are present as natural minerals and high CO_2 absorption capacity at moderate working temperatures [13–16]. In this study, we review the CO_2 capture capacity of these different potential oxides.

Selection criteria for metal oxides

Recently, researchers proposed a theoretical screening methodology to identify the most promising CO_2 sorbent candidates from the vast array of possible solid materials [17–19]. The methodology uses both the thermodynamic database searching with first principles density functional theory and phonon lattice dynamics calculations. The thermodynamic properties of solid materials are used for computing the thermodynamic reaction equilibrium

Fig. 2 Cyclic CO_2 capture process for metal oxides (MO) and metal carbonates (MCO_3)

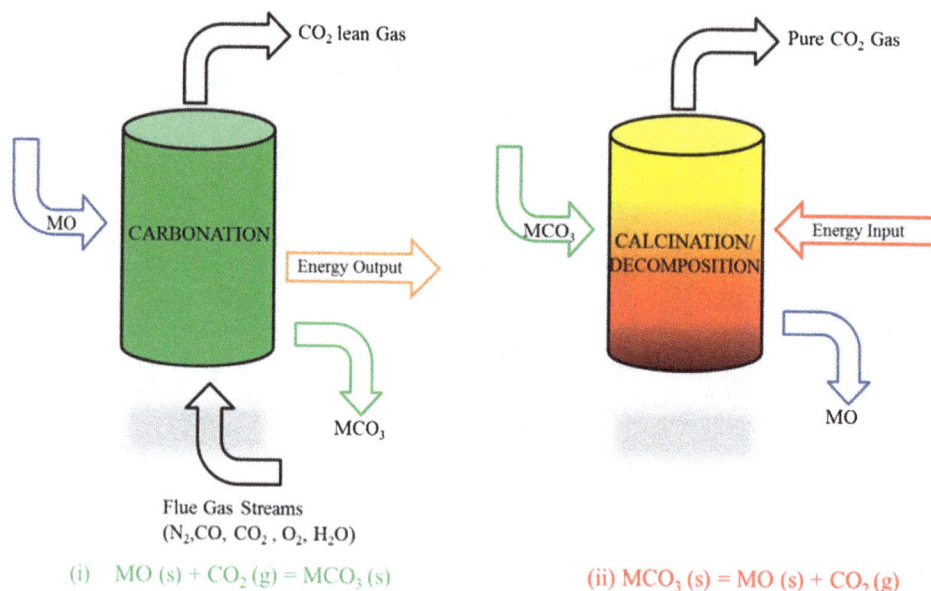

(i)   MO (s) + CO_2 (g) = MCO_3 (s)   (ii) MCO_3 (s) = MO (s) + CO_2 (g)

properties of CO_2 absorption/desorption cycle based on the chemical potential and heat of reaction analysis. The selection is based on the pre- and post-combustion technologies and conditions in power plants. This includes only those solid materials that may be suitable as CO_2 sorbent candidates and further be considered for experimental validation, if they meet the following criteria: lead to lowering in energy cost for the capture and regeneration process and could operate at desired conditions of CO_2 pressure and temperature. Figure 3 illustrates the thermodynamic properties and CO_2 wt% absorbed by different metal oxides. However, the thermodynamic properties and CO_2 wt% absorbed by any sorbent could not be the only factors to select them for experimental studies. Various other important parameters such as availability or cost, regeneration temperature, kinetics, reversibility and durability must also be carefully considered before selecting any metal oxides for a large-scale application.

Based on the above figure, it can deduce that BeO has a favorable thermodynamics, but this system is not a viable candidate for CO_2 capture because of the health issues related to beryllium powder or dust. It should be noted that a high exothermic forward reaction (MO(s) + CO_2 (g) = MCO_3(s)) generally requires a large regeneration temperature for the oxide (MCO_3(s) = MO (s) + CO_2 (g)). Also, a low CO_2 wt% ($\sim$40–50) will require a large amount of the solid materials to deal with a huge volume of flue gas stream at any fossil fuel burning plants. Thus, a high regeneration temperature and low CO_2 wt% can cause a vast energy or carbon emission penalty. Moreover, such oxides can eventually increase the electricity cost by a significant factor and thus may not be suitable for CO_2 capture at any plants.

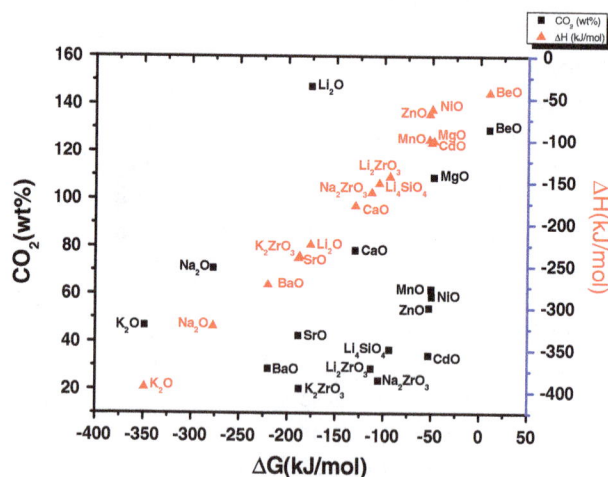

Fig. 3 Thermodynamic properties and CO_2 wt% for metal oxides MO(s) + CO_2 (g) = MCO_3(s) at 300 K

The turnover temperature represents the value above which sorbents cannot absorb CO_2 and the release of CO_2 begins. The flue gas conditions are different for pre- and post-combustion CO_2 capture at power plants. Under pre-combustion conditions, after water–gas shift reaction, the gas stream mainly contains CO_2, H_2O and H_2. The partial CO_2 pressure in this case is around 20–25 bar, and the temperature is about 300–350 °C. But for post-combustion conditions, the flue gas stream mainly contains CO_2 and N_2. In this case, the partial pressure of CO_2 is around 0.1–0.2 bar, while the temperature range can vary around 27–77 °C. The goal set up by the US Department of Energy is to capture at least 90 % CO_2 with an increase in the electricity cost of no more than 10 and 35 % for pre- and post-combustion technologies, respectively [20]. To minimize the energy consumption, the ideal sorbents should work in the above-indicated pressure and temperature ranges to separate CO_2 from H_2. A list of turnover temperatures for different oxides can be found elsewhere [17–19].

Based on the above-discussed limitations, exclusively few oxides seem to be promising and thus chosen to study for the experimental validation. The two alkaline-earth metal oxides (such as CaO and MgO) draw the tremendous attention because of accessibility and favorable thermodynamics. Recently, lithium-, sodium- and potassium-based silicates or zirconates also gained interest due to their high CO_2 absorption capacity. Moreover, FeO can also be seen as a promising material for CO_2 capture at both power and non-power sectors (iron and steel industry). In the following sections, we review the experimental studies for the most promising CO_2 sorbent candidates.

Metal oxides

Alkali metal-based oxides

Lithium zirconate (Li_2ZrO_3)

The reaction between lithium carbonate (Li_2CO_3) and zirconium dioxide (ZrO_2) is well known to synthesize lithium zirconate (Li_2ZrO_3). However, the reverse reaction was not considered for CO_2 capture until 1998 [21]. Nakagawa and Ohashi first investigated the capture of CO_2 using Li_2ZrO_3 at high temperatures (400–600 °C) [21]. Li_2ZrO_3 has great potential because it has an excellent CO_2 sorption capacity (28 wt%) as well as a small volume change during the CO_2 sorption/desorption cycles [22]. The reaction (1) occurs mainly due to the Li ion mobility in the ceramics [23, 24]. As can be seen from Fig. 4, during CO_2 sorption, the Li ions diffuse from the core of the particles to the surface and react with CO_2 to form Li_2CO_3. The diffusion of CO_2 in the solid Li_2CO_3 is recognized as the rate-limiting step.

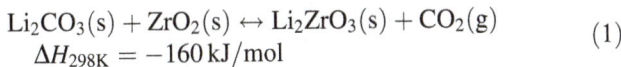

$$Li_2CO_3(s) + ZrO_2(s) \leftrightarrow Li_2ZrO_3(s) + CO_2(g) \quad (1)$$
$$\Delta H_{298K} = -160\,kJ/mol$$

The separation of CO_2 with aid of Li_2ZrO_3 works in a temperature range of 450–700 °C. The main advantage of this new material with respect to calcium-based absorbents is its unique thermal stability that allows achieving many absorption/regeneration cycles (over 20) without loss of capacity. Previous reports suggest that only 11–13 wt% change after absorption of CO_2 using Li_2ZrO_3 at 500–700 °C in more than an hour is possible. Thus, it is evident that Li_2ZrO_3 has extremely slow absorption kinetics, and hence, recently, the addition of dopants was proposed as a solution. Hence, the addition of dopants such as K_2CO_3 and Y_2O_3 to Li_2ZrO_3 could potentially accelerate its CO_2 sorption kinetics.

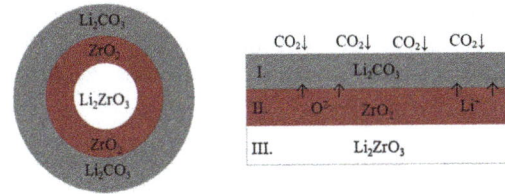

Reaction at
I. $CO_2 + O^{2-} + 2Li^+ \rightarrow Li_2CO_3$
III. $Li_2ZrO_3 \rightarrow ZrO_2 + O^{2-} + 2Li^+$

Fig. 4 Proposed mechanisms for CO_2 sorption on Li_2ZrO_3

In the case of the pure Li_2ZrO_3 (without K_2CO_3 additive), formation of an impervious shell of Li_2CO_3 on the surface leads to a residual unreactive Li_2ZrO_3 core (Fig. 4). This observation suggests that the diffusion resistance of CO_2 through the solid Li_2CO_3 surface layer is the rate-limiting step. At 500 °C, Li_2CO_3 and K_2CO_3 form a eutectic mixture. During the absorption process, the eutectic mixture helps to diffuse more CO_2 toward the inner unreacted particles [25]. K_2CO_3- and Li_2CO_3-containing Li_2ZrO_3 sample reacts with gaseous CO_2 at 500 °C to form an outer layer of a molten liquid and a solid interior ZrO_2 core. Consequently, CO_2 diffuses through the molten carbonate layer at a much faster rate. Such difference in CO_2 sorption rate of doped and undoped Li_2ZrO_3 can be partially attributed to the molten carbonate layer.

Another dopant, Y_2O_3, can improve the O^{-2} ionic diffusion in the ZrO_2 layer that is formed during the absorption. Unfortunately, Y_2O_3 failed to increase the CO_2 absorption kinetics of Li_2ZrO_3 [26]. Other findings reveal that solid solutions of lithium and potassium meta zirconates ($Li_{2-x}K_xZrO_3$) increase both the kinetics and CO_2 absorption capacity compared to undoped Li_2ZrO_3 [27]. In the same line, solid solutions of lithium and sodium zirconates were also explored [28]. It was concluded that the higher the Na content in the samples the faster the absorption kinetics. On the contrary, the higher the Li content in the mixture the faster the regeneration kinetics.

Lithium orthosilicate (Li_4SiO_4)

Lithium silicate-based (Li_4SiO_4) sorbent material has been considered as a promising candidate for high-temperature CO_2 removal in the recent times [29, 30]. Li_4SiO_4 could adsorb different concentrations of CO_2 in the temperature range of 450–700 °C, but it has relatively (compared to amine solutions) higher decomposition temperature (>800 °C). A high decomposition temperature implies that more heat input and costly equipment would be required to perform the process. However, compared to other solid sorbents such as CaO, the decomposition temperature is low and thus less supply of energy will be needed [31]. As can be calculated from reaction (2), the theoretical CO_2 adsorption capacity of Li_4SiO_4 is 36.7 wt% ($\sim$8.34 mmol CO_2/g Li_4SiO_4).

$$Li_4SiO_4(s) + CO_2(g) \leftrightarrow Li_2SiO_3(s) + Li_2CO_3(s) \tag{2}$$
$$\Delta H_{298K} = -141.97\,kJ/mol$$

In a study, Wang et al. [32] reported an interesting way to synthesize Li_4SiO_4 using rice husk, which were calcined at 800 °C in the presence of Li_2CO_3. This material is ideally suited for CO_2 removal from synthesis gas (syngas) derived from gasification of carbonaceous fuels (coal, coke, natural gas, biomass, etc.) and is hence considered as a pre-combustion CO_2 sorbent.

Li_4SiO_4 has shown the ability to remove more than 90 % of the CO_2 from simulated syngas in both fixed-bed and fluidized-bed process configuration. The lithium silicate-based sorbent is highly effective at temperatures of 250–550 °C, pressures of 0–20 atm, CO_2 concentrations of 2–20 % and in the presence of contaminants such as hydrogen sulfide. Yamachui et al. [33] reported a carbonation yield of 87.2 % under 100 cm^3/min flow rate of pure CO_2. The water vapor plays a major role in increasing the CO_2 absorption rates of Li_4SiO_4. About 22–26 wt% (capacities of $\sim$5–6 mmol/g) were obtained using a gas stream containing about 15 % CO_2 and 10 % H_2O in N_2 at 550 °C [34]. Moreover, the sorbent has shown excellent regenerability and attrition resistance in thermal cycling tests. Recent analysis has shown that the lithium silicate-based sorbent has the capability not only to separate CO_2 from syngas, but also to promote the water–gas shift reaction [35]. It is reported that dopants such as Al, Fe, Na, K and Cs increase the CO_2 uptake of lithium orthosilicate [36, 37].

Researchers have attempted to improve the reaction kinetics of CO_2 adsorption by altering the synthesis routes [38]. The new methods are aimed at reducing the particle size of Li_4SiO_4. The experimental results reveal that a smaller particle size yields higher CO_2 adsorption efficiency and adsorption rate. Further studies focused on decreasing the precursor particle size [39] or choosing more sintering-resistant precursor to result in the smaller product grain size [40]. Similarly, based on the particle size effects on Li_4SiO_4, the reactivity of Li_4SiO_4 was modeled as a reaction of CO_2 at the solid surface followed by the rate-limiting diffusion of lithium [41].

A calculation based on the most optimistic assumptions illustrates that despite the fact that Li_4SiO_4-based absorbent demonstrates high CO_2 uptake capacity, it would be impractical to perform the process at the scale of a fixed-bed temperature swing process at a 500-MW coal-fired power plant [34]. However, efforts have been made to modify the Li_4SiO_4 synthesis methods such as solid-state

reactions [30, 42], sol–gel method [38, 43], impregnated suspension [44], precipitation method [45], amorphous silica [40], Quartz [39], fly ashes [46], rice husk ashes [47] and diatomite [48].

Sodium zirconate (Na_2ZrO_3)

Sodium zirconate has gained attention due to its favorable thermodynamics and high theoretical CO_2 sorption capacity [23, 49, 50]. Na-based sorbents can be used as solid CO_2 acceptors at high temperatures. The use of sodium-based oxides is more economical than lithium-based oxides. Moreover, previous reports suggest that Na_2ZrO_3 can have a better reaction rate than the synthetic sorbents such as Li_2ZrO_3 and Li_4SiO_4 [51].

Na_2ZrO_3 has a lamellar structure, where sodium ions are located among the zirconate layers, which permits for sodium diffusion. The mechanism for CO_2 chemisorption on Na_2ZrO_3 has already been proposed. Initially, the surface of the alkaline ceramic particles reacts with CO_2 and forms carbonates on the periphery. Once the external layer of carbonate is completely formed, sodium atoms have to diffuse throughout this external carbonate shell to reach the surface and be able to react with CO_2.

$$Na_2ZrO_3(s) + CO_2(g) \leftrightarrow Na_2CO_3(s) + ZrO_2(s) \tag{3}$$

The maximum theoretical CO_2 absorption should correspond to an increase of weight equal to 23.75 wt%. The reaction (3) is controlled by two different processes—CO_2 absorption over the surface of Na_2ZrO_3 particles and sodium diffusion. Alcerreca-Corte et al. [52] performed the kinetic analysis for chemisorption of CO_2 on the Na_2ZrO_3 particles in the temperature range of 150–700 °C. A fast kinetics was observed between 550 and 700 °C. However, at low temperatures, kinetics was relatively low and was attributed to the sintering effect as well as diffusion problems. The optimum temperature for the reaction (3) was 600 °C. All the isotherms were fitted to a double exponential model. The activation energies for the CO_2 absorption and the diffusion of sodium were calculated to be 33.87 and 48.01 kJ/mol, respectively. Therefore, it was concluded that sodium diffusion is the limiting step of the process. Similarly, Jimenez et al. reported the CO_2 sorption kinetics of Na_2ZrO_3 at similar temperatures and $P_{CO_2} = 0.4$–0.8 atm. Contrary to previous study, the surface reaction is identified to be the rate-limiting step for the reaction and attributed to the fast CO_2 sorption kinetics of Na_2ZrO_3 [53].

In general, the addition of steam enhances the capture kinetics and also favors the regeneration. It is believed that the presence of steam increases the mobility of alkaline ions and therefore accelerates the reactions. Therefore, several studies have been performed to evaluate the effect of steam addition on the CO_2 capture properties of Na_2ZrO_3 [54, 55]. One such study was performed by Santillán-Reyes et al. in the presence of water vapor and low temperature range (30–70 °C). They synthesized Na_2ZrO_3 by solid-state reaction using the reagents Na_2CO_3 and ZrO_2 in the molar ratio of 1.1:1.0, respectively. Further, the reaction mechanism of Na_2ZrO_3–CO_2–H_2O was proposed and illustrated in (4). Na_2ZrO_3 reacts with water vapor to produce hydroxyl species at the particle surfaces and then absorbs 2 mol of CO_2 and 1 mol of H_2O to eventually form $NaHCO_3$ and ZrO_2.

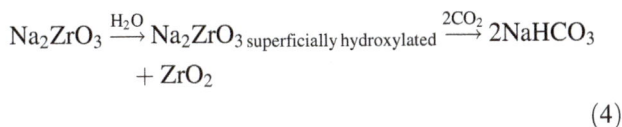

$$Na_2ZrO_3 \xrightarrow{H_2O} Na_2ZrO_3 \text{ superficially hydroxylated} \xrightarrow{2CO_2} 2NaHCO_3 + ZrO_2$$

(4)

Na_2ZrO_3 was able to absorb 5.8 mmol/g of CO_2 (which has theoretical capacity is 10.8 mmol/g). Therefore, Na_2ZrO_3 can also be used as CO_2 sorbent at low temperatures and in the presence of water vapor. The reported absorption results for Na_2ZrO_3 are still higher than for other materials in the same temperature conditions (≤6 mmol/g of material) [56].

Nanosized materials are expected to increase the CO_2 capture/regeneration rates due to their improved characteristics, such as high surface area and higher surface reactivity because of unsaturated bonds on their pristine surfaces [57–59]. Therefore, Zhao et al. synthesized nanosized Na_2ZrO_3 using a soft-chemical route starting with the formation of a complex from zirconoxy nitrate and sodium citrate, followed by highly exothermic reaction between citrate and nitrate during calcination in a controlled atmosphere [49]. It was found that much faster CO_2 capture rates can be achieved on the monoclinic Na_2ZrO_3 compared to the hexagonal one, even at a very low CO_2 pressure (0.025 bar). Thus, a higher CO_2 capture rate is obtained for nanosized Na_2ZrO_3 due to the dual effect of its crystal size and structure.

Moreover, the doping effect of Li to Na_2ZrO_3 was investigated. The change in sorption/regeneration kinetics was evident from the experimental data. The greater the Li content, the lower the CO_2 sorption kinetics. This observation can be attributed to the fact that Li substitutes some of the Na in the Na_2ZrO_3 structure and results in the formation of small amount of Li_2ZrO_3 [60]. Lopez-Ortiz et al. concluded that the highest activity toward CO_2 sorption was exhibited by Na_2ZrO_3, followed by sodium antimoniate (Na_3SbO_4) and finally sodium titanate (Na_2TiO_3). Other applications for Na_2ZrO_3 may involve its use for the development of inorganic membranes for CO_2 separations [24, 61].

Sodium metasilicate (Na_2SiO_3)

Sodium metasilicate has not yet been extensively examined. One of the first studies illustrates the synthesis of Na_2SiO_3 using solid-state and precipitation methods [62]. It was noticed that Na_2SiO_3 can absorb only a small quantity (∼1–2 wt%) of CO_2 in the range of room temperature and 130 °C. Further, the kinetic analysis of CO_2 absorption was also performed, and the results indicate that the sorption mechanism proceeds through a two-step process: (1) superficial chemical sorption and (2) sodium diffusion process. The calculated activation energies for surface reaction and sodium diffusion were 17.48 and 23.97 kJ/mol, respectively. The sodium diffusion is recognized as the limiting step for the carbonation process. Moreover, the CO_2 absorption increases with the decrease in the particle size of Na_2SiO_3 and can be attributed to the subsequent increase in the surface area.

Another study was performed at much lower temperatures (30–60 °C) for the Na_2SiO_3–CO_2–H_2O system [63]. Na_2SiO_3 was synthesized using two methods—(1) solid-state reaction (Na_2CO_3 + SiO_2); and (2) combustion method (NaOH, SiO_2 and urea (CO$(NH_2)_2$). The reactions were performed under the flow of two different carrier gasses—N_2 and CO_2. In the presence of N_2 as a carrier gas, Na_2SiO_3 traps water physically and chemically, where the water vapor adsorption and/or absorption depend on temperature and relative humidity. When CO_2 was used as the carrier gas, Na_2SiO_3 traps both water vapor and CO_2. Interestingly, the presence of water vapor promotes the higher CO_2 chemisorption (>1 wt%). It can be seen in reaction series (5) that at first, Na_2SiO_3 reacts with water superficially, producing Na–OH and Si–OH species. Further, superficially activated Na_2SiO_3 combines with CO_2 to form $NaHCO_3$.

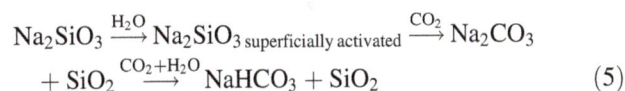

$$Na_2SiO_3 \xrightarrow{H_2O} Na_2SiO_3 \text{ superficially activated} \xrightarrow{CO_2} Na_2CO_3 + SiO_2 \xrightarrow{CO_2+H_2O} NaHCO_3 + SiO_2$$

(5)

It has also been observed that under the condition of thermal humidity, Na_2SiO_3 absorbs up to 16.39 mmol of CO_2/gm of ceramic, twice more CO_2 than the quantity absorbed under dry conditions. The synthesis method of Na_2SiO_3 influences its CO_2 capture capacity. Higher the surface area, the better is the CO_2 absorption. Here, the surface areas were 0.5 and 1.6 m^2/g for solid-state and combustion method, respectively. Hence, the latter has about 8.6 mmol/g CO_2 capture efficiency, almost twice of that of the former sample. Thus, CO_2 absorption capacity of Na_2SiO_3 is governed by the combined effect of water vapor and surface area. These recent results support the potential of Na_2SiO_3 as a CO_2 capture sorbents at moderate or environmental temperatures.

Table 1 CO_2 uptake capacity of MgO at different conditions

	Sorbent	Gas stream	Carbonation temperature (°C)	Pressure (bar) or flow rate	Particle size	Regeneration temperature (°C)	CO_2 capture capacity (mmol/g) conversion (%)	References
1	MgO	Pure CO_2	50–1,000	100 mL/min (flow rate)	–	–	0.99	[66]
2	MgO/Al$_2$O$_3$ (10 wt% MgO)	(13 v % H$_2$O, 13v % CO$_2$)	30,150	1	(20–40) mesh size	350	1.36	[67]
3	MgO	(11 v % H$_2$O, 1v % CO$_2$)	50–100	0.01	–	150–400	1.05	[68]
4	K$_2$CO$_3$/MgO	(11 v % H$_2$O, 1 v % CO$_2$)	50–100	–	–	150–400	2.98	[56]
5	MgO	330/660 ppm in air	0,100	0.2	–	–	0.64,0.43	[56, 69]
6	MgO/MCM-41	Pure CO_2	25	1	–	–	1.06	[64]
7	Mesoporous MgO	Pure CO_2	25,100	1	–	–	1.82,2.27	[70]
8	Non-porous MgO	Pure CO_2	25	1	–	–	0.45	[70]
9	MgO–ZrO$_2$	Pure CO_2	30,150	1	–	–	1.15,1.01	[71, 72]
10	MgO (31.7 wt%)/Al$_2$O$_3$ (22.4 wt%)	Pure CO_2	20,200,300	1	–	–	0.13,0.24,0.5	[73]
11	MgO (33.8 wt%)/Al$_2$O$_3$ (20.8 wt%)	Pure CO_2	20,200,300	1	–	–	0.08,0.12,0.5	[73]
12	K$_2$CO$_3$/MgO/ Al$_2$O$_3$	Flue gas	60	1	–	480	2.49	[74]
13	K$_2$CO$_3$/MgO	(9 v % H$_2$O, 1 v % CO$_2$)	60	40 mL/min (flow rate)	–	400	2.7	[75]
14	MgO nanocrystal	Flue gas	60	25sccm (flow rate)	5 nm	60–600	6.4	[76]
15	MgO	Pure CO_2	350	1.33,3.33	–	–	0.089,0.091	[77]
16	MgO	Pure CO_2	300–500	9–36	<44 μm	–	70–80 % (~200 min)	[78]
17	MgO	Pure CO_2	–	20–40	<44 μm	–	100 % (~120 min)	[79]

Alkali-earth metal oxides

Magnesium oxide (MgO)

MgO combines with CO_2 to form $MgCO_3$. $MgCO_3$ is thermodynamically stable at ambient conditions. A simple calculation shows that to capture 1 ton of CO_2, about 0.92 ton of MgO is required (assuming 100 % conversion). Thus, it is certain that to capture CO_2 using MgO at any power plant site, a large quantity of MgO will be needed. Moreover, the kinetics and recycling efficiency of MgO–CO_2 reaction are also the major issues to use MgO as a sorbent for practical applications. MgO is an attractive candidate for both pre- and post-combustion capture technologies due to its low regenerating temperatures: $T_2 = 287$ °C (post-combustion) and $T_1 = 447$ °C (pre-combustion) [64]. One should note that MgO when recycled between naturally occurring magnesite and dolomite can cause relatively lesser energy and carbon emission penalty, respectively [65].

Many researchers studied the heterogeneous reaction between magnesium species and CO_2 in the presence or absence of H_2O. Although a majority of research was not directly aimed at carbon capture, still the results are valuable additions to the mineral carbonation research. The CO_2 uptake capacity on different MgO sorbents at various conditions is listed in Table 1. Most of the experiments are restricted to low-temperature and ambient or low-pressure conditions. Hence, the reported sorption capacities of CO_2 on MgO are not very high. In contrast, recent work of Highfield et al. reports a conversion of about 70–80 % MgO to $MgCO_3$ in about 2 h with steam pressure between 5 and 10 % of the total pressure 20–40 bar and temperature around 300–350 °C.

One of the key factors in gas–solid reaction is the presence of water. There have been several observations when

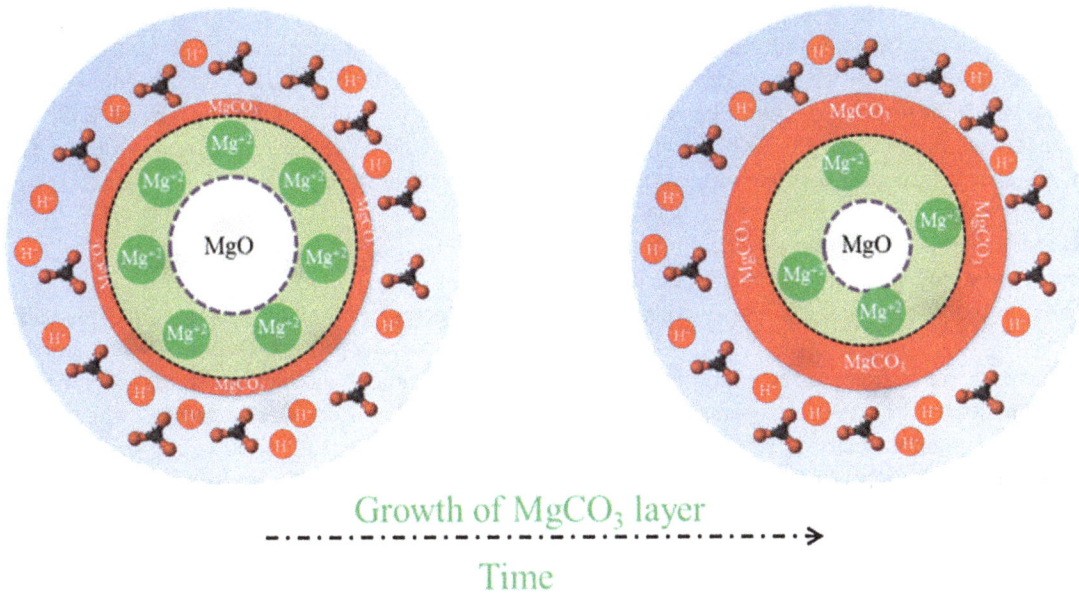

Fig. 5 Adsorption/absorption model for MgO–CO_2–H_2O reaction

water catalyzed the reaction [80, 81]. Figure 5 depicts the absorption/adsorption model of MgO–CO_2–H_2O.

MgO particles are surrounded by water vapor where CO_2 reacts to form CO_3^{2-} ions and H^+ ions. Free Mg^{+2} ions could further react with the CO_3^{2-} ions to form $MgCO_3$. Recently, Fagerlund et al. [78] proposed the reaction mechanism for MgO carbonation in the presence of steam:

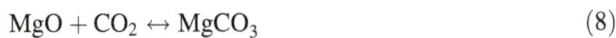

$$MgO + H_2O \leftrightarrow MgO \cdot H_2O^* \qquad (6)$$

$$MgO \cdot H_2O^* + CO_2 \leftrightarrow MgCO_3 + H_2O \qquad (7)$$

$$MgO + CO_2 \leftrightarrow MgCO_3 \qquad (8)$$

However, $MgCO_3$ forms an impervious layer around unreacted MgO particles and hinders the further diffusion of CO_2 molecules. Thus, only increasing the amount of water vapor cannot lead to the complete carbonate conversion of MgO. Therefore, besides the amount of steam, surface properties of MgO (such as surface area, particle size and porosity) are also very crucial parameters for the carbonation process. It is observed that in the absence or low partial pressure of water vapor, $MgO \cdot 2MgCO_3$ is formed. It can be easily calculated that the CO_2 capture capacity of $MgO \cdot 2MgCO_3$ is about two-thirds that of $MgCO_3$. Hence, formation of $MgO \cdot 2MgCO_3$ can significantly reduce the further uptake of CO_2 by MgO particles. But at a high temperature (>350 °C), $MgO \cdot 2MgCO_3$ decomposes to MgO.

MgO–CO_2 reaction has been studied to some extent. The role of water vapor in increasing the activity of MgO is now well established. The direct relation of porosity, surface area and particle size with the carbonation yield has been examined. However, several factors such as carbonation kinetics, sorbent reversibility and durability are still not completely resolved and thus need to be investigated.

Calcium oxide (CaO)

In 1995, Silaban and Harrison proposed the method that involves the separation of CO_2 at high temperatures (>600 °C) using the reversible reaction of CaO [82].

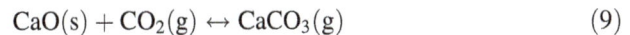

$$CaO(s) + CO_2(g) \leftrightarrow CaCO_3(g) \qquad (9)$$

The low price (because of naturally occurring mineral, limestone) and favorable thermodynamics of CaO have attracted a lot of interest in the past one decade, and thus, the CaO–CO_2 system has been extensively studied. A great advantage of calcium-based sorbents is that the absorption of CO_2 can occur at temperatures above 600 °C (possibility of heat recovery) [83]. The efficiency of the CO_2 capture depends on many parameters such as diffusion resistance, which depends on the size of sorbent particles, volume and the pore structure, as well as surface size and reaction kinetics [84]. These parameters influence carbonation, but many of them also affect calcination. A detailed list of recent methods used to enhance the CO_2 uptake by CaO-based sorbents can be found elsewhere [85].

Bulk CaO, as a CO_2 sorbent, has three severe limitations: (1) kinetics of the carbonation reaction, despite being highly exothermic, becomes slow after the first layer of carbonate formation because the reaction is limited by the diffusion of CO_2 through the thin surface layer of $CaCO_3$ formed on CaO [86, 87]. The uptake kinetics of CaO-based sorbents was improved by innovative synthesis methods or precursors [88, 89]. (2) Regeneration step is very energy intensive (>800°C for CaCO$_3$ decomposition), so excessive sintering and mechanical failure of the oxide occur. This leads to a drastic loss in adsorption activity after a few sorption and regeneration cycles [90]. (3) To be an effective sorbent, CaO, that is suitable for high surface area exposure, such as a powder. However, powders can be used in fluidized beds but the pressure drop associated with them is very large. Also, use of fine powders can be problematic due to entrainment in the process flow and attrition of the material [91].

The improvement in the sorbent reversibility can be seen as a major challenge for extended operation purpose. Decrease in absorption capacity is associated with sintering of the sorbent surface because of the influence of temperature and the reduction in porosity and active surface [92–95]. Other factors that reduce the activity of sorbents are the attrition of sorbent grains during the process and chemical inactivation [96]. It should be noted that the reaction with sulfur oxide (SO$_2$) is competitive to the carbonation reaction. Sulfurization and carbonation reactions are similar because of being heterogeneous and occurring in the porous structure of the sorbent. However, sulfurization is irreversible in the temperature range 500–900 °C, occurs in small pores and covers the surface of the sorbent [97–100]. Decrease in the absorption capacity after 45 cycles was 60 % for limestone, 40 % for dolomite and less than 20 % for huntite, and can be attributed to the differences in the sorbent structure [101]. The decrease in absorption capacity by CaO can be limited in several ways. The parameters that have impact are dopant (NaCl, Na$_2$CO$_3$, KCl, KMnO$_4$, MgCl$_2$, CaCl$_2$, Mg (NO$_3$)$_2$), sorbent grain size, raw material for calcium-based sorbent, thermal pretreatment of the sorbent and steam reactivation. The properties of CaO obtained from calcium acetate are better than the properties of CaO obtained from natural limestone because it has a much better conversion and greater durability [102].

Calcium silicate (CaSiO$_3$)

Calcium silicate received attention because it is inexpensive and has lower sorption temperature compared to CaO. Wang et al. [103] explored the sorption–desorption property of CaSiO$_3$. CaSiO$_3$ is synthesized through solid chemical reaction of CaCO$_3$ and SiO$_2$ at 800 °C. CaSiO$_3$ starts to absorb CO$_2$ at 400 °C with about 28.72 % sorption efficiency, using 15 % CO$_2$ and rest N$_2$. The observed regeneration temperature for CaSiO$_3$ was 800 °C. However, the CO$_2$ capture capacity drastically decreases from large number of cycles and could be attributed to sintering of the materials which leads to the loss of specific surface area. Similarly, Gupta and Fan reported that CaSiO$_3$ captured 27.85 % CO$_2$ at 700 °C and 1 atm [104]. Under the same conditions, Tilekar et al. reported the CO$_2$ capture capacity of CaSiO$_3$ to be 14.19 wt% [105]. The water molecules in the CaSiO$_3$ facilitate the capture of CO$_2$.

Transition metal-based oxides

Transition metal-based oxides act as an oxygen carrier in the chemical looping process. In 1983, Richter and Knoche introduced chemical looping for heat generation using metal oxides as chemical ingredients, which are now considered as a potential solution for the simultaneous power generation and CO$_2$ capture [106]. Metal oxide (Me$_x$O$_y$) is reduced by the carbonaceous fuel to generate CO$_2$ and/or H$_2$O in the fuel reactor. The reduced metal (Me$_x$O$_{y-1}$) is then oxidized back to Me$_x$O$_y$ by oxygen in the air reactor, which is an exothermic reaction. In the process, transition metal oxides (oxygen carriers) are transported within and between the fuel reactor and the air reactor.

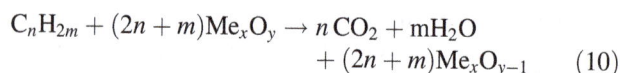

$$C_nH_{2m} + (2n + m)Me_xO_y \rightarrow n\,CO_2 + mH_2O \\ + (2n + m)Me_xO_{y-1} \quad (10)$$

$$2Me_xO_{y-1} + O_2 \rightarrow 2Me_xO_y \quad (11)$$

The widely developed oxygen carriers include Ni-, Cu- and Fe-based metal oxides and their composited particles. Despite the great reactivity of Ni- or Cu-based oxygen carriers, their development is restricted because of their relatively high cost. Fe-based oxides are the strong candidates to be commercialized because of their economic feasibility. A number of reviews about the current status of the chemical looping process have been published in recent times [107–111]. Here, the focus is on the direct interaction of CO$_2$ with transition metal-based oxides via carbonation–calcination cycle.

Iron-based oxides

Iron oxide-based materials can be used for CO$_2$ capture purpose in both the power and non-power sectors. The main advantages of iron-based sorbent are as follows: (1) accessibility, (2) favorable thermodynamics and (3) slow degradation of sorption capacity.

Recently, Ohio State University developed an iron oxide-based oxygen carrier for particles suitable for operation in the coal direct chemical looping (CDCL) process. CDCL is an efficient power generation process. The process consists of a unique moving bed reactor, namely the reducer, where pulverized coal is fully converted using iron oxide.

$$Fe_2O_3(s) + C(s) \rightarrow FeO(s) + Fe(s) + CO_2(g) \quad (12)$$

$$2FeO(s) + 0.5O_2(g) \rightarrow Fe_2O_3(s) \quad (13)$$

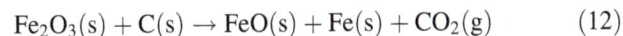

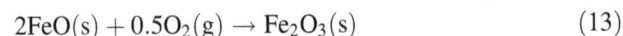

Fe_2O_3, an oxygen carrier, is reduced to FeO and Fe while the exhaust gas is a stream of pure CO_2 that can be sequestered or used for enhanced oil recovery [eq. (12)]. The reduced FeO and Fe mixture can then be oxidized to Fe_2O_3 using air in the combustor reactor, liberating heat to produce steam [112]. Another recent study presents the CO_2 capture technique for iron and steel industries using iron-based raw materials easily available at the iron-making sites [113].

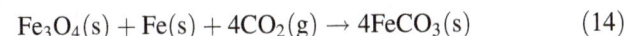

$$Fe_3O_4(s) + Fe(s) + 4CO_2(g) \rightarrow 4FeCO_3(s) \quad (14)$$

$$3FeCO_3(s) + 0.5O_2(g) \rightarrow Fe_3O_4(s) + 3CO_2(g) \quad (15)$$

The global steel production is growing rapidly, from 1,248 Mt in 2006 to 1,490 Mt in 2011—an increase of about 16 % in a short span of 5 years [114]. Steel production is an energy intensive process and thus consumes huge amount of fossil fuels. On world average, each ton of steel production emits 2.2 ton of CO_2. Therefore, iron-making industry is one of the biggest targets next to power plants to curb vast emission of greenhouse gasses.

A mixture of magnetite and iron combines with CO_2 and forms siderite. The reaction generates heat and thus can be utilized to produce steam. Siderite once formed can further decompose back to magnetite in the presence of oxygen at 350 °C. Once the absorption capacity of magnetite degrades, it can be sent back to blast furnace for further processing. For example, the sorbents can be processed in a blast furnace for the production of iron or steel. Thus, the proposed method can advantageously reduce or even eliminate the loss of raw materials. This system and method can be more thermodynamically favorable and can save energy. The reversibility and suitable kinetics can make the system highly favorable for CO_2 capture for iron and steel industries.

Other transition metal oxides

It is considered that MnO can be used for both post- and pre-combustion conditions. At $P_{CO_2} = 0.1$ bar, the driving force for MnO to convert in $MnCO_3$ was high even at a low temperature 77 °C. The CO_2 could be released at about 377°C at partial $P_{CO_2} = 10$ bar. For pre-combustion, MnO will capture CO_2 at a partial pressure of about 10 bar below 327 °C and then could be regenerated by heating above 377 °C to produce a stream of pure CO_2. According to the thermodynamics, CdO can be used for both pre- and post-combustion conditions. However, the toxicity of Cd system would be a big hurdle in its practical applications.

ZnO can be used for post-combustion but not for pre-combustion. The equilibrium temperature required for carbonate formation is 77 °C at $P_{CO_2} = 0.1$ bar and around 167 °C for $P_{CO_2} = 10$ bars [18]. Therefore, it may be essential to cool post-combustion gasses prior to carbonation for high driving force and thus make the sorbent unfit for post-combustion capture. Few researcher groups studied the heterogeneous reaction between ZnO and CO_2 in the presence or absence of H_2O under mild conditions. Previous experiments were performed under mild conditions, mainly room temperature and P_{CO_2} varied from 0.01 atmospheres to 1 atmosphere (atm) [115–117]. Basic zinc carbonate with chemical formula $Zn_3CO_3 (OH)_4 \cdot 2H_2O$ was reported as a predominant phase at these conditions. Chen et al. synthesized the compound, zinc carbonate hydroxide hydrate (JCPDS #11-0287), using chemical precipitation method [5.0 % Na_2CO_3 (aq.) added to 0.1 mol/L $ZnSO_4$(aq.)] [118]. A direct relation of heating rate of the sample to its decomposition temperature was noticed. Interestingly, only one endothermic peak for zinc carbonate hydroxide hydrate was observed and was thus attributed to the combined evolution of water vapor and CO_2. The decomposition temperature was observed in the range of 197–257 °C for various heating rate (5–20 °C/min).

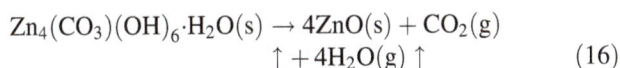

$$Zn_4(CO_3)(OH)_6 \cdot H_2O(s) \rightarrow 4ZnO(s) + CO_2(g)$$
$$\uparrow + 4H_2O(g) \uparrow \quad (16)$$

ZnO–CO_2–H_2O process is favored by low operating and regeneration temperature and could be ideal for post-combustion CO_2 capture. However, its application is limited due to its rare occurrence in the nature, and this could be the reason for the sparse study of ZnO–CO_2 system.

Similarly, NiO can be considered for post-combustion capture, but not for pre-combustion capture. According to thermodynamics, post-combustion absorption will occur below 77 °C and NiO could be regenerated above 187 °C. However, NiO availability is still a significant concern for its practical application. Therefore, the catalytic potential of NiO leads to its huge application in the CO_2 utilization field. Under sunlight, CO_2 can be converted into methanol in the presence of $NiO/lnTaO_4$ [119–121].

Perovskites

Perovskites have a stable structural framework, despite the movement of active ions into and out of the structure. Such materials have the potential to obviate the challenges associated with the other oxides. Currently, the main issues that most of the oxides face are as follows: (1) rapid degradation of their CO_2 capture capacity over repeated cycles and (2) a large volume change when oxide is converted into carbonate. Recent reported experimental results have substantially attracted the interest toward perovskites, as potential CO_2 capture materials.

One such study by Galven et al. investigates the CO_2 adsorption behavior of $Li_2SrTa_2O_7$, a layered perovskite family of the Ruddlesden–Popper (RP) phase (general formula, $A_{n-1} A'_2 B_n X_{3n+1}$, where A, A' and B are cations, X is an anion, and n is the number of the layers of octahedra in the perovskite-like stack) [122]. Under humid CO_2 environment, $Li_2SrTa_2O_7$ is transformed into $LiHSrTa_2O_7$ releasing LiOH, which then locks CO_2 in the form of Li_2CO_3.

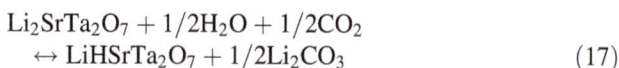

$$Li_2SrTa_2O_7 + 1/2H_2O + 1/2CO_2 \\ \leftrightarrow LiHSrTa_2O_7 + 1/2Li_2CO_3 \qquad (17)$$

It is interesting to note that the operating temperature of $Li_2SrTa_2O_7$ is 140 °C compared to ~ 400 °C of most of the other inorganic solid compounds. On heating, $Li_2SrTa_2O_7$ can be fully recovered at increased temperature (~ 700 °C) showing that the CO_2 capture is reversible and can be performed for a number of cycles. Similar CO_2 absorption capacity was observed for $Li_2SrNb_2O_7$. These materials, $Li_2SrTa_2O_7$ and $Li_2SrNb_2O_7$, have phases consisting of layers of corner-connected TaO_6 or NbO_6 octahedra separated by a layer of corner-connected LiO_4 tetrahedra [123]. In the presence of water at room temperature, H^+ shifts Li^+ in the layered structure and allows the shifted Li^+ to react with CO_2 to form Li_2CO_3. During regeneration, Li^+ is reinserted into the Ruddlesden–Popper structure, as in alkali garnet materials [124–127].

Another perovskite material, $Ba_4Sb_2O_9$, can be synthesized using a solid-state reaction in air from $BaCO_3$ and Sb_2O_3 [128]. $6H-Ba_4Sb_2O_9$ reacts with CO_2 to form $BaSb_2O_6$ and $BaCO_3$.

$$6H-Ba_4Sb_2O_9 + 3CO_2 \leftrightarrow BaSb_2O_6 + 3BaCO_3 \qquad (18)$$

$6H-Ba_4Sb_2O_9$, was found to be able to capture CO_2, mixed with N_2, at 600 °C while the regeneration temperature was around 950 °C under N_2 atmosphere [129]. No significant reduction in the CO_2 absorption capacity was observed for 100 cycles. After 100 cycles, the capacity [~ 0.1 g (CO_2)/g sorbent] is stable and translates to 73 % of the total molar capacity. The reaction between $6H-Ba_4Sb_2O_9$ and CO_2 has similar equilibrium constant to that of CaO-based sorbents. However, the CO_2 absorption capacity was stable for a large number of cycles, unlike CaO that degrades rapidly over cycles under same operating conditions.

Miscellaneous

Aluminosilicate zeolite is attractive for post-combustion CO_2 capture due to its selective CO_2 adsorption and high CO_2 capacities [130]. An extensive research has been carried out on aluminosilicate zeolite, and the results exhibit high uptake capacity than pure-silica zeolites [131–133]. In general, adsorption kinetics of CO_2 on zeolites is relatively fast and can achieve equilibrium capacity within a few minutes. The temperature and pressure strongly govern the CO_2 adsorption on zeolites. As the temperature increases, the adsorption of CO_2 decreases but as the gas-phase partial pressure of CO_2 increases, CO_2 adsorption increases accordingly. However, the presence of water vapor may limit the application of zeolite sorbents by decreasing its capacity. It is clear that optimizing different factors, such as basicity, pore size of zeolites and electric field strength (due to the exchangeable cations in the cavities), can significantly influence the CO_2 adsorption capacities of zeolites.

Conclusion

It is certain that an improved and cost-efficient process for CO_2 capture obtained from flue gas streams from power plants burning fossil fuels is needed to mitigate the huge emission of greenhouse gasses. Among several different proposed processes, metal oxides are promising options for suitable sorbent and have a great potential in the future. However, these sorbents also have limitations and challenges which must be solved before these can be employed commercially for CO_2 capture purpose. Based on the current scenario, following suggestions are proposed for the future research on CO_2 capture using metal oxides:

(a) Improvement in the potential sorbent performance, in terms of capture capacity, reversibility rate, carbonation kinetics, multi-cycle durability and sorbent reversibility;

(b) In-depth analysis of performance check of potential sorbent under actual flue gas conditions;

(c) A detailed techno-economic assessment of the potential sorbents;

Acknowledgments S. Kumar would like to acknowledge the financial support from Florida International University Graduate School Dissertation Year Fellowship (DYF).

References

1. Aaron, D., Tsouris, C.: Separation of CO_2 from flue gas: a review. Sep. Sci. Technol. **40**, 321–348 (2005)

2. White, C.M., Strazisar, B.R., Granite, E.J., Hoffman, J.S., Pennline, H.W.: Separation and capture of CO_2 from large stationary sources and sequestration in geological formations. J. Air Waste Manag. Assoc. **53**, 645–715 (2003)

3. Bryant, E.: Climate Process and Change. Cambridge University Press, Cambridge (1997)

4. Oreskes, N.: The scientific consensus on climate change. Science **306**, 1686 (2004)

5. Merkel, T.C., Lin, H., Wei, X., Baker, R.: Power plant carbon dioxide capture: an opportunity for membranes. J. Membr. Sci. **359**, 126–139 (2010)

6. Little, R.J., Versteeg, G.F., Van Swaaij, W.P.M.: Physical absorption into non-aqueous solutions in a stirred cell reactor. Chem. Eng. Sci. **46**, 3308–3313 (1991)

7. Chiesa, P., Consonni, S.P.: Shift reactors and physical absorption for low-CO_2 emission IGCCs. J. Eng. Gas Turbines Power **121**, 295–305 (1999)

8. Bishnoi, S., Rochelle, G.T.: Absorption of carbon dioxide into aqueous piperazine: reaction kinetics, mass transfer and solubility. Chem. Eng. Sci. **55**, 5531–5543 (2000)

9. Rochelle, G.T.: Amine scrubbing for CO_2 capture. Science **325**, 1652–1654 (2009)

10. Harlick, P.J.E., Tezel, F.H.: An experimental adsorbent screening study for CO_2 capture from N_2. Microporous Mesoporous Mater. **76**, 71–77 (2004)

11. Powell, C.E., Qiao, G.G.: Polymeric CO_2/N_2 gas separation membranes for the capture carbon dioxide from power plant flue gases. J. Membr. Sci. **279**, 1–49 (2006)

12. Benson, S.M., Cole, D.R.: CO_2 sequestration in deep sedimentary formations. Elements **4**, 325–331 (2008)

13. Guoxin, H., Huang, H., Li, Y.H.: The gasification of wet biomass using $Ca(OH)_2$ as CO_2 absorbent: the microstructure of char and absorbent. Int. J. Hydrogen Energy **33**, 5422–5429 (2008)

14. Siriwardane, R., Poston, J., Chaudhari, K., Zinn, A., Simonyi, T., Robinson, C.: Chemical-looping combustion of simulated synthesis gas using nickel oxide oxygen carrier supported on bentonite. Energy Fuels **21**, 1582–1591 (2007)

15. Lee, S.C., Kim, J.C.: Dry potassium-based sorbents for CO_2 capture. Catal. Surv. Asia **11**, 171–185 (2007)

16. Abanades, J.C., Anthony, E.J., Wang, J., Oakey, J.E.: Fluidized bed combustion systems integrating CO_2 capture with CaO. Environ. Sci. Technol. **39**, 2861–2866 (2005)

17. Duan, Y., Sorescu, D.C.: CO_2 capture properties of alkaline earth metal oxides and hydroxides: a combined density functional theory and lattice phonon dynamics study. J. Chem. Phys. **133**, 074508 (2010)

18. Zhang, B., Duan, Y., Johnson, K.: Density functional theory study of CO_2 capture with transition metal oxides and hydroxides. J. Chem. Phys. **136**, 064516 (2012)

19. Duan, Y., Luebke, D., Pennline, H.: Efficient theoretical screening of solid sorbents for CO_2 capture applications. Int. J. Clean Coal Energy **1**, 1–11 (2012)

20. DOE-NETL, Cost and performance baseline for fossil energy plants, Volume 1: Bituminous Coal and Natural Gas to Electricity Final Report. August 2007

21. Nakagawa, K., Ohashi, T.J.: A novel method for CO_2 capture from high temperature gases. J. Electrochem. Soc. **145**, 1344–1346 (1998)

22. Ochoa-Fernández, E., Rusten, H.K., Jakobsen, H.A., Rønning, M., Holmen, A., Chen, D.: Sorption enhanced hydrogen production by steam methane reforming using Li_2ZrO_3 as sorbent: sorption kinetics and reactor simulation. Catal. Today **106**, 41–46 (2005)

23. Pfeiffer, H., Vázquez, C., Lara, V.H., Bosch, P.: Thermal behavior and CO_2 absorption of $Li_{2-x}Na_xZrO_3$ solid solutions. Chem. Mater. **19**, 922–926 (2007)

24. Ida, J., Lin, Y.S.: Mechanism of high-temperature CO_2 sorption on lithium zirconate. Environ. Sci. Technol. **37**, 1999–2004 (2003)

25. Ida, J., Xiong, R., Lin, Y.S.: Synthesis and CO_2 sorption properties of pure and modified lithium zirconate. Sep. Purif. Technol. **36**, 41–51 (2004)

26. Pannocchia, G., Puccini, M., Seggiani, M., Vitolo, S.: Experimental and modeling studies on high-temperature capture of CO_2 using lithium zirconate based sorbents. Ind. Eng. Chem. Res. **46**, 6696–6706 (2007)

27. Veliz-Enriquez, M., González, G., Pfeiffer, H.: Synthesis and CO_2 capture evaluation of $Li_{2-x}K_xZrO_3$ solid solutions and crystal structure of a new lithium–potassium zirconate phase. J. Solid State Chem. **180**, 2485–2492 (2007)

28. Hernández, L.O.G., Gutiérrez, D.L., Martínez, V.C., Ortiz, A.L.: Synthesis, characterization and high temperature CO_2 capture evaluation of Li_2ZrO_3-Na_2ZrO_3 mixtures. J. New Mater. Electrochem. Syst. **11**, 137–142 (2008)

29. Essaki, K., Nakagawa, K., Kato, M., Uemoto, H.: CO_2 absorption by lithium silicate at room temperature. J. Chem. Eng. Jpn. **37**, 772–777 (2004)

30. Kato, M., Nakagawa, K., Essaki, K., Maezawa, Y., Takeda, S., Kogo, R., Hagiwara, Y.: Novel CO_2 absorbents using lithium containing oxide. Int. J. Appl. Ceram. Technol. **2**, 467–475 (2005)

31. Puccini, M., Seggiani, M., Vitolo, S.: Lithium silicate pellets for CO_2 capture at high temperature. Chem. Eng. Trans. **35**, 373–378 (2003)

32. Wang, K., Guo, X., Zhao, P., Wang, F., Zheng, C.: High temperature capture of CO_2 on lithium-based sorbents from rice husk ash. J. Hazard. Mater. **189**, 301–307 (2011)

33. Yamauchi, K., Murayama, N., Shibata, J.: Absorption and release of carbon dioxide with various metal oxides and hydroxides. Mater. Trans. **48**, 2739–2742 (2007)

34. Quinn, R., Kitzhoffer, R.J., Hufton, J.R., Golden, T.C.: A high temperature lithium orthosilicate-based solid absorbent for post combustion CO_2 capture. Ind. Eng. Chem. Res. **51**, 9320–9327 (2012)

35. Figueroa, J.D., Fout, T., Plasynski, S., McIlvried, H., Srivastava, R.D.: Advances in CO_2 capture technology—the U.S. Department of Energy's Carbon Sequestration Program. Int. J. Greenhouse Gas Control **2**, 9–20 (2008)

36. Korake, P.V., Gaikwad, A.G.: Capture of carbon dioxide over porous solid adsorbents lithium silicate, lithium aluminate and magnesium aluminate at pre-combustion temperatures. Front. Chem. Sci. Eng. **5**, 215–226 (2011)

37. Gauer, C., Heschel, W.: Doped lithium orthosilicate for absorption of carbon dioxide. J. Mater. Sci. **41**, 2405–2409 (2006)

38. Venegas, M.J., Fregoso-Israel, E., Escamilla, R., Pfeiffer, H.: Kinetic and reaction mechanism of CO2 sorption on Li4SiO4: study of the particle size effect. Ind. Eng. Chem. Res. **46**, 2407–2412 (2007)

39. Xu, H., Cheng, W., Jin, X., Wang, G., Lu, H., Wang, H., Chen, D., Fan, B., Hou, T., Zhang, R.: Effect of the particle size of quartz powder on the synthesis and CO_2 absorption properties of Li_4SiO_4 at high temperature. Ind. Eng. Chem. Res. **52**, 1886–1891 (2013)

40. Seggiani, M., Puccini, M., Vitolo, S.: High-temperature and low concentration CO_2 sorption on Li_4SiO_4 based sorbents: study of the used silica and doping method effects. Int. J. Greenhouse Gas Control **5**, 741–748 (2011)

41. Mosqueda, R.R., Pfeiffer, H.: Thermo-kinetic analysis of the CO_2 chemisorption on Li_4SiO_4 by using different gas flow rates and particle sizes. J. Phys. Chem. A **114**, 435–454 (2010)

42. Lopez-Ortiz, A., Perez Rivera, N.G., Rojas, R.A., Gutierrez, D.L.: Novel carbon dioxide solid acceptors using sodium containing oxides. Sep. Sci. Technol. **39**, 3559–3572 (2005)

43. Wu, X., Wen, Z., Xu, X., Wang, X., Lin, J.: Synthesis and characterization of Li_4SiO_4 nano-powders by a water-based sol–gel process. J. Nucl. Mater. **392**, 471–475 (2009)

44. Bretado, M.E., Velderrain, V.G., Gutierrez, D.L., Collins-Martinez, V., Ortiz, L.A.: A new synthesis route to Li_4SiO_4 as CO_2 catalytic/sorbent. Catal. Today **107–08**, 863–867 (2005)

45. Shan, S., Li, S., Jia, Q., Jiang, L., Wang, Y., Peng, J.: Impregnation precipitation preparation and kinetic analysis of Li_4SiO_4-based sorbents with fast CO_2 adsorption rate. Ind. Eng. Chem. Res. **52**, 6941–6945 (2013)

46. Olivares-Marin, M., Drage, T.C., Maroto-Valer, M.M.: Novel lithium-based sorbents from fly ashes for CO_2 capture at high temperatures. Int. J. Greenhouse Gas Control **4**, 623–629 (2010)

47. Wang, K., Zhao, P., Guo, X., Li, Y., Han, D., Chao, Y.: Enhancement of reactivity in Li_4SiO_4-based sorbents from the nano-sized rice husk ash for high-temperature CO_2 capture. Energy Convers. Manag. **81**, 447–454 (2014)

48. Shan, S., Jia, Q., Jiang, L., Li, Q., Wang, Y., Peng, J.: Novel Li_4SiO_4-based sorbents from diatomite for high temperature CO_2 capture. Ceram. Int. **39**, 5437–5441 (2013)

49. Zhao, T., Rønning, M., Chen, D.: Preparation and high-temperature CO_2 capture properties of nanocrystalline Na_2ZrO_3. Chem. Mater. **19**, 3294–3301 (2007)

50. Pfeiffer, H., Lima, E., Bosch, P.: Lithium–sodium metazirconate solid solutions, $Li_{2-x}Na_xZrO3$ ($0 \leq x \leq 2$): a hierarchical architecture. Chem. Mater. **18**, 2642–2647 (2006)

51. López-Ortiz, A., Perez-Rivera, N.G., Reyes-Rojas, A., Lardizabal-Gutierrez, D.: Novel carbon dioxide solid acceptors using sodium containing oxides. Sep. Sci. Technol. **39**, 3559–3572 (2004)

52. Alcérreca-Corte, I., Fregoso-Israel, E., Pfeiffer, H.: CO2 absorption on Na2ZrO3: a kinetic analysis of the chemisorption and diffusion processes. J. Phys. Chem. C **112**, 6520–6525 (2008)

53. Jimenez, D.B., Bretado, M.A.E., Gutierrez, D.L., Guiterrez, J.M.S., Ortiz, A.L., Martinez, V.C.: Kinetic study and modeling of the high temperature CO_2 capture by Na_2ZrO_3 solid absorbent. Int. J. Hydrogen Energy **38**, 2557–2564 (2013)

54. Ochoa-Fernández, E., Zhao, T., Ronning, M., Chen, D.: Effects of steam addition on the properties of high temperature ceramic CO_2 acceptors. J. Environ. Eng. **37**, 397–403 (2009)

55. Santillán-Reyes, G.G., Pfeiffer, H.: Analysis of the CO_2 capture in sodium zirconate (Na_2ZrO_3): effect of water vapor addition. Int. J. Greenhouse Gas Control **5**, 1624–1629 (2011)

56. Choi, S., Drese, J.H., Jones, W.C.: Adsorbent materials for carbon dioxide capture from large anthropogenic point sources. ChemSusChem **2**, 796–854 (2009)

57. Cushing, B.L., Kolesnichenko, V.L., O'Connor, C.J.: Recent advances in the liquid-phase syntheses of inorganic nanoparticles. Chem. Rev. **104**, 3893–3946 (2004)

58. Vasylkiv, O., Sakka, Y.: Nanoexplosion synthesis of multimetal oxide ceramic nanopowders. Nano Letter **5**, 2598–2604 (2005)

59. Pfeiffer, H., Bosch, P.: Thermal stability and high-temperature carbon dioxide sorption on hexa-lithium zirconate ($Li_6Zr_2O_7$). Chem. Mater. **17**, 1704–1710 (2005)

60. Jimenez, B.D., Reyes Rojas, C.M., López-Ortiz, G.V.: Novel developments in adsorption; the effect of Li as a dopant in Na2ZrO3 high temperature CO2 acceptor. In: AIChE Annual Meeting (2004)

61. Yoshida, T.: Development of molten carbonate fuel cell (MCFC) power generation technology. Ishikawajima-Harima Eng. Rev. **4**, 125–134 (1994)

62. Rodriguez, M.T., Pfeiffer, H.: Sodium metasilicate (Na_2SiO_3): a thermo-kinetic analysis of its CO_2 chemical sorption. Thermochim. Acta **473**, 92–95 (2008)

63. Rodríguez-Mosqueda, R., Pfeiffer, H.: High CO2 capture in sodium metasilicate (Na2SiO3) at low temperatures (30–60 C)

64. Bhagiyalakshmi, M., Lee, J.Y., Jang, H.T.: Synthesis of mesoporous magnesium oxide: its application to CO_2 chemisorption. Int. J. Greenhouse Gas Control **4**, 51–56 (2010)

65. Feng, B., An, H., Tan, E.: Screening of CO_2 adsorbing materials for zero emission power generation systems. Energy Fuels **21**, 426–434 (2007)

66. Li, L., Wen, X., Fu, X., Wang, F., Zhao, N., Xiao, F., Wei, W., Sun, Y.: MgO/Al2O3 sorbent for CO_2 capture. Energy Fuels **24**, 5773–5780 (2010)

67. Lee, S.C., Chae, H.J., Lee, S.J., Choi, B.Y., Yi, C.K., Lee, J.B., Ryu, C.K., Kim, J.C.: Development of regenerable MgO-based sorbent promoted with K_2CO_3 for CO_2 capture at low temperatures. Environ. Sci. Technol. **42**, 2736–2741 (2008)

68. Ward, S.M., Braslaw, J., Gealer, R.L.: Carbon dioxide sorption studies on magnesium oxide. Thermochim. Acta **64**, 107–114 (1983)

69. Fu, X., Zhao, N., Li, J., Xiao, F., Wei, W., Sun, Y.: Carbon dioxide capture by MgO-modified MCM-41 materials. Adsorpt. Sci. Technol. **27**, 593–601 (2009)

70. Liu, S., Zhang, X., Li, J., Zhao, N., Wei, W., Sun, Y.: Preparation and application of stabilized mesoporous MgO–ZrO_2 solid base. Catal. Commun. **9**, 1527–1532 (2008)

71. Li, B., Wen, X., Zhao, N., Wang, X.Z., Wei, W., Sun, Y., Ren, Z.H., Wang, Z.J.: Preparation of high stability MgO-ZrO_2 solid base and its high temperature CO_2 capture properties. J. Fuel Chem. Technol. **38**, 473–477 (2010)

72. Yong, Z., Mata, V., Rodrigues, A.E.: Adsorption of carbon dioxide onto hydrotalcite-like compounds (HTlcs) at high temperatures. Ind. Eng. Chem. Res. **40**, 204–209 (2001)

73. Li, L., Li, Y., Wen, X., Wang, F., Zhao, N., Xiao, F., Wei, W., Sun, Y.: CO_2 capture over K_2CO_3/MgO/Al_2O_3 dry sorbent in a fluidized bed. Energy Fuels **25**, 3835–3842 (2011)

74. Lee, S.C., Choi, B.Y., Lee, T.J., Ryu, C.K., Soo, Y.S., Kim, J.C.: CO_2 absorption and regeneration of alkali metal-based solid sorbents. Catal. Today **111**, 385–390 (2006)

75. Ruminski, A.M., Jeon, K.J., Urban, J.J.: Size- and surface-dependent CO_2 capture in chemically synthesized magnesium oxide nanocrystals. J. Mater. Chem. **21**, 11486–11491 (2011)

76. Beruto, D., Botter, R., Searcy, A.W.: Thermodynamics of two, two-dimensional phases formed by carbon dioxide chemisorption on magnesium oxide. J. Phys. Chem. **91**, 3578–3581 (1987)

77. Lu, W., Yuan, D., Sculley, J., Zhao, D., Krishna, R., Zhou, H.C.: Sulfonate-grafted porous polymer networks for preferential CO_2 adsorption at low pressure. J. Am. Chem. Soc. **133**, 18126–18129 (2011)

78. Fagerlund, J., Highfield, J., Zevenhoven, R.: Kinetics studies on wet and dry gas–solid carbonation of MgO and $Mg(OH)_2$ for CO_2 sequestration. RSC Adv. **2**, 10380–10393 (2012)

79. Highfield, J., Bu, J., Fagerlund, J., Zevenhoven, R. ICCDU-XI (2011)

80. Butt, D.P., Pile, S.P., Park, Y., Vaidya, R., Lackner, K.S., Wendt, C.H., Nomura, K.: Report LA-UR-98-7631 (1998)

81. Hassanzadeh, A., Abbasian, J.: Regenerable MgO-based sorbents for high-temperature CO_2 removal from syngas: 1. Sorbent development, evaluation, and reaction modeling. Fuel **89**, 1287–1297 (2010)

82. Silaban, A., Harrison, D.P.: High temperature capture of carbon dioxide: characteristics of the reversible reaction between CaO (s) and CO_2 (g). Chem. Eng. Commun. **137**, 177–190 (1995)

83. Salvador, C., Lu, D., Anthony, E., Abanades, J.: Enhancement of CaO for CO_2 capture in an FBC environment. Chem. Eng. J. **96**, 187–195 (2003)

84. Hughes, R., Lu, D., Anthony, E., Wu, Y.: Improved long-term conversion of limestone-derived sorbents for in situ capture of

through the CO2–H2O chemisorption process. J. Phys. Chem. C **117**, 13452–13461 (2013)

CO_2 in a fluidized bed combustor. Ind. Eng. Chem. Res. **43**, 5529–5539 (2004)

85. Moranska, M.K., Tomaszewicz, G., Labojko, G.: Comparison of different methods for enhancing CO_2 capture by CaO-based sorbents-review. Physicochem. Prob. Miner. Process. **48**, 77–90 (2012)

86. Barker, R.: The reactivity of calcium oxide towards carbon dioxide and its use for energy storage. J. Appl. Chem. Biotech. **24**, 221–227 (1974)

87. Bhatia, S.K., Perlmutter, D.D.: Effect of the product layer on the kinetics of the CO_2-lime reaction. Am. Instit. Chem. Eng. J. **29**, 79–86 (1983)

88. Lu, H., Smirniotis, P.G., Ernst, F.O., Pratsinis, S.E.: Nanostructured Ca-based sorbents with high CO_2 uptake efficiency. Chem. Eng. Sci. **64**, 1936–1943 (2009)

89. Lu, H., Reddy, E.P., Smirniotis, P.G.: Calcium oxide based sorbents for capture of carbon dioxide at high temperatures. Ind. Eng. Chem. Res. **45**, 3944–3949 (2006)

90. Abanades, J.C., Alvarez, D.: Conversion limits in the reaction of CO_2 with lime. Energy Fuels **17**, 308–315 (2003)

91. Gruene, P., Belova, A.G., Yegulalp, T.M., Farrauto, R.J., Castaldi, M.J.: Dispersed calcium oxide as a reversible and efficient CO_2—sorbent at intermediate temperatures. Ind. Eng. Chem. Res. **50**, 4042–4049 (2011)

92. Wang, J., Manovic, V., Wu, Y., Anthony, E.: A study on the activity of CaO-based sorbents for capturing CO_2 in clean energy processes. Appl. Energy **87**, 1453–1458 (2010)

93. Bouquet, E., Leyssens, G., Schönnenbeck, C., Gilot, P.: The decrease of carbonation efficiency of CaO along calcination–carbonation cycles: experiments and modeling. Chem. Eng. Sci. **64**, 2136–2146 (2009)

94. Lysikov, A., Trukhan, S., Okunev, A.: Sorption enhanced hydrocarbons reforming for fuel cell powered generators. Int. J. Hydrogen Energy **33**, 3061–3066 (2008)

95. Stanmore, B., Gilot, P.: Review—calcination and carbonation of limestone during thermal cycling for CO_2 sequestration. Fuel Process. Technol. **86**, 1707–1743 (2005)

96. Li, Y., Zhao, Ch., Chen, H., Liang, C., Duan, L., Zhou, W.: Modified CaO-based sorbent looping cycle for CO_2 mitigation. Fuel **88**, 697–704 (2009)

97. Adanez, J., de Diego, L., Garcia-Labiano, F.: Calcination of calcium acetate and calcium magnesium acetate: effect of the reacting atmosphere. Fuel **78**, 583–592 (1999)

98. Nimmo, W., Patsias, A., Hampartsoumian, E., Gibbs, B., Fairweather, M., Williams, P.: Calcium magnesium acetate and urea advanced reburning for NO control with simultaneous SO_2 reduction. Fuel **83**, 1143–1150 (2004)

99. Patsias, A., Nimmo, W., Gibbs, B., Williams, P.: Calcium-based sorbents for simultaneous NOx/Sox reduction in a down-fired furnace. Fuel **84**, 1864–1873 (2005)

100. Silaban, A., Narcida, M., Harrision, D.: Characteristics of the reversible reaction between CO_2 (g) and calcined dolomite. Chem. Eng. Commun. **146**, 149–162 (1996)

101. Bandi, A., Specht, M., Sichler, P., Nicoloso, N.: In-situ gas. Conditioning in fuel reforming for hydrogen generation. 5th International Symposium on Gas Cleaning at High Temperature, Morgantown (2002)

102. Manovic, V., Anthony, E., Grasa, G., Abanades, J.: CO_2 looping cycle performance of a high-purity limestone after thermal activation/doping. Energy Fuels **22**, 3258–3264 (2008)

103. Wang, M., Lee, C.G.: Absorption of CO_2 on $CaSiO_3$ at high temperatures. Energy Convers. Manag. **50**, 636–638 (2009)

104. Gupta, H., Fan, L.S.: Carbonation-calcination cycle using high reactivity calcium oxide for carbon dioxide separation from flue gas. Ind. Eng. Chem. Res. **41**, 4035–4042 (2002)

105. Tilekar, G., Shinde, K., Kale, K., Raskar, R., Gaikwad, A.: The capture of carbon dioxide by transition metal aluminates, calcium aluminate, calcium zirconate, calcium silicate and lithium zirconate. Front. Chem. Sci. Eng. **5**, 477–491 (2011)

106. Richter, H.J., Knoche, K. F.: Reversibility of combustion processes. In: Gaggioli, R.A. (ed.) Efficiency and Costing: Second Law Analysis of Processes. ACS Symposium Series, vol. 235, pp 71–85 (1983)

107. Fan, L.S., Zeng, L., Wang, W., Luo, S.: Chemical looping processes for CO_2 capture and carbonaceous fuel conversion—prospect and opportunity. Energy Environ. Sci. **5**, 7254–7280 (2012)

108. Adanez, J., Abad, A., Garcia-Labiano, F., Gayan, P., de Diego, L.: Progress in chemical-looping combustion and reforming technologies. Prog. Energy Combust. Sci. **38**, 215–282 (2012)

109. Chiu, P.C., Ku, Y.: Chemical looping process—a novel technology for inherent CO_2 capture. Aerosol Air Qual. Res. **12**, 1421–1432 (2012)

110. Imtiaz, Q., Hosseini, D., Müller, C.R.: Review of oxygen carriers for the chemical looping with oxygen uncoupling: thermodynamics, material development, and synthesis. Energy Technol. **1**, 633–647 (2013)

111. Mattisson, T.: Materials for chemical-looping with oxygen Uncoupling. ISRN Chem. Eng. **2013**, 1–19 (2013)

112. Kim, H.R. Coal-direct chemical looping combustion process for in situ carbon dioxide capture—Operational Experience of Integrated 25-kWth Sub-Pilot Scale. Unit Document no. osu1352996758

113. Kumar, S., Drozd, V., Durygin, A., Saxena, S.K.: Method and system for sequestering carbon dioxide and producing hydrogen gas. U.S. Provisional Patent Application No. 61/733,297 (2012)

114. www.worldsteel.org

115. Habashi, F.: Can. Instit. Min. Bull. **94**, 71–76 (2001)

116. Ohkuma, N., Funayama, Y., Ito, H., Mizutani, N., Kato, M.: Nihon kagaku kaishi 802 (1987)

117. Galvez, J., Arana, R.: Univ. Murcia Scie. **38**, 153 (1979)

118. Chen, J., Zhao, R., Jiang, H.: Trans. Nonferrous Metals Soc. **8**, 149–153 (1988)

119. Wang, Z.Y., Chou, H.C., Wu, J.C.S., Tsai, D.P., Mul, G.: CO_2 photoreduction using $NiO/InTaO_4$ in optical-fiber reactor for renewable energy. Appl. Catal. A **380**, 172–177 (2010)

120. Chen, H.C., Chou, H.C., Wu, J.C.S., Lin, H.Y.: Sol–gel prepared $InTaO_4$ and its photocatalytic characteristics. J. Mater. Res. **23**, 1364–1370 (2008)

121. Pan, P.W., Chen, Y.W.: Photocatalytic reduction of carbon dioxide on $NiO/InTaO_4$ under visible light irradiation. Catal. Commun. **8**, 1546–1549 (2007)

122. Galven, C., Fourquet, J.L., Suard, E., Crosnier-Lopez, M.P., Le Berre, F.: Mechanism of a reversible CO_2 capture monitored by the layered perovskite $Li_2SrTa_2O_7$. Dalton Trans. **39**, 4191–4197 (2010)

123. Pagnier, T., Rosman, N., Galven, C., Suard, E., Fourquet, J.L., Le Berre, F., Crosnier-Lopez, M.P.: Phase transition in the Ruddlesden-Popper layered perovskite Li2SrTa2O7. J. Solid State Chem. **182**, 317–326 (2009)

124. Galven, C., Dittmer, J., Suard, E., Le Berre, F., Crosnier-Lopez, M.P.: Instability of lithium garnets against moisture. Structural characterization and dynamics of $Li_{7-x}H_xLa_3Sn_2O_{12}$ and $Li_{5-x}H_xLa_3Nb_2O_{12}$. Chem. Mater. **24**, 3335–3345 (2012)

125. Galven, C., Fourquet, J.L., Crosnier-Lopez, M.P., Le Berre, F.: Instability of the lithium garnet $Li_7La_3Sn_2O_{12}$: Li^+/H^+ exchange and structural study. Chem. Mater. **23**, 1892–1900 (2011)

126. Narayanan, S., Ramezanipour, F., Thangadurai, V.: Enhancing Li ion conductivity of garnet-type $Li_5La_3Nb_2O_{12}$ by Y-and Li-codoping: synthesis, structure, chemical stability, and transport properties. J. Phys. Chem. C **116**, 20154–20162 (2012)

127. Nyman, M., Alam, T.M., McIntyre, S.K., Bleier, G.C., Ingersoll, D.: Alternative approach to increasing li mobility in Li-La-Nb/Ta garnet electrolytes. Chem. Mater. **22**, 5401–5410 (2010)

128. Dunstan, M.T., Pavan, A.F., Kharton, V.V., Avdeev, M., Kimpton, J.A., Koltygin, V.A., Tsipis, E.V., Ling, C.D.: Phase behavior and mixed ionic–electronic conductivity of $Ba_4Sb_2O_9$. Solid State Ionics **235**, 1–7 (2013)

129. Dunstan, M.T., Liu, W., Pavan, A.F., Kimpton, J.A., Ling, C.D., Scott, S.A., Dennis, J.S., Grey, C.P.: Reversible CO_2 absorption by the 6H Perovskite $Ba_4Sb_2O_9$. Chem. Mater. **25**, 4881–4891 (2013)

130. Yaghi, O.M., Keeffe, M., Ockwig, N., Hee, C., Eddaoudi, M., Kim, J.: Reticular synthesis and the design of new materials. Nature **423**, 705–714 (2003)

131. Hudson, M., Queen, W., Mason, J., Fickel, D., Lobo, R., Brown, C.: Unconventional, highly selective CO_2 adsorption in Zeolite SSZ-13. J. Am. Chem. Soc. **134**, 1970–1973 (2012)

132. Pires, J., de Carvalho, M.B., Ribeiro, F.R., Derouane, E.G.: Carbon dioxide in Y and ZSM-20 zeolites: adsorption and infrared studies. J. Mol. Catal. **85**, 295–303 (1993)

133. Samanta, A.K., Zhao, A., Shimizu, G.K.H., Sarkar, P., Gupta, R.: Post-combustion CO_2 capture using solid sorbents: a review. Ind. Eng. Chem. Res. **51**, 1438–1463 (2012)

Influence of calcination temperature on the morphology and energy storage properties of cobalt oxide nanostructures directly grown over carbon cloth substrates

R. B. Rakhi · Wei Chen · Dongkyu Cha ·
H. N. Alshareef

Abstract Nanostructured and mesoporous cobalt oxide (Co_3O_4) nanowire in flower-like arrangements have been directly grown over flexible carbon cloth collectors using solvothermal synthesis for supercapacitor applications. Changes in the morphology and porosity of the nanowire assemblies have been induced by manipulating the calcination temperature (200–300 °C) of the one-dimensional (1-D) structures, resulting in significant impact on their surface area and pseudocapacitive properties. As the calcination temperature increases from 200 to 250 °C, the flower morphology gradually modifies to the point where the electrolyte could access almost all the nanowires over the entire sample volume, resulting in an increase in specific capacitance from 334 to 605 Fg^{-1}, depending on the nanowire electrode morphology. The 300 °C calcination results in the breakdown of the mesoporous morphology and decreases the efficiency of electrolyte diffusion, resulting in a drop in pseudocapacitance after 300 °C. A peak energy density of 44 Wh kg^{-1} has been obtained at a power density of 20 kW kg^{-1} for the 250 °C calcined sample.

Keywords Supercapacitor · Pseudocapacitance · Calcination · Cobalt oxide · Mesoporous · Specific capacitance

Introduction

Significant research interest has been dedicated over the past decade in the development of lightweight and environmentally friendly efficient energy storage devices to meet the various requirements of modern portable consumer electronics and hybrid electric vehicles [1–5]. Supercapacitors, with their higher power densities, fast charge–discharge rates and excellent cycle stabilities than batteries, and higher energy densities than conventional dielectric capacitors, offer a promising approach to satisfy the increasing power demands of energy storage systems in the twenty-first century [6–8]. Depending on the charge storage mechanisms, supercapacitors are broadly classified into two types: (1) electric double-layer capacitors which are based on high surface area carbon nanomaterials with the capacitance arising from the charge separation at the electrode/electrolyte interface and (2) pseudocapacitors based on transition metal oxides (TMOs) or conducting polymers with an additional Faradaic capacitance arising from fast, reversible redox reactions occurring at or near the solid electrode surface [9, 10]. Performance of a supercapacitor strongly depends on the morphology and properties of its electrodes [11]. An ideal electrode material should have large surface area, high conductivity, suitable pore size distribution, long-term cycle stability and electrochemical reproducibility [11]. Carbon materials possess high conductivity and long-term stability; but have lower specific capacitance as compared to pseudocapacitance materials [12–18]. On the other hand, TMOs exhibit higher specific capacitances, but have lower conductivity than the carbon-based systems [19–24]. Conductivity of TMO-based electrodes can be enhanced by the direct growth of TMOs over conducting substrates; this process also eliminates the use of ancillary conducting material and binder in the electrode fabrication process [5, 25–30].

R. B. Rakhi · W. Chen · D. Cha · H. N. Alshareef (✉)
Materials Science and Engineering, King Abdullah University of Science and Technology (KAUST), Thuwal 23955-6900, Saudi Arabia
e-mail: husam.alshareef@kaust.edu.sa

Among the different TMOs, cobalt oxide (Co_3O_4) is a promising candidate for pseudocapacitors due to its easy availability, existence of variable oxidation states of metal ions which facilitate redox transitions, good pseudocapacitive behavior with high pseudocapacitance and stable cycling performance and cost-effectiveness as compared to the popular pseudocapacitor material, ruthenium oxide [21, 31–33]. As the electrochemical performance of Co_3O_4 nanostructures largely depends on their morphology and surface area, the development of controlled synthesis techniques of Co_3O_4 nanostructures with high electroactive area is of great importance [34]. There have been a variety of reports on the synthesis of Co_3O_4 nanostructures with different morphologies by different synthetic routes. He et al. reported the size- and shape-controlled synthesis of monodispersed Co_3O_4 nanocrystals by solubility-controlled liquid-phase reaction process. Cao et al. reported the preparation and growth kinetics of highly ordered Co_3O_4 superstructures with tunable morphologies from simple nanoplates to well-organized cabbage-like structures, and then to microspherical composites using a polyol process. Wang et al. reported the synthesis of mesoporous Co_3O_4 nanorods by a simple hydrothermal method [35]. Reports are also available on the direct growth of Co_3O_4 on conducting substrates. Electrodeposition of Co_3O_4 oxide thin films on different substrates are reported by Kandalkar et al. [36] (copper substrates) and wu et al. [37] (Ni foam substrate). Qing et al. [27] reported the facile synthesis of Co_3O_4 nanoflowers on Ni foam by a simple solvothermal method. Tummala et al. [38] reported the deposition of nanostructured Co_3O_4 on stainless steel current collectors by plasma spray technique.

Among the various possible morphologies of Co_3O_4, mesoporous one-dimensional (1-D) nanostructures are best suited for supercapacitor applications as they have the advantages of large surface area, improved electrolyte accessibility and reduced mass and charge diffusion distances [31, 34, 39–42]. Very few reports are available on the direct growth of one-dimensional Co_3O_4 nanostructures over different conducting substrates for supercapacitor applications. Gao et al. [43] reported a specific capacitance of 754 F/g at a current density of 5 mA cm^{-2} for Co_3O_4 nanowire arrays directly grown over nickel foam by a simple template-free method. Zhang et al. [33] reported the facile synthesis of freestanding Co_3O_4 nanowire arrays on Ni foam substrate with a remarkable specific capacitance of 1,160 F/g at a current density of 2 A/g. However, it is well known that the use of Ni foam as a current collector can bring about substantial errors to the specific capacitance values of electrode materials [44]. Carbon-coated flexible textiles can be used as an alternative to the Ni foam [45]. Yang et al. [26] reported the direct growth of Co_3O_4 nanowire network (nanonet) on a carbon fiber paper with a mass loading of 0.4 mg/cm^2 with a large specific capacitance (1,124 F/g) at a high discharge current density of 25.34 A/g. In a recent work, we reported remarkable pseudocapacitance for self-organized microstructures of mesoporous Co_3O_4 nanowires grown over two different types of carbon paper substrates by a solvothermal method [25]. The work was focused on how the surface texture of paper substrate influenced the morphology and pseudocapacitance of Co_3O_4. We could achieve Co_3O_4 nanowires which are organized into flower-like morphology over planar graphitized carbon paper and brush-like morphology over fibrous non-graphitized carbon paper substrate. The work dealt with samples calcined at 250 °C. But during the course of study we came across some interesting observations which revealed that the calcination temperature played a crucial role in tuning the pseudocapacitance, morphology and pore size of the sample with flower-like morphology [25].

As the carbon substrate is sensitive to high temperature, extreme care needs to be taken in selecting the suitable calcination temperature, which can dramatically enhance the supercapacitor performance of Co_3O_4 without damaging the carbon substrate. To the best of our knowledge, no detailed reports are available on the influence of calcination temperature on the morphology, pore structure and hence pseudocapacitance of 1-D Co_3O_4 nanostructures grown directly over carbon cloth substrate. We could reproduce the flower-like morphology over graphitized carbon cloth substrate using a solvothermal method. In the present study, we report the effect of calcination temperature on the pseudocapacitive properties of self-organized micro flowers of 1-D Co_3O_4 nanostructures directly grown over graphitized carbon cloth.

Experimental

Synthesis of 1-D Co_3O_4 nanostructures over carbon cloth

One-dimensional Co_3O_4, nanostructures were directly grown over graphitized carbon cloth substrate by a simple solvothermal method. 2 g $Co(NO_3)_2 \cdot 6H_2O$, 1 g hexa decyl trimethyl ammonium bromide (CTAB) and 6 ml water were dissolved in 30 ml absolute methanol by vigorous magnetic stirring. The resulting solution was then transferred into a 40 ml Teflon-lined stainless steel autoclave. A piece of graphitized carbon cloth (1 cm × 2 cm)was immersed in the growth solution in the autoclave followed by heating the autoclave in an oven at 180 °C for 24 h to allow growth of Co_3O_4 nanoneedles. The substrate covered with products was then washed with H_2O and ethanol, dried in a vacuum at 120 °C for 10 h, and the final product was denoted as as-prepared sample. The as-prepared samples were then calcined at different temperatures, −200, 225, 250 and 300 °C, for 4 h. Calcination temperatures were selected on the basis of thermogravimetric analysis (NETZSCH TG 209 F1 Iris) of as-prepared nanoneedles. The carbon cloth substrate was not able to survive the calcination temperature above 300 °C. The weight of Co_3O_4 deposit was accurately calculated from the difference in the weight of the substrate before the solvothermal process and with Co_3O_4 loading after calcination using an analytical micro balance (XP 26 Mettler Toledo, max 22 g, 0.001 mg of resolution).

General characterization

Co_3O_4 nanostructures were characterized by a powder X-ray diffraction system (XRD, Bruker, D8 ADVANCE) equipped with Cu K_α radiation ($\lambda = 0.15406$ nm). BET surface area of the samples were determined using surface area and porosimetry system 'Micromeritics' (ASAP 2420) at 77 K. Before measurements, the samples were dried at 70 °C for 10 h in a vacuum oven and then degassed at 200 °C for 12 h until the vacuum was <2 μm Hg. The surface morphology and microstructure of the samples were investigated by a scanning electron microscopy (SEM, FEI Helios NanoLab) and transition electron microscopy (TEM, FEI Titan).

Electrochemical characterization

Electrochemical measurements were carried out in symmetric two-electrode configurations using a Model 660D electrochemical workstation (CH Instruments). Assembled coin cells for the two-electrode configuration were prepared by sandwiching two identical pieces of Co_3O_4 nanostructures over carbon cloth (each with an area of 0.5×1 cm^2) calcined at a particular temperature by a monolayer polypropylene separator (25 μm thick, Celgard 3501), inside a coin cell (CR2032, MTI), with 30 wt% KOH (Sigma-Aldrich) as electrolyte. The mass of Co_3O_4 in each electrode was approximately 2 mg. The approximate thickness of the active materials on the electrodes was nearly 20 μm. The area of the electrodes was nearly 0.25 cm^2.

The electrochemical properties of the supercapacitor electrodes were studied by cyclic voltammetry (CV), galvanostatic charge–discharge (CD) and electrochemical impedance spectroscopy (EIS). The CV measurements were conducted in a voltage window between 0 and 0.8 V at a wide range of scan rates, ranging from 1 mV/s to 100 V/s. The CD measurements were also carried out in the same voltage window under a wide range of current densities, from 0.25 to 25 A/g. The EIS was performed in the frequency range from 100 kHz to 10 mHz at open circuit voltage by applying a 5 mV signal. All measurements were carried out at room temperature.

The specific capacitance (C_{sp}) of symmetric supercapacitors was calculated from the cyclic voltammogramms and charge–discharge curves according to Eqs. (1) and (2).

$$C_{sp} = \frac{2i}{fm} \qquad (1)$$

where 'i' is the average cathodic current of the CV loop and 'f' is the scan rate.

$$C_{sp} = \frac{2}{m} \times \frac{I}{\left(\Delta V/\Delta t\right)} \qquad (2)$$

where 'I' is the constant current for charge–discharge, $\Delta V/\Delta t$ is the slope of the discharge curve and 'm' is the mass of one electrode. In the present study, for specific capacitance calculation, electrode mass was taken as 2 mg (mass of the Co_3O_4 deposit only).

Results and discussion

1-D Co_3O_4 structures with nanocrystalline and mesoporous nature directly grown over conducting substrates can be considered as ideal candidates for supercapacitor electrodes. Morphology, crystallinity, porosity and capacitive performance of these structures can be varied to a great extent by changing the calcination temperature. Based on the above considerations, in the present work a systematic investigation was conducted to identify the influence of calcination temperature on the morphology and hence the pseudocapacitance properties of 1-D Co_3O_4 nanostructures directly grown over flexible carbon cloth substrate. Self-organization of these nanostructures into flower-like morphologies over flexible carbon cloth collectors was achieved by a simple solvothermal process. Changes in the morphology and porosity of the nanostructure assemblies were induced by manipulating the calcination temperature (200–300 °C) of the one-dimensional (1-D) structures, resulting in significant impact on their surface area and pseudocapacitive properties.

The thermal behaviors of the Co_3O_4 samples were determined by thermogravimetric analysis (TGA) as shown in Fig. 1a. The typical TGA curves of the two precursors clearly display two-step weight loss each, due to dehydration and decomposition of the cobalt carbonate nitrate hydroxide hydrate and β-$Co(OH)_2$ phases. The first weight loss at the lower temperature (<250 °C) is due to the elimination of adsorbed and intercalated water, while the second weight loss from 250 to 300 °C corresponds to the loss of structural water and thermal oxidative decomposition of the initial products as well as desorption of residual nitrate and carbonate anions. There is no major weight loss at higher temperature in the TGA curve indicating the absence of additional structural changes in those regions. The XRD patterns of the as-prepared and calcined Co_3O_4 samples are shown in Fig. 1b. The as-prepared sample contains a mixture of β-$Co(OH)_2$ (JCPDS 30-0443), cobalt carbonate nitrate hydroxide hydrate (JCPDS50-1891) and Co_3O_4 (JCPDS Card no. 43-1003) as identified from the XRD pattern (Fig. 1i). The sample calcined at 200 °C for 4 h retains (101) peak of the β-$Co(OH)_2$ phase along with the peaks from Co_3O_4 (Fig. 1ii). This residual phase can be removed by increasing the calcination temperature. Samples subjected to calcination temperatures of 225, 250 and 300 °C show identical crystal structures of FCC-type Co_3O_4 nanowires with space group of Fd3 m (Fig. 1iii, iv and v). The peaks observed in the XRD pattern can be indexed as (111), (220), (311), (400), (422), (511), (440), (620) and (533) planes of the nano-structured FCC-type Co_3O_4. Absence of any secondary peaks in the XRD pattern clearly illustrates the complete decomposition of precursors to Co_3O_4 samples after annealing the samples above a temperature of 225 °C.

The nitrogen adsorption and desorption isotherms of the Co_3O_4 samples calcined at different temperatures are shown in Fig. 2a. The Brunauer–Emmett–Teller (BET) surface area values calculated for the as-prepared sample and the samples calcined temperatures of 200, 225, 250 and 300 °C are, respectively, 9.27, 12.89, 13.27, 75.92 and 52.31 m²/g. All the calcined samples exhibit distinct hysteresis loops in the range of 0.45–1.0 P/P₀, which suggests the presence of a mesoporous structure. This loop is absent in the case of the as-prepared sample. The hysteresis loop is prominent in the sample calcined at 250 °C. This can be attributed to the well-formed mesoporous nature with the maximum BET surface area. Xiong et al. [46] reported a BET surface area of 25.12 m²/g for mesoporous Co_3O_4 nanosheets. Meher and Rao [32] reported the synthesis of mesoporous Co_3O_4 nanowires with a BET surface area of 60 m²/g. BET surface area value obtained for the Co_3O_4 nanowires in the present study is higher than these reported values, which is favorable for better supercapacitor performance of the composites. Calcination at 300 °C leads to reduction in BET surface area as the smaller particles aggregate to form large crystals in Co_3O_4 structure. Adsorption average pore width (4 V/A by BET) for the as-prepared sample and the samples calcined temperatures of 200, 225, 250 and 300 °C are, respectively, 9.20, 16.39, 17.42, 18.14 and 18.89 nm. The pore size distributions of the samples calculated by desorption isotherms using Barret–Joyner–Halenda (BJH) method are shown in Fig. 2b. BJH Desorption cumulative volume of pores between 1.7000 and 300.0000 nm diameter for the as-prepared sample and the samples calcined temperatures of 200, 225, 250 and 300 °C are 0.04570, 0.05174, 0.06449, 0.10361 and 0.17601 cm³/g, respectively.

SEM and TEM images of as-prepared samples are shown in Fig. 3. SEM images (Figs. 2a and b) show self-organization of nanoneedles into flower-like morphology on the surface of the graphitized carbon cloth. The nano-needles have varying lengths of 1–15 μm and diameters <100 nm. From TEM (Fig. 3c) and HRTEM (Fig. 3d) images, it is clear that the nanoneedle exhibits a smooth texture and single crystalline feature.

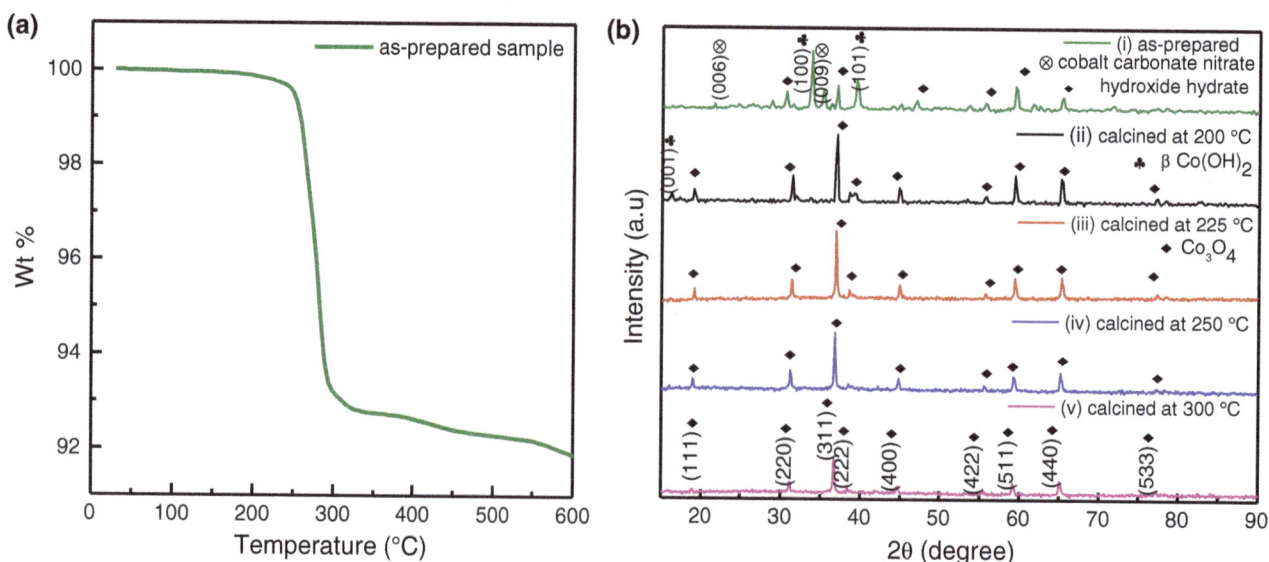

Fig. 1 **a** TGA curve of the as-prepared sample; **b** powder X-ray diffraction pattern of (i) as-prepared sample, (ii) sample calcined at 200 °C, (iii) sample calcined at 225 °C, (iv) sample calcined at 250 °C and (v) sample calcined at 300 °C

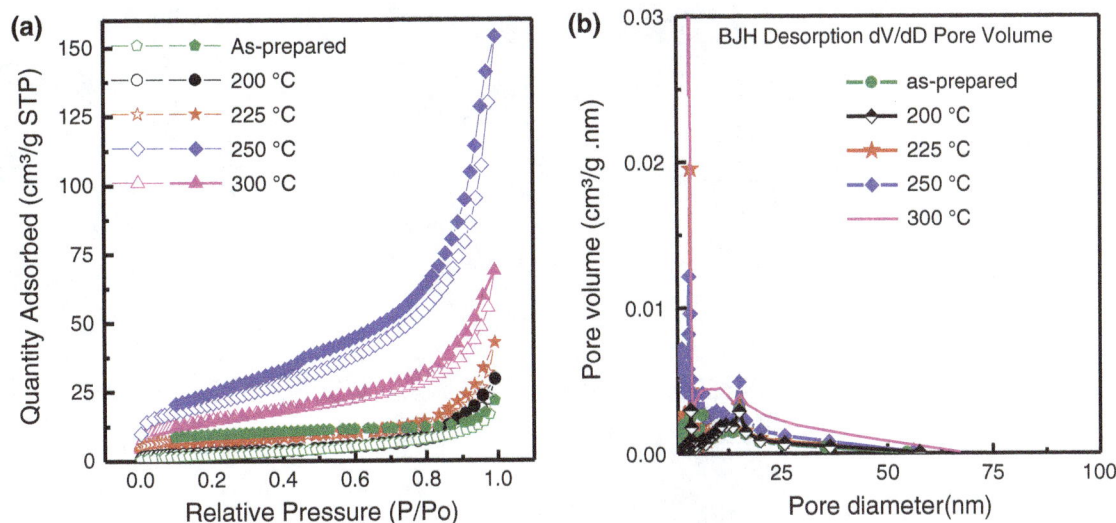

Fig. 2 **a** Nitrogen adsorption and desorption isotherms and **b** pore size distributions of as-prepared and calcined samples

Fig. 3 **a** Low and **b** high magnification SEM images of the as-prepared sample. **c** Low and **d** high magnification TEM images of the as-prepared sample

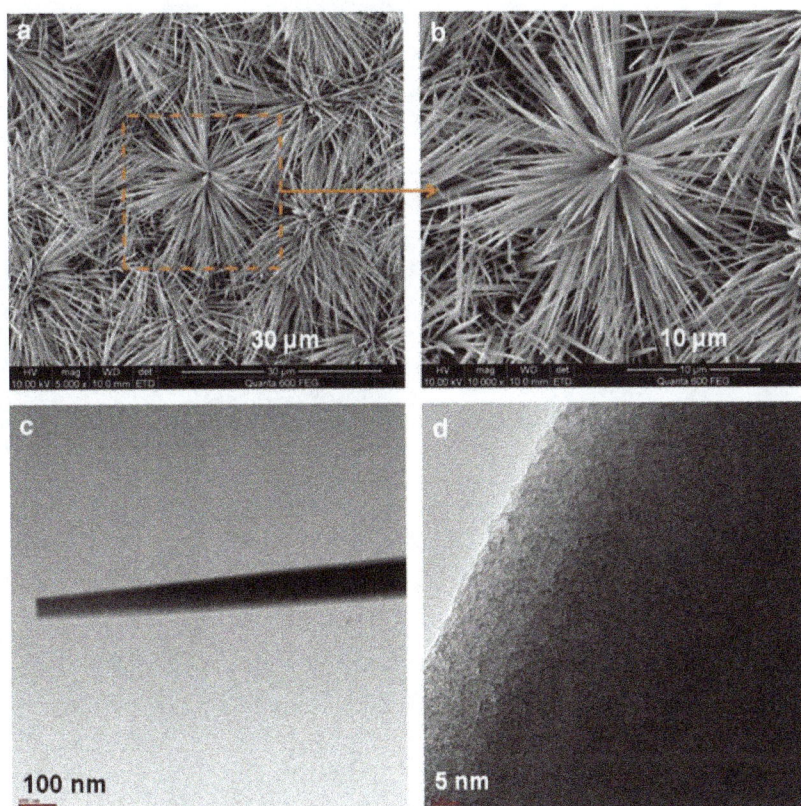

The effect of calcination temperature on the morphology of Co_3O_4 nanostructures is clearly evident from the low and high magnification SEM images of calcined samples as shown in Fig. 4. Figure 4a and b indicates that calcination at 200 °C resulted in the flattening of rigid sharp cylindrical nanoneedles. With further increase in calcination temperature (at 225 °C), the flowers grow bigger with the center changing into flexible nanowires as shown in Fig. 4c, d. Well-organized flower-like morphologies of flexible Co_3O_4 nanowires are achieved by calcination of as-prepared samples at 250 °C (Fig. 4e and f). High magnification images indicate that these flower-like structures are composed of cobalt oxide nanowires, with each having a length ranging from 1 to 10 μm. Further increase in calcination temperature (at 300 °C) results in the breaking down of the Co_3O_4 nanowire, destroying the flower-like morphology.

Fig. 4 SEM images of Co_3O_4 nanostructures showing the surface morphology. Low and high magnification SEM images of samples calcined at 200 °C (**a** and **b**), 225 °C (**c** and **d**), 250 °C (**e** and **f**) and 300 °C (**g** and **h**)

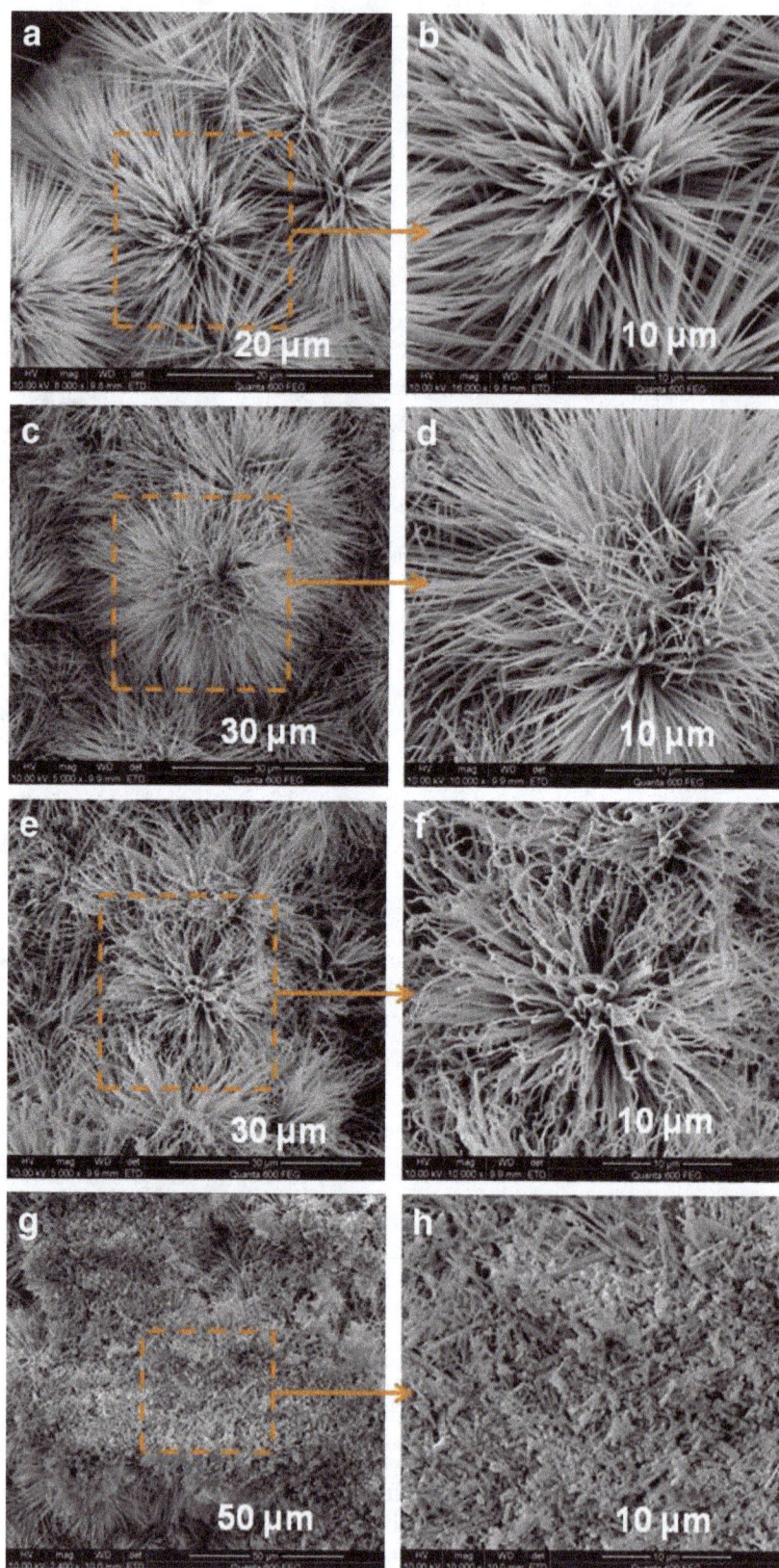

To investigate the microstructure of Co_3O_4 nanostructures in detail, TEM and HRTEM images are used (Fig. 5). Nanowire calcined at 200 °C (Fig. 5a and b) exhibits the presence of numerous nanocrystallites of Co_3O_4 having size <3 nm as compared to the smooth texture of the as-prepared sample. The size of these nanocrystals increases further with increase in the calcination temperature. The formation of such large crystal structures can probably be attributed to the re-crystallization and aggregation of the particles during calcination, in which the small nanocrystals grow into an interconnected porous structure. From Fig. 4c–h, it is clear that Co_3O_4 nanowires calcined above 225 °C consist of numerous interconnected nanoparticles and presents a rough appearance due to a large number of mesopores. This porous structure can be ascribed to the successive release and loss of H_2O and different gases during the oxidation/decomposition of the intermediate products $Co(OH)_2$ or cobalt carbonate nitrate hydroxide hydrate. The diameter of the nanowires varies from nearly 160 to 20 nm from the point of contact to the carbon cloth to the tip. For the samples calcined at 250 °C, the mesoporous structure is clearly formed.

The individual nanocrystals have a size in the range of 8–12 nm and the pore sizes are in the range of a few nanometers. Further increase in the calcination temperature to 300 °C leads to the formation of large crystallites having cuboid shape with average size in the range of 16–20 nm, which in turn leads to a reduction in the BET surface area and increase in pore volume.

As the Co_3O_4 nanowires composed of nanocrystallites possess high surface area as well as mesoporosity and are self-organized into flower-like morphologies on the surface of the current collectors, they can be used as efficient electrode materials for supercapacitors. However, the charge storage efficiency of the samples calcined at different temperatures can vary due to the observed differences in orientation and morphology of the 1-D nanostructures. To identify which morphology is favorable for high-rate capacitive energy storage, cyclic voltammetry (CV) and galvanostatic charge–discharge (CD) measurements are conducted in symmetric two-electrode configurations. It has been reported that as compared to three-electrode configuration, measurements using two-electrode configuration is more suited for evaluating the performance of a supercapacitor test cell as it mimics the physical configuration, internal voltages and charge transfer that occurs in a real supercapacitor application and thus provides the best indication of an electrode material's performance [47].

Figure 6a–d, respectively, shows CV loops obtained for symmetric button cell supercapacitors based on Co_3O_4 samples calcined at 200, 225, 250 and 300 °C at different scan rates of 1, 2, 5, 10, 20, 50 and 100 mV/s in a fixed potential range of 0–0.8 V. Test cells of samples calcined above 225 °C retain nearly rectangular CV loops up to a scan rate of 100 mV/s, which are characteristics for supercapacitors with excellent capacitance behavior and low contact resistance. Supercapacitor based on the sample calcined at 200 °C retains symmetric rectangular shape

only up to a scan rate of 20 mV/s. Above 20 mV/s, the CV loops exhibit quasi rectangular shape indicating lower rate performance of the sample as compared to the samples calcined above 225 °C.

Improvement in the performance of samples can be attributed to the formation of mesoporous structure above 225 °C. For samples annealed at 200, 225 and 300 °C, the CV loops at different scan rates show no peaks, which indicate that the electrodes are charged and discharged at a pseudo-constant rate over the complete voltammetric cycle. The CV loops of the supercapacitor based on the sample calcined at 250 °C, shows the presence of a pair of redox peaks indicating the existence of a strong Faradic reaction of Co_3O_4 according to Eq. (3).

$$Co_3O_4 + H_2O + OH^- \Leftrightarrow 3CoOOH + e^- \qquad (3)$$

A comparison of CV loops of symmetric supercapacitors based on Co_3O_4 samples calcined at different temperatures and carbon cloth substrate at a scan rate of 20 mV/s is shown in Fig. 7a. For the same mass loading, the CV curves show different areas indicating different levels of stored charge. From the CV loops, specific capacitances of 256, 341, 549 and 435 F/g, respectively [using Eqs. (1) and (3)], are obtained for the samples with calcination temperatures 200, 225, 250 and 300 °C. From Fig. 7a, it is also evident that the capacitive contribution from the carbon cloth substrate to the electrodes is negligibly small. Figure 7b shows the comparison of galvanostatic charge discharge curves for the samples at a constant current density of 1 A/g. The constant current charge discharge curves of 1-D Co_3O_4 nanostructures are nearly triangular, with reduced internal resistance at the beginning of the discharge curve. The reduction in internal resistance may be attributable to the direct contact of nanowires and the conducting carbon cloth substrate. The specific capacitance values for the supercapacitors based on samples with calcination temperatures 200, 225, 250 and 300 °C, respectively, are 334, 435, 605 and 509 F/g. Even though the samples calcined at 200 and 250 °C have flower-like morphology, they have less mesopores as compared to the samples calcined at 250 and 300 °C. Hence the electrode–electrolyte accessibility will be less, leading to lower capacitance. The tremendous improvement capacitance for sample calcined at 250 °C can be attributed to the mesoporous nanowire morphology of Co_3O_4 as well as their self-organization into flower-like microarchitecture on the surface of the carbon cloth, capable of holding a large quantity of electrolyte and thereby providing more electrode–electrolyte accessibility. The decrease in the specific capacitance of the sample calcined at 300 °C may be due to the damage in the flower-type morphology, which reduces the quantity of electrolyte trapped in between the

Fig. 5 TEM images of Co_3O_4 nanowires showing a porous strcture. Low and high magnification TEM images of samples calcined at 200 °C (**a** and **b**), 225 °C (**c** and **d**), 250 °C (**e** and **f**) and 300 °C (**g** and **h**)

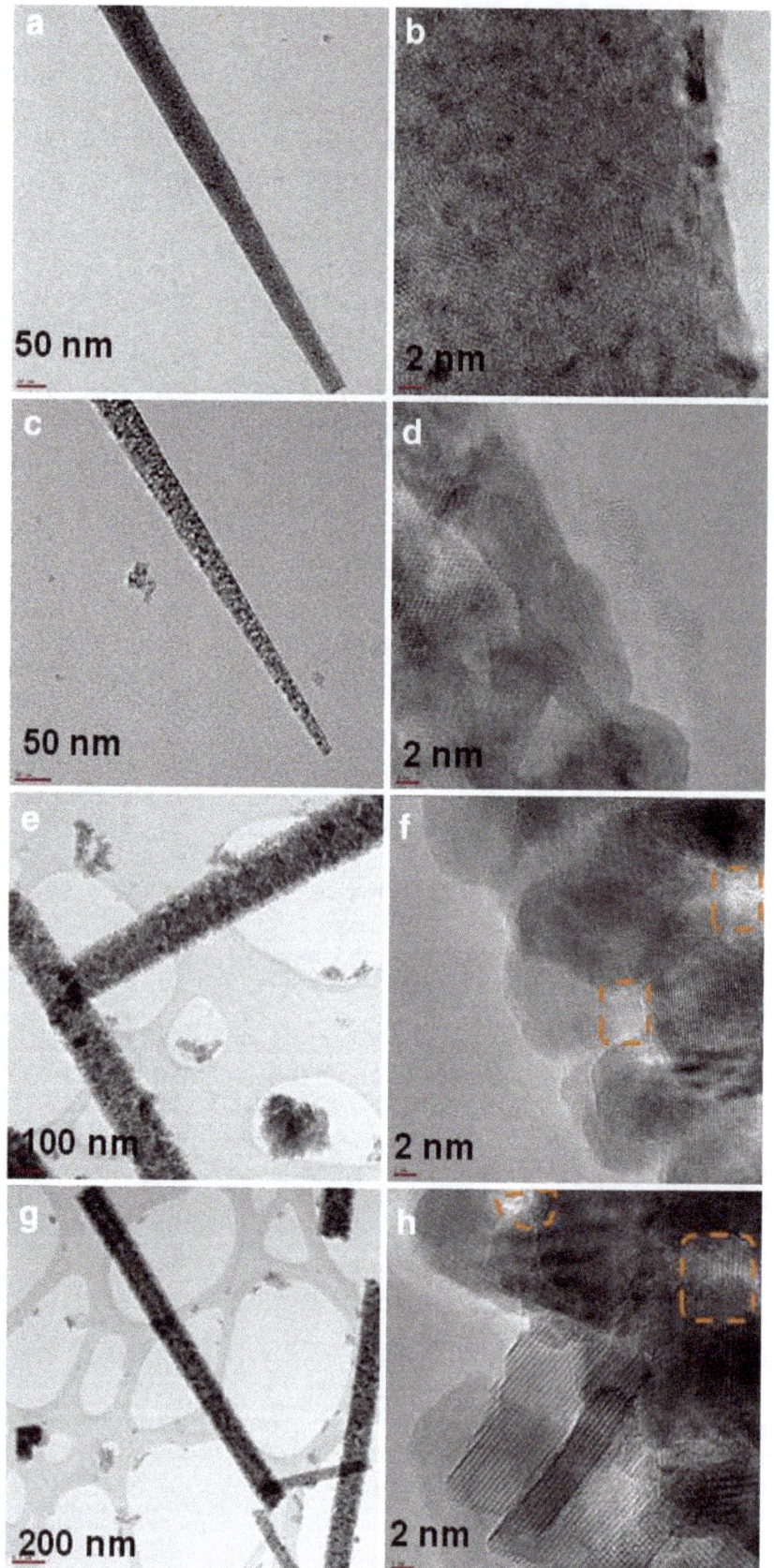

Fig. 6 Cyclic voltammograms of Co_3O_4 samples calcined at **a** 200 °C, **b** 225 °C, **c** 250 °C and **d** 300 °C at different scan rates

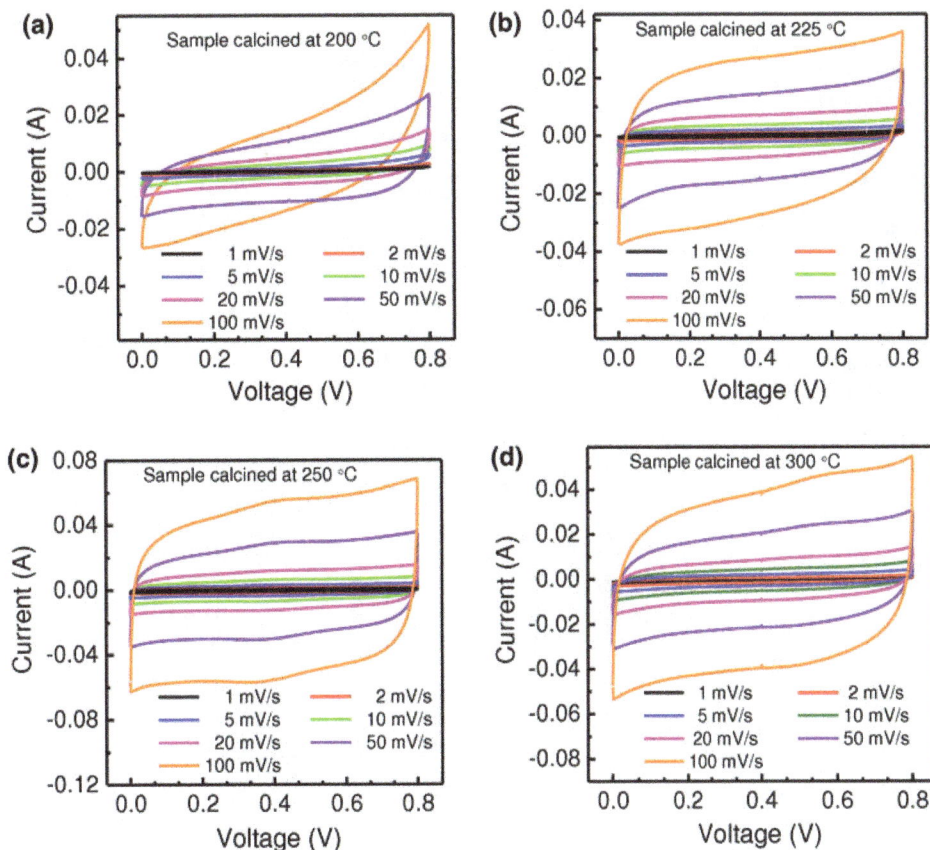

nanowires. But due to the well-formed mesoporous structure, the sample still exhibits higher specific capacitance than the samples calcined at 200 and 225 °C.

High-performance supercapacitor electrodes should have high-rate performance. Variations in specific capacitances of symmetric supercapacitors based on the calcined samples with increase in current density are shown in Fig. 7c. In general, at lower current densities (from 0.25 to 2 A/g), the specific capacitance decreases with the increase in discharge current density; it may be caused by the internal resistance of the electrode. At an extremely low current density of 0.25 A/g, the maximum specific capacitance values for the supercapacitors based on samples with calcination temperatures 200, 225, 250 and 300 °C, respectively, are 372, 524, 716 and 621 F/g. At lower current densities, ions can penetrate into the inner structure of electrode material having access to almost all available pores of the electrode, but at higher current densities an effective utilization of the material is limited only to the outer surface of electrodes. It results in the reduction of specific capacitance values. At current densities above 2 A/g, specific capacitance tends to stabilize. The sample calcined at 250 °C retains more than 83 % of its capacitance (605 F/g at a low current density of 1 A/g) retention even when the current density increases as high as 25 A/g, indicating the relatively good high-rate capability. The

sample calcined at 300 °C is also able to retain 79 % of its capacitance at a high current density of 25 A/g. But samples calcined at 200 and 225 °C are able to retain only 60 and 65 % of their capacitance, due to the incomplete formation of nanocrystallites and mesopores. Cycle life is another important factor for evaluating the performance of a supercapacitor. Cyclic stability curves for symmetric supercapacitors at a constant current density of 1 A/g are illustrated in Fig. 7d. For samples annealed at 250 and 300 °C, the specific capacitance increases gradually at the beginning of cycles and then there is a slight decrease. This phenomenon indicates that there is an initial activation process for Faradaic pseudocapacitance of mesoporous Co_3O_4. During the first 500 cycles, the specific capacitance of the sample annealed at 250 °C increases from 605 to 625 F/g and that of Co_3O_4 nanowires with flower-like morphology increases from 509 to 520 F/g. For the samples annealed at 250 and 300 °C, the specific capacitance slowly decreases with increase in cycle numbers. At the end of 5,000 cycles, supercapacitors based on samples with calcination temperatures 200, 225, 250 and 300 °C retain, respectively, 80, 85, 97 and 93 % of their maximum capacitance. Direct growth of 1-D Co_3O_4 nanostructures on carbon substrates and the self-organization of nanostructures into flower-like microarchitectures also have dramatic impact on supercapacitor performance, as each

Fig. 7 a Comparison of cyclic voltammograms of Co_3O_4 samples and carbon cloth substrate at a scan rate of 20 mV/s. **b** Comparison of charge–discharge curves of calcined Co_3O_4 samples at a constant current density of 1 A/g. **c** Specific capacitances of calcined Co_3O_4 samples at different current densities. **d** Cycling performance of calcined Co_3O_4 samples at a constant current density of 1 A/g (5,000 charge–discharge cycles)

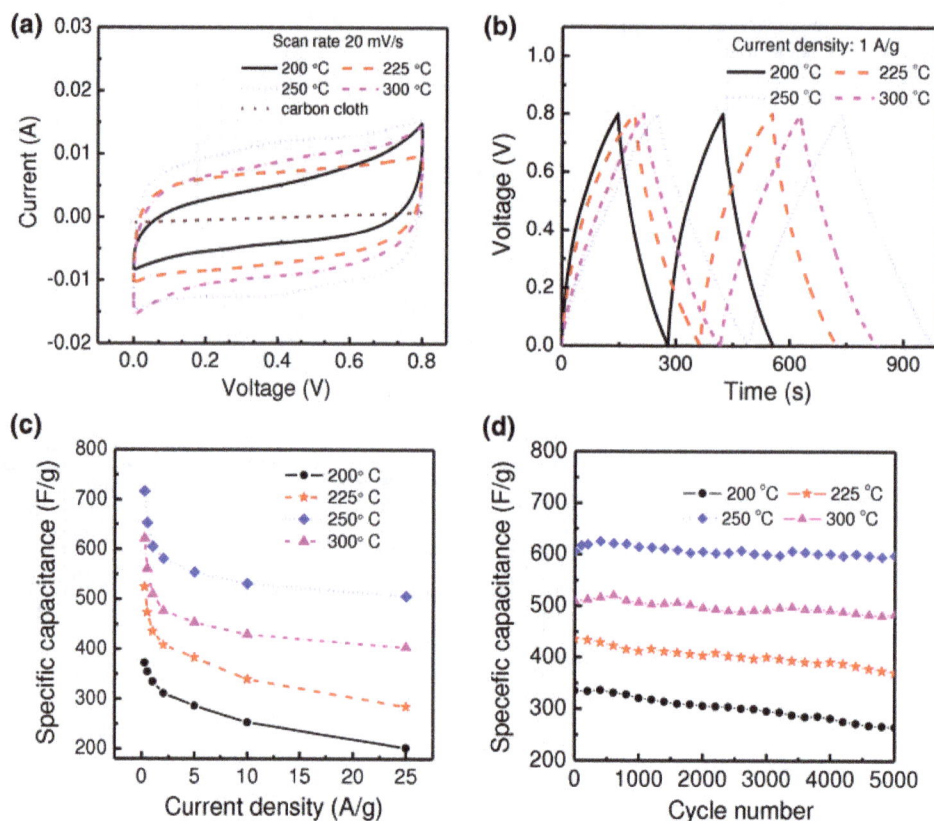

nanostructure will be electronically attached to the current collector which in turn will reduce the internal resistance of the electrode facilitating the rapid transport of the electrolyte ions and increasing the electrochemical utilization and pseudocapacitive performance of Co_3O_4.

Electrochemical impedance spectroscopy (EIS) can be applied to investigate electrical conductivity and ion transfer of the supercapacitor test cells. Figure 8a shows the experimental Nyquist impedance spectra for symmetric supercapacitor test cells of Co_3O_4 samples prepared at different calcination temperatures. The impedance spectra can be divided into two regions by the so-called knee frequency, with a semicircle arc in the high-frequency

region and a straight line in the low-frequency region. The real axis intercept at high frequency corresponds to the uncompensated resistance of the bulk electrolyte solution (R_s) and it is also known as equivalent series resistance (ESR). The diameter of semicircle in the high-frequency range gives the value of charge transfer resistance (R_{ct}). R_{ct} is a surface property of the porous electrode which is related to the electroactive surface area. It is a combination of electrolyte accessible area and electrical conductivity of the electrode material. The larger the electroactive surface area, the lower is the charge transfer resistance. In the present study all the samples exhibit very low value for R_{ct}, indicating high conductivity of the electrodes with

Fig. 8 a Nyquist plots for supercapacitors based on calcined Co_3O_4 samples. **b** Ragone plot (power density vs. energy density) of calcined Co_3O_4 samples based on symmetric supercapacitors. The energy densities and power densities were derived from the charge–discharge curves at different current densities

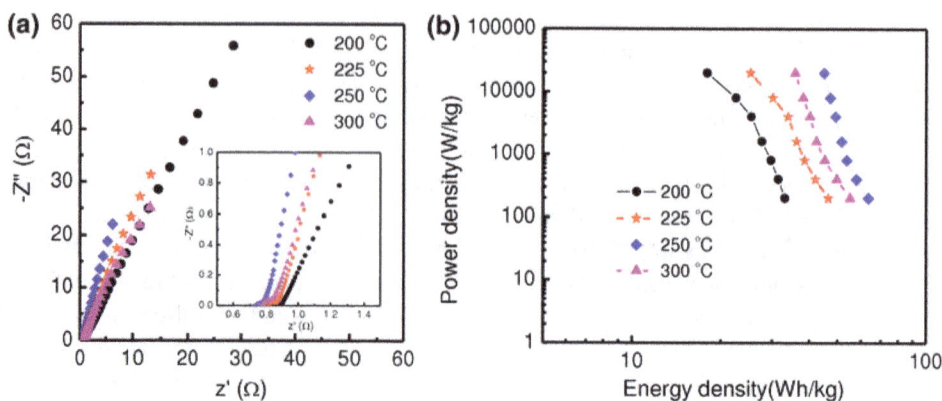

excellent electrolyte accessibility. The line at the low-frequency region making an angle 45° with the real axis is the Warburg line, which is a result of the frequency dependence of ion diffusion in the electrolyte to the electrode interface. Experimental results show that the magnitude of ESR obtained from the x-intercept of the impedance spectra for supercapacitors based on Co_3O_4 samples prepared at calcination temperatures 200, 225, 250 and 300 °C are 0.88, 0.81, 0.73 and 0.76 Ω, respectively. These lower values indicate consistent interfacial contact between the cobalt oxide nanowires and the carbon substrates. ESR data are an important factor in determining the maximum possible power density of a supercapacitor. The maximum power density (P_{max}) of the supercapacitor devices are calculated from the low-frequency data of the impedance spectra, according to the Eq. (4).

$$P_{max} = \frac{V_i^2}{4MR} \qquad (4)$$

where V_i is the initial voltage (here it is 1 V), R is ($R_s + R_{ct}$) and M is the total mass of two electrodes with a cell voltage of 0.8 V. Maximum power densities of 31.9, 34.6, 38.5 and 35.5 kW/kg are obtained, respectively, for supercapacitors based on Co_3O_4 samples prepared at calcination temperatures of 200, 225, 250 and 300 °C.

Figure 8b shows the Ragone plot (power density vs. energy density) of the symmetric supercapacitors based on Co_3O_4 calcined at different temperatures. The energy (E) and power densities (P) for the supercapacitors were calculated from charge–discharge curves at different current densities using Eq. (5) and (6), respectively.

$$E = \frac{1}{2} C_{sp} \Delta V^2 \qquad (5)$$

where 'ΔV' is the potential window of the discharge process.

$$P = E/\Delta t \qquad (6)$$

At a constant power density of 20 kW/kg, the energy densities obtained for supercapacitors based on samples with calcination temperatures 200, 225, 250 and 300 °C are 17, 25, 44 and 35 Wh/kg, respectively. At a low power density of 0.2 kW/kg, the energy densities reach as high as 33, 47, 63 and 55 Wh/kg, respectively, for the samples calcined at 200, 225, 250 and 300 °C. From the analysis of Ragone plot, it is evident that mesoporous Co_3O_4 nanowires prepared by the calcination of the as-prepared sample at 250 °C works as a very promising electrode material for high-performance supercapacitors. This sample maintains a high power density without much reduction in energy density.

The present study clearly demonstrates that calcination temperature plays a significant role in tuning the

morphology, porosity and capacitive performance of 1-D Co_3O_4 nanostructures. By changing calcination temperature from 200 °C (flattened nanoneedles) to 250 °C (mesoporous nanowires), we could nearly double the maximum capacitance of the Co_3O_4. The flower-like self-organization of mesoporous Co_3O_4 nanowires obtained by the calcination of as-prepared sample at 250 °C exhibits superior supercapacitor performance as compared to the other three samples. This can be attributed to the mesoporous nanowire morphology of Co_3O_4 retaining the flower-like microarchitecture, which accelerates the Faradaic charge transfer reactions, by providing easy diffusion of KOH electrolyte between different nanowires, accessing almost all the nanowires over the entire sample. In a recent study we reported a maximum pseudocapacitance of 620 F/g at a current density of 0.25 A/g for Co_3O_4 nanowires with flower-like morphology directly grown over carbon paper. Improvement in capacitive performance of a similar sample grown over carbon cloth substrate in the present study can be attributed to the improved wettability of the carbon cloth. It is interesting to note that even in two-electrode configuration with a high mass loading of 2 mg per electrode, the sample calcined at 250 °C yields higher pseudocapacitance and rate performance behavior than most of the previously reported Co3O4 electrodes.

Conclusions

In summary, we have demonstrated that the calcination temperature plays a crucial role in tuning the nanostructure and pseudocapacitive performance of mesoporous Co_3O_4 nanowires grown directly over carbon cloth substrate. The Co_3O_4 nanostructure sample calcined at 250 °C exhibited the best capacitive performance with high specific capacitance values (716 F/g at 0.25 A/g and 605 F/g at 1 A/g), high-rate performance (83 %) and excellent cycling stability (97 % after 5,000 charge discharge cycles) when used in symmetric two-electrode configuration. This improvement in the performance was attributed to the mesoporous nature of the Co_3O_4 nanowires retaining the flower-like morphology and providing maximum access of aqueous electrolyte to the electrodes.

Acknowledgments The authors thank the 'Advanced nanofabrication, Imaging and Characterization Laboratory' and 'Analytic Core Laboratory', KAUST. R. B. R. acknowledges the financial support from SABIC Postdoctoral Fellowship. W. C. acknowledges support from KAUST Graduate Fellowship. H. N. A. acknowledges the generous support from KAUST Baseline Fund.

References

1. Simon, P., Gogotsi, Y.: Materials for electrochemical capacitors. Nat Mater 7(11), 845–854 (2008).

2. Conway, B.E.: Electrochemical supercapacitors: scientific fundamentals and technological applications. Kluwer Academic/Plenum, New York (1999)

3. Jayalakshmi, M., Balasubramanian, K.: Simple capacitors to supercapacitors—an overview. Int J Electrochem Sci 3(11), 1196–1217 (2008)

4. Xu, J.J., Wang, K., Zu, S.Z., Han, B.H., Wei, Z.X.: Hierarchical nanocomposites of polyaniline nanowire arrays on graphene oxide sheets with synergistic effect for energy storage. ACS Nano 4(9), 5019–5026 (2010).

5. Benson, J., Boukhalfa, S., Magasinski, A., Kvit, A., Yushin, G.: Chemical vapor deposition of aluminum nanowires on metal substrates for electrical energy storage applications. ACS Nano 6(1), 118–125 (2012).

6. Jacob, G.M., Yang, Q.M., Zhitomirsky, I.: Composite electrodes for electrochemical supercapacitors. J Appl Electrochem 39(12), 2579–2585 (2009).

7. Naoi, K.: 'Nanohybrid capacitor': the next generation electrochemical capacitors. Fuel Cells 10(5), 825–833 (2010).

8. Pumera, M.: Graphene-based nanomaterials for energy storage. Energy Environ Sci 4(3), 668–674 (2011)

9. Chen, W., Rakhi, R.B., Hu, L.B., Xie, X., Cui, Y., Alshareef, H.N.: High-performance nanostructured supercapacitors on a sponge. Nano Lett 11(12), 5165–5172 (2011).

10. Rakhi, R.B., Chen, W., Cha, D., Alshareef, H.N.: Nanostructured ternary electrodes for energy-storage applications. Adv Energy Mater 2(3), 381–389 (2012).

11. Wang, G.P., Zhang, L., Zhang, J.J.: A review of electrode materials for electrochemical supercapacitors. Chem Soc Rev 41(2), 797–828 (2012).

12. Zhang, H., Cao, G.P., Yang, Y.S.: Carbon nanotube arrays and their composites for electrochemical capacitors and lithium-ion batteries. Energy Environ Sci 2(9), 932–943 (2009).

13. Pandolfo, A.G., Hollenkamp, A.F.: Carbon properties and their role in supercapacitors. J Power Sour 157(1), 11–27 (2006).

14. Liu, C.G., Liu, M., Wang, M.Z., Cheng, H.M.: Research and development of carbon materials for electrochemical capacitors—II—The carbon electrode. New Carbon Mater 17(2), 64–72 (2002)

15. Lota, G., Fic, K., Frackowiak, E.: Carbon nanotubes and their composites in electrochemical applications. Energy Environ Sci 4(5), 1592–1605 (2011)

16. Biener, J., Stadermann, M., Suss, M., Worsley, M.A., Biener, M.M., Rose, K.A., Baumann, T.F.: Advanced carbon aerogels for energy applications. Energy Environ Sci 4(3), 656–667 (2011)

17. Li, X., Rong, J.P., Wei, B.Q.: Electrochemical behavior of single-walled carbon nanotube supercapacitors under compressive stress. ACS Nano 4(10), 6039–6049 (2010).

18. Izadi-Najafabadi, A., Yamada, T., Futaba, D.N., Yudasaka, M., Takagi, H., Hatori, H., Iijima, S., Hata, K.: High-power supercapacitor electrodes from single-walled carbon nanohorn/nanotube composite. ACS Nano 5(2), 811–819 (2011).

19. Zheng, J.P., Cygan, P.J., Jow, T.R.: Hydrous ruthenium oxide as an electrode material for electrochemical supercapacitors. J Electrochem Soc 142(8), 2699–2703 (1995)

20. Xu, C.J., Kang, F.Y., Li, B.H., Du, H.D.: Recent progress on manganese dioxide based supercapacitors. J Mater Res 25(8), 1421–1432 (2010).

21. Guan, C., Liu, J.P., Cheng, C.W., Li, H.X., Li, X.L., Zhou, W.W., Zhang, H., Fan, H.J.: Hybrid structure of cobalt monoxide nanowire @ nickel hydroxidenitrate nanoflake aligned on nickel foam for high-rate supercapacitor. Energy Environ Sci 4(11), 4496–4499 (2011).

22. Mai, L.Q., Yang, F., Zhao, Y.L., Xu, X., Xu, L., Luo, Y.Z.: Hierarchical MnMoO(4)/CoMoO(4) heterostructured nanowires with enhanced supercapacitor performance. Nature Commun 2, 381 (2011).

23. Dong, S., Chen, X., Gu, L., Zhou, X., Li, L., Liu, Z., Han, P., Xu, H., Yao, J., Wang, H., Zhang, X., Shang, C., Cui, G., Chen, L.: One dimensional MnO2/titanium nitride nanotube coaxial arrays for high performance electrochemical capacitive energy storage. Energy Environ Sci 4(9), 3502–3508 (2011)

24. Yan, J.A., Khoo, E., Sumboja, A., Lee, P.S.: Facile coating of manganese oxide on tin oxide nanowires with high-performance capacitive behavior. ACS Nano 4(7), 4247–4255 (2010).

25. Rakhi, R.B., Chen, W., Cha, D.Y., Alshareef, H.N.: Substrate dependent self-organization of mesoporous cobalt oxide nanowires with remarkable pseudocapacitance. Nano Lett 12(5), 2559–2567 (2012).

26. Yang, L., Cheng, S., Ding, Y., Zhu, X.B., Wang, Z.L., Liu, M.L.: Hierarchical network architectures of carbon fiber paper supported cobalt oxide nanonet for high-capacity pseudocapacitors. Nano Lett 12(1), 321–325 (2012).

27. Qing, X.X., Liu, S.Q., Huang, K.L., Lv, K.Z., Yang, Y.P., Lu, Z.G., Fang, D., Liang, X.X.: Facile synthesis of Co(3)O(4) nanoflowers grown on Ni foam with superior electrochemical performance. Electrochim Acta 56(14), 4985–4991 (2011).

28. Xia, X.H., Tu, J.P., Zhang, Y.Q., Wang, X.L., Gu, C.D., Zhao, X.B., Fan, H.J.: High-quality metal oxide core/shell nanowire arrays on conductive substrates for electrochemical energy storage. ACS Nano 6(6), 5531–5538 (2012).

29. Sassin, M.B., Chervin, C.N., Rolison, D.R., Long, J.W.: Redox deposition of nanoscale metal oxides on carbon for next-generation electrochemical capacitors. Acc Chem Res 46(5), 1062–1074 (2013).

30. Yu, G.H., Xie, X., Pan, L.J., Bao, Z.N., Cui, Y.: Hybrid nanostructured materials for high-performance electrochemical capacitors. Nano Energy 2(2), 213–234 (2013).

31. Xia, X.H., Tu, J.P., Mai, Y.J., Wang, X.L., Gu, C.D., Zhao, X.B.: Self-supported hydrothermal synthesized hollow Co(3)O(4) nanowire arrays with high supercapacitor capacitance. J Mater Chem 21(25), 9319–9325 (2011).

32. Meher, S.K., Rao, G.R.: Ultralayered Co(3)O(4) for high-performance supercapacitor applications. J Phys Chem C 115(31), 15646–15654 (2011).

33. Zhang, F.Z.C.Y.X.L.L.Z.Q.C.X.: Facile growth of mesoporous Co3O4 nanowire arrays on Ni foam for high performance electrochemical capacitors. J Power Sources 203, 250–256 (2012)

34. Zhang, L.C.J.L.X.G.: Preparation and properties of Co3O4 nanorods as supercapacitor material. J Appl Electrochem 39, 1871–1876 (2009)

35. Wang, H.Z.L., Tan, X., Hol, C.M.B., Zahiri, B., Olsen, B.C., Mitlin, D.: Supercapacitive properties of hydrothermally synthesized Co3O4 nanostructures. J Phys Chem C 115, 17599–17605 (2011)

36. Kandalkar, S.G., Lee, H.M., Chae, H., Kim, C.K.: Structural, morphological, and electrical characteristics of the electrodeposited cobalt oxide electrode for supercapacitor applications.

Mater Res Bull **46**(1), 48–51 (2011).

37. Wu, J.B., Lin, Y., Xia, X.H., Xu, J.Y., Shi, Q.Y.: Pseudocapacitive properties of electrodeposited porous nanowall Co(3)O(4) film. Electrochim Acta **56**(20), 7163–7170 (2011).

38. Tummala, R., Guduru, R.K., Mohanty, P.S.: Nanostructured Co3O4 electrodes for supercapacitor applications from plasma spray technique. J Power Sources **209**, 44–51 (2012).

39. Asano, Y., Komatsu, T., Murashiro, K., Hoshino, K.: Capacitance studies of cobalt compound nanowires prepared via electrodeposition. J Power Sources **196**(11), 5215–5222 (2011).

40. Xu, J.A., Gao, L., Cao, J.Y., Wang, W.C., Chen, Z.D.: Preparation and electrochemical capacitance of cobalt oxide (Co(3)O(4)) nanotubes as supercapacitor material. Electrochim Acta **56**(2), 732–736 (2010).

41. Xia, X.H., Tu, J.P., Zhang, Y.Q., Mai, Y.J., Wang, X.L., Gu, C.D., Zhao, X.B.: Freestanding Co3O4 nanowire array for high performance supercapacitors. Rsc Advances **2**(5), 1835–1841 (2012).

42. Lu, X., Zheng, D., Zhai, T., Liu, Z., Huang, Y., Xie, S., Tong, Y.: Facile synthesis of large-area manganese oxide nanorod arrays as a high-performance electrochemical supercapacitor. Energy Environ Sci **4**(8), 2915–2921 (2011)

43. Gao, Y.Y., Chen, S.L., Cao, D.X., Wang, G.L., Yin, J.L.: Electrochemical capacitance of Co(3)O(4) nanowire arrays supported on nickel foam. J Power Sources **195**(6), 1757–1760 (2010).

44. Xing, W., Qiao, S.Z., Wu, X.Z., Gao, X.L., Zhou, J., Zhuo, S.P., Hartono, S.B., Hulicova-Jurcakova, D.: Exaggerated capacitance using electrochemically active nickel foam as current collector in electrochemical measurement. J Power Sources **196**(8), 4123–4127 (2011).

45. Jost, K., Perez, C.R., McDonough, J.K., Presser, V., Heon, M., Dion, G., Gogotsi, Y.: Carbon coated textiles for flexible energy storage. Energy Environ Sci **4**(12), 5060–5067 (2011)

46. Xiong, S., Yuan, C., Zhang, X., Xi, B., Qian, Y.: Controllable synthesis of mesoporous Co3O4 nanostructures with tunable morphology for application in supercapacitors. Chem Eur J **15**, 5320–5326 (2009)

47. Stoller, M.D., Ruoff, R.S.: Best practice methods for determining an electrode material's performance for ultracapacitors. Energy Environ Sci **3**(9), 1294–1301 (2010).

12

Nano-nickel catalytic dehydrogenation of ammonia borane

Dileep Kumar · H. A. Mangalvedekar ·
S. K. Mahajan

Abstract The complex chemical hydride, ammonia borane (NH_3BH_3, AB) is a hydrogen rich compound. It is a promising hydrogen source for applications using proton exchange membrane fuel cells (PEMFCs) due to hydrogen content. It has reasonably lower operating temperatures compared with other solid-state hydriding materials. At present, AB is an expensive disposable source which in its pure form releases 1 mol of hydrogen at around 110 °C. This temperature is much higher than the operating temperature of PEMFC (~ 80 °C). At the operating temperatures of the fuel cell, the slow kinetics of pure AB is a deterrent which provides enough scope for experimentation. The paper is the result of experimental thermolysis effort by using nano-nickel as a catalyst with pure AB. The neat and catalyzed AB isothermal decomposition and kinetic behavior are illustrated through the experimental results obtained under various conditions. The focus of experimentation is to increase the rate and extent of release of hydrogen at lower temperatures. The experimental results indicated that the use of nickel as catalyst reduced the induction period with significant improvement in hydrogen release compared with neat AB.

Keywords Ammonia borane · Nano-nickel · Dehydrogenation · Thermolysis · Kinetics · Decomposition

D. Kumar (✉) · H. A. Mangalvedekar
VJTI, Matunga (E), Mumbai 400019, MS, India
e-mail: dknayak@vpmthane.org

S. K. Mahajan
Directorate of Technical Education, 3, Mahapalika Marg,
Mumbai 400001, MS, India

Introduction

Ammonia borane (NH_3BH_3, AB) is a promising hydrogen source material due to its high hydrogen content (19.6 wt%). It is a white crystalline solid compound which was first synthesized and characterized by Shore and Parry [1] in the year 1955. The primary elements that make ammonia borane are nitrogen, hydrogen and boron. The elements nitrogen and hydrogen exist widely in the nature, e.g., in air and water [2]. Ammonia borane is a non-volatile material with appreciable degree of stability in air and water under ambient conditions. It can prove out to be an energy carrier for low power applications using proton exchange membrane fuel cell (PEMFC) at lower temperatures. Release of H_2 from amino boranes is a difficult and complex process. Efforts are made to study the phenomenon in the experimentation. The molecular description of NH_3BH_3 indicates that it is a donor–acceptor adduct formed as a result of the dative bond between a Lewis acid (BH_3) and a Lewis base (NH_3) [3]. The compound is a solid at room temperature primarily due to di-hydrogen bonding and dipole–dipole interactions. Ammonia borane and diammoniate of diborane (DADB) are chemically similar with varying stability characteristics. It may be inferred that AB is more readily applicable than DADB to hydrogen storage for automotive use [4].

Hydrogen release studies

Ammonia borane can release more than 2 mol of H_2 with heating to modest temperatures. The reactions of hydrogen evolution are summarized in Table 1.

The optimum thermal decomposition reaction of ammonium borohydride, $NH_4BH_4 \rightarrow BN + 4H_2$ occurs by a four-step process with H_2 yields of 24 wt%. This far exceeds the US DOE set ambitious and stiff target which is 9 wt% for the year 2015. However, this will not make it an immediate option as many issues about its practical usage are yet to be addressed. The strict hydrogen purity requirements for fuel cell applications demand minimization of side reactions [6]. The possibility of toxic gaseous boranes in the evolved H_2 is likely to affect the fuel cell performance. Above 500 °C, AB can be completely decomposed to form boron nitride (BN). The residue BN is not preferred for regeneration due to high chemical and thermal stability and hence ammonia borane is treated as a disposable source [7]. The main hurdle in hydrogen release is the slow kinetics at lower temperatures leading to long induction period [8, 9]. Neat AB thermally decomposes initially at 70 °C and reaches a maximum at 112 °C with the observed melting of AB to yield 1 mol of H_2 and the by-product is polyaminoborane $(NH_2BH_2)_n$. [2, 10, 11].

Use of catalyst is one of the methods explored by many researchers in the past to improve kinetics as well as hydrogen release at lower temperatures. Yao et al. [12] used lithium (Li) catalysis and mesoporous carbon (CMK-3) for thermal decomposition which released over 7 wt% hydrogen at 60 °C. Chen et al. [13] used Co- and Ni-based catalysts and observed a release of 1 mol of hydrogen at 59 °C. Burrell et al. [14] used Pt-catalyzed hydrogen release from AB with 4 wt% at 70 °C. Kalidindi et al. [15] used Cu and Ni nanoparticles and observed higher kinetics and hydrogen release. Sun et al. [16] performed monodisperse nickel particles catalysis in hydrolytic dehydrogenation of AB with the goal of preparation of non-noble metal catalyst. Gangal et al. [8] used silicon nanoparticles as catalyst and noted substantial reduction in activation energy and absence of induction period. Manners et al. [17] reported metals catalysis using Rh, Pd, Ru that could dehydrogenate ammonia borane at lower temperatures. Baker et al. [18] used Ni to develop unprecedented ability of hydrogen release from ammonia borane. Most of the above works included the use of organic or inorganic solvents and hydrolysis method to obtain improvements in the performance. After reviewing the performance and experimental processes adopted by earlier works, we decided to use nickel in its nano form as catalyst which holds enough promise to improve the dehydrogenation process. The goal of our work is to use the low cost and abundantly available nickel, a non-noble metal catalyst to optimize the hydrogen release from ammonia borane.

Experimental

Ammonia borane complex (97 % pure) purchased from Aldrich was used as received. Nano-nickel purchased from Laboratory Chemical Co. was used in the required quantity

for mechanical mixing. Neat AB sample fourier transform infra-red (FTIR) analysis was done before dehydrogenation using Bruker Germany model 3000 Hyperion Microscope with Vertex 80 FTIR imaging system at the Sophisticated Analytical Instrumentation Facility (SAIF) in IIT Bombay.

The neat AB sample and catalyst Ni sample are characterized for XRD spectrum using XPERT-PRO diffractometer system with Cu anode and K_α having wave length of 1.554060°A, in the 2Θ range of 5.0214°–99.9834°. The TEM imaging of the both the samples was done using Philips TEM model CM200.

The experiments of isothermal decomposition were conducted in an indigenously developed Sievert's type apparatus at 80, 100, 120, 140 and 160 °C, with different samples of AB and ABNi. The catalyzed sample was prepared by mechanical milling technique using nano-nickel in 10 % quantity i.e., 2 mg of Ni for 20 mg of AB before each experiment. The volume of the evolved gas with respect to time and pressure is recorded. The fine mix turned into gray colour and is used to perform isothermal decomposition test at five different temperatures starting from 80 °C. Each sample was loaded in a crucible of cylindrical shape within the reactor. By using different valves, sample holder is connected at separate instants to vacuum pump, hydrogen gas cylinder, and the apparatus. In each set of experiment and monitoring, the sample holder was evacuated and filled with hydrogen gas before finally connecting to the measuring setup of apparatus. A thermocouple is placed very close to the sample holder to measure the accurate temperature to which the sample is subjected. The temperature and pressure are recorded digitally at intervals of 1 min using a data logger. The change in pressure of the apparatus was used to calculate the amount of hydrogen released in wt%. The FTIR analysis was performed before and after dehydrogenation for each of the catalyzed sample to study the bond structures of reaction products.

Results and discussion

XRD and TEM of Ni samples

The XRD pattern of the Ni sample shown in Fig. 1 matches with the ICSD 64989, JCPDS reference code:00-004-0850. The presence of sharp peaks in the spectrum indicates the crystalline nature of the sample. The average crystal size calculated for the dominant peak (111) using Scherrer's is found to be 55 nm. Hence, it can be confirmed that the Ni is nano sized and the catalyst being used is nano-nickel.

The TEM images of Ni shown in Fig. 2 indicate nano particles of average size 40–65 nm. The diffraction rings indicate the crystalline nature of the material matching with the XRD pattern.

The crystallite size found using the Scherrer's formula is estimated to be 46 nm and the dominant peak (110) appearing at the 2Θ value of 24.1831°.

The XRD pattern of the AB sample shown in Fig. 3 matches well with JCPDS reference no: 01-074-0894. In AB, the presence of sharp peaks indicates crystalline nature of the sample. This is in well accordance with the details reported in literature where it is mentioned that AB shows orthorhombic structure at lower temperatures. At higher than ambient temperatures, AB shows tetragonal structure [19].

The TEM image of the considered AB sample in Fig. 4 confirms the crystalline structure in the range 34–70 nm.

Hence, it is found through XRD characterization and TEM imaging that both ammonia borane and nickel used are nano-sized crystalline structure materials.

Isothermal decomposition

Thermolysis is a method which requires heating with temperature control. The Sievert's type apparatus is a setup which permits this type of test on small quantities of material. Apparatus is designed to deal with challenges associated with solid-state reactions at elevated temperatures for the dehydrogenation in various stages from the material.

The isothermal decomposition of neat AB at lower temperatures is an extremely slow process with very long induction period. Strong chemical bond structure of neat AB is responsible for prolonged induction period and less hydrogen release at this temperature. The max release was achieved beyond 250 min which remained steady thereafter. Considerable reduction in the induction period is observed when isothermal decomposition was performed at 100 and 120 °C. Beyond induction period, 6–7 wt% gas is released. The literature provides the details of melting point of AB as 112–114 °C and at this temperature 1 mol equivalent of gas is liberated [20].

During 120 °C isothermal cycle, the hydrogen released is over 9 wt% which remains steady after 100 min with nominal induction period. The observations indicate that the increase in temperature decreases the induction period as well as expedites dehydrogenation process. The isothermal volumetric hydrogen release measurements presented the features that the neat AB sample released first mole of hydrogen at 110 °C and the second mole at 160 °C, respectively, in agreement with past results, respectively [19].

Catalytic dehydrogenation

The dehydrogenation results of AB and ABNi and subsequent characterization are indicated graphically in Fig. 5,

6, 7, 8, 9. This helps to compare the pristine AB and catalyzed AB thermolysis performances at each stage of experiment. The amount of catalyst used is around 10 % of the mass of neat AB. Mechanical mixing ensures proper dispersion of the catalyst with the base material to increase material surface area. It can be observed from Fig. 5 that the ABNi dehydrogenation started with reduced induction period of 8 min compared to over 125 min for neat AB at 80 °C. The reduction in induction period may be due to nickel particles assisting in breaking of di-hydrogen bonds.

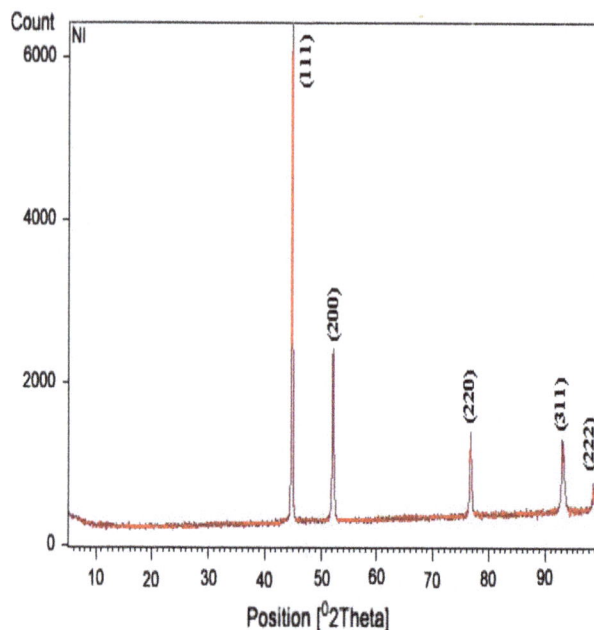

Fig. 1 XRD of Ni sample

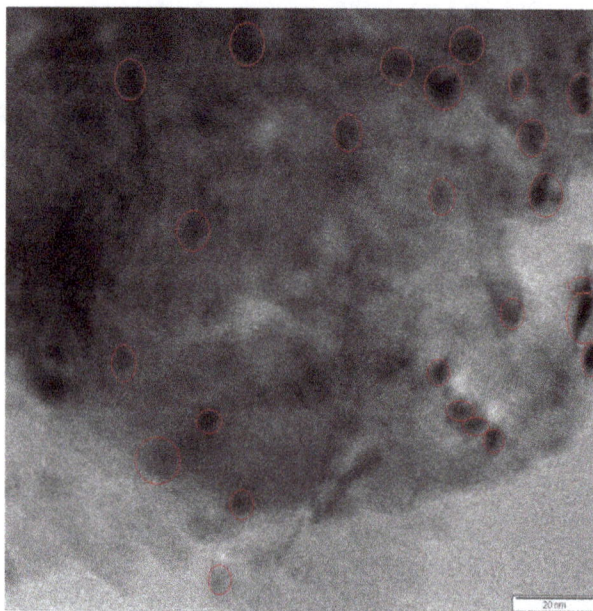

Fig. 2 TEM of Nickel sample

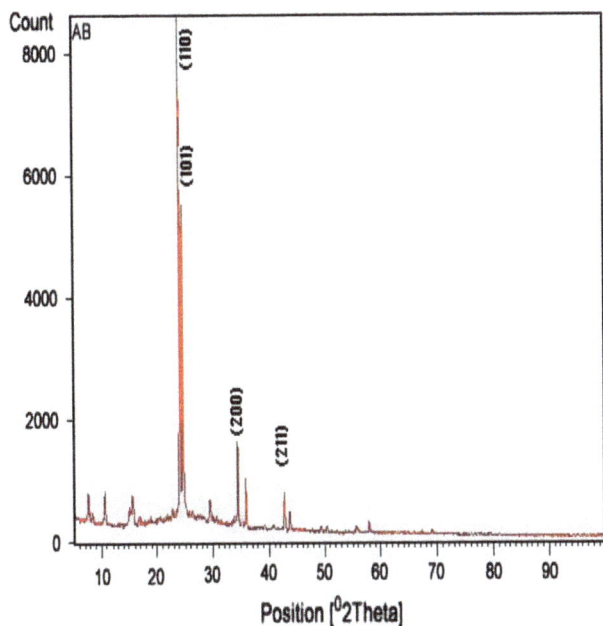

Fig. 3 XRD of ammonia borane sample

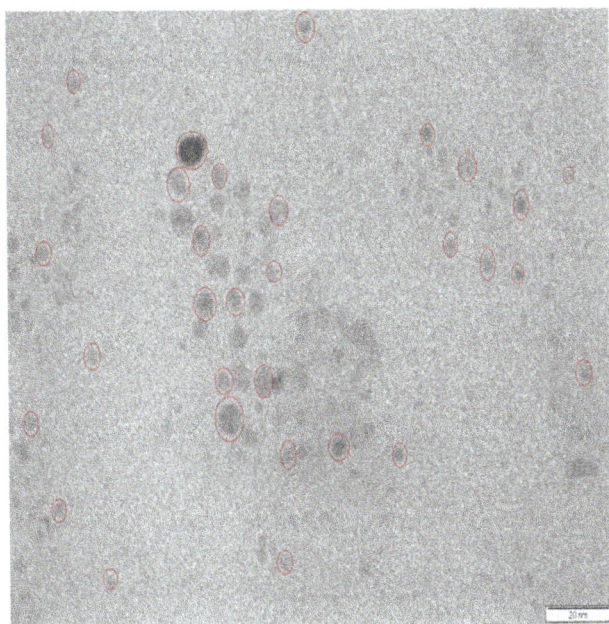

Fig. 5 The curve indicating the hydrogen release through isothermal decomposition of neat AB and catalyzed AB at 80 °C

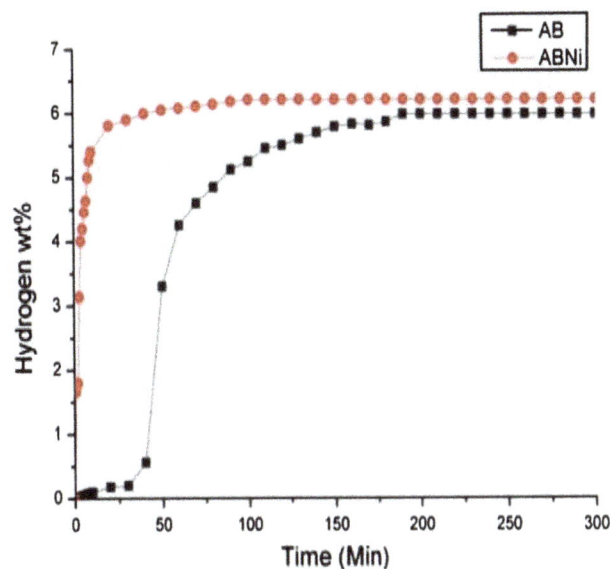

Fig. 4 TEM of ammonia borane sample

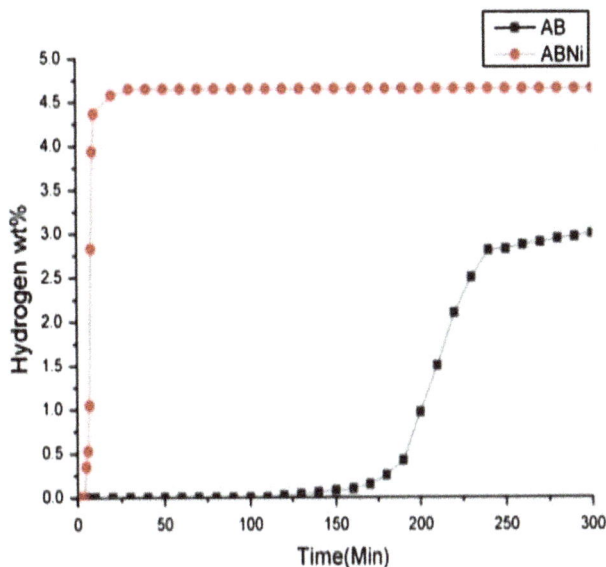

Fig. 6 The curve indicating the hydrogen release through isothermal decomposition of neat AB and catalyzed AB at 100 °C

The ABNi maximum hydrogen release is 4.5 wt% which is much higher compared to 3 wt% of neat AB at this temperature. It was also noted that the catalyzed samples did not foam during dehydrogenation.

The induction period and hydrogen release wt% values for various temperatures are summarized in Table 2.

It could be noted from Fig. 6 that the first mole of hydrogen is released at around 100 °C. Figure 7 indicates that the second mole at 120 °C in ABNi samples with very negligible induction period. Figures 8 and 9 show that the

ABNi decomposition at 140 and 160 °C also have negligible induction period with substantial increase in the hydrogen gas release. The neat AB hydrogen release at these temperatures was in the range of 12 wt%, whereas catalyzed decomposition resulted in release of 13.5 wt%.

Transmittance spectrum of neat AB and Ni catalyzed AB

FTIR transmittance spectrum provides information about the chemical bonds in the sample under study [8]. The

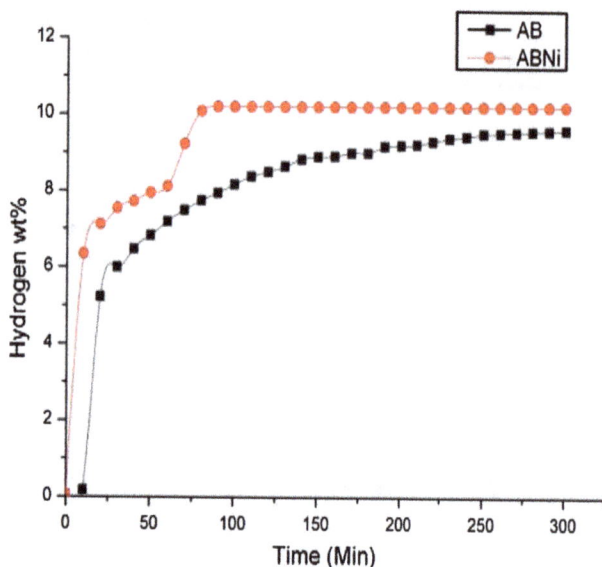

Fig. 7 The curve indicating the hydrogen release through isothermal decomposition of neat AB and catalyzed AB at 120 °C

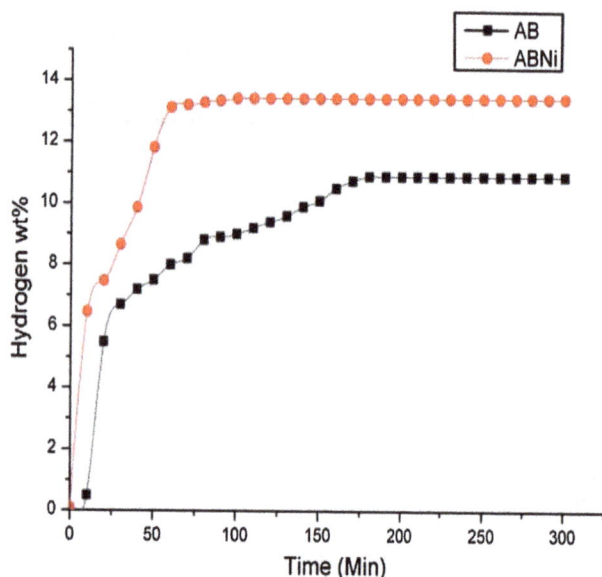

Fig. 9 The curve indicating the hydrogen release through isothermal decomposition of neat AB and catalyzed AB at 160 °C

Transmittance spectrum of dehydrogenated catalyzed ABNi

The FTIR analysis of catalyzed dehydrogenated sample solid residue after each experiment with 80, 100, 120, 140 and 160 °C was performed to understand the effect of catalyst on chemical bond structures. The transmittance spectrum in each case is indicated in Fig. 12. The spectrum in (a)–(e) is almost identical with the broadening of peaks in the N–H stretch or B–H stretch which indicates the material decomposition and hydrogen release. The release of hydrogen in terms of wt% has showed improvement in comparison with neat AB with considerable reduction in the induction period. The second mole release in the experiment at 160 °C is visible in the transmittance curve with the broadening of N–H and B–H stretch.

The characteristic frequencies corresponding to N–H, N–H_2 and N–H_3 near 3,200 cm^{-1} go on broadening as the decomposition temperature increases and finally they almost disappear. The same phenomenon is observed for the frequency bands corresponding to B–H–B–H_2 stretch observed near 2,200 cm^{-1}. Shifting and broadening of the peaks corresponding to B–H_2 torsion and bend in the range 1,000–1,500 cm^{-1} are also seen clearly in the Fig. 12.

The peak near 700 cm^{-1} corresponding B–N also appears to be slightly affected. It can be inferred that the hydrogen attached to borane and nitrogen is getting dissociated and the borane–nitrogen bond in its place, ruling out the possibility of formation of ammonia.

Broadening of peaks indicates the disruption of bonds resulting in the release of hydrogen. It is seen that the B–N

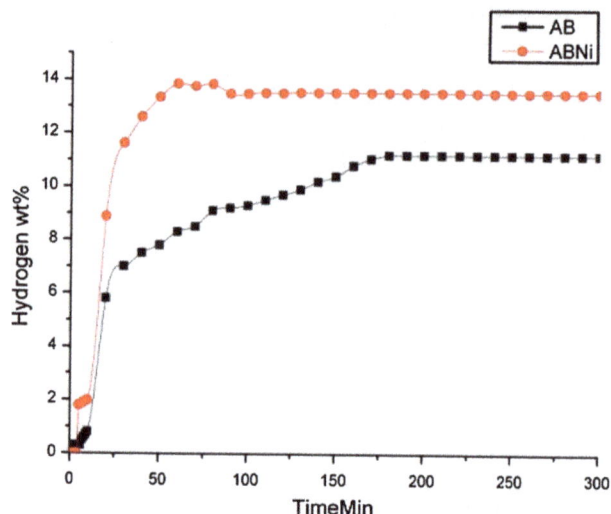

Fig. 8 The curve indicating the hydrogen release through isothermal decomposition of neat AB and catalyzed AB at 140 °C

FTIR analysis of neat AB and Ni catalyzed AB was conducted before and after decomposition.

The FTIR of neat AB before and after decomposition is shown in Fig. 10. The FTIR of ABNi before and after decomposition is shown in Fig. 11. It is observed that various bonds are resembling in both the samples indicating the B–N bend at 700 cm^{-1}, B–H_2 torsion at 1,500 cm^{-1}, B–H_2 bend at 1,300 cm^{-1}, B–H, B–H_2 stretch at 2,300 cm^{-1} and N–H, N–H_2, N–H_3 stretch above 3,000 cm^{-1}. This justifies the use of catalyst in the catalyzed sample is just facilitating the dehydrogenation process.

Table 1 Decomposition reactions for a thermal chemical hydride [5]

Decomposition reaction	Storage density[a], wt% H_2	Temperature[b], °C	Product
$NH_4BH_4 \rightarrow NH_3BH_3 + H_2$	6.1	<25	Ammonia borane
$NH_3BH_3 \rightarrow NH_2BH_2 + H_2$	6.5	<120	Polyaminoborane
$NH_2BH_2 \rightarrow NHBH + H_2$	6.9	>120	Polyaminoborane
$NHBH \rightarrow BN + H_2$	7.3	>500	Borane nitride/borazine

[a] Theoretical maximum

[b] Decomposition temperature

Table 2 AB and ABNi hydrogen release statistics

Temperature (°C)	Induction period in Min		Maximum release duration in Min		Max wt%	
	AB	ABNi	AB	ABNi	AB	ABNi
80	125	8	275	60	3	4.75
100	15	0	200	50	5.75	6.00
120	10	0	300	75	9.25	10.5
140	5	0	175	60	11	13.0
160	0	0	175	50	12	13.5

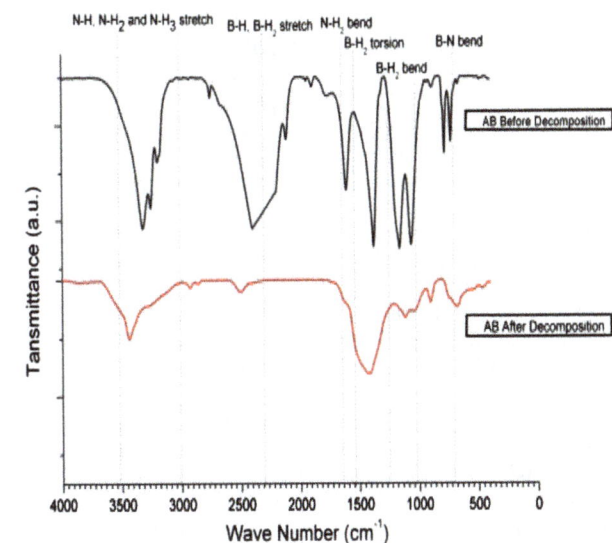

Fig. 11 FTIR curves for ABNi before and after decomposition at 80 °C

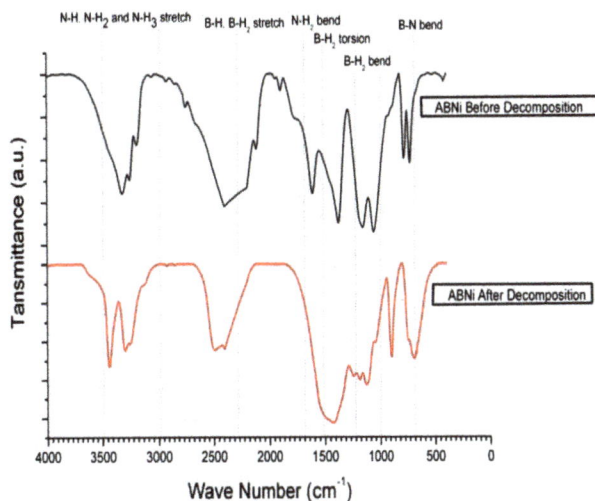

Fig. 10 FTIR curves for neat AB before and after decomposition at 160 °C

bend is intact and has not got affected either due to catalysis or due to elevated temperatures. The non-formation of borazine is an indication that there are no unwanted reactions and by-products.

The FTIR study has indicated that there is gradual release of the hydrogen gas as the bonds are getting affected and the reactions are taking place at lower temperatures. With the rise in temperature, significant improvement in hydrogen release is observed at every stage with fast reaction kinetics and faster release of hydrogen.

Conclusion

XRD and TEM of Ni and AB confirmed the nano-sized crystalline structure of samples used.

Isothermal decomposition of ABNi clearly indicated significant improvement in hydrogen release with reduced induction period compared to AB.

The FTIR of neat and catalyzed AB is indicating no much difference in the transmittance characteristics. This helps in concluding that the presence of nickel as a catalyst is just facilitating the dehydrogenation process.

The FTIR of dehydrogenated sample indicates the release of hydrogen from various B–H and N–H bonds. The mechanical mixing of Ni with neat AB has resulted in considerable reduction in the induction time and improvement in hydrogen release rate.

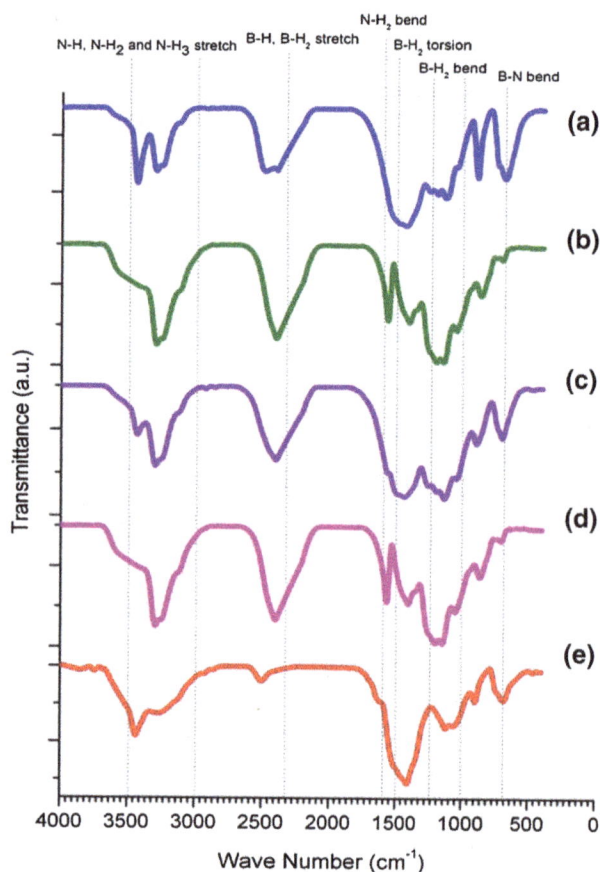

Fig. 12 FTIR curves for dehydrogenated sample *a* 80 °C *b* 100 °C *c* 120 °C *d* 140 °C *e* 160 °C

The presence of BN bands in the FTIR of dehydrogenated ABNi sample near 700 cm^{-1} suggests that this band is not ruptured and the possibility of ammonia formation can be ruled out.

Acknowledgments The authors acknowledge the assistance of COE in CNDS Lab, VJTI Mumbai, SAIF Lab, IIT Bombay and Advanced Materials Lab at V.P.M.'s Polytechnic, Thane for the facilities offered for the experimental investigation work.

References

1. Shore, S.G., Parry, R.W.: The crystalline compound ammonia borane. J. Am. Chem. Soc. **77**, 6084 (1955)
2. Peng, B., Chen, J.: Ammonia borane as an efficient and lightweight hydrogen storage medium. Energy Environ. Sci. **1**, 479–483 (2008)
3. Weaver, J.R., Shore, S.G., Parry, R.W.: The dipole moment of ammonia–borane. J. Chem. Phys. **29**, 1 (1958)
4. Karkamkar, A., Aardahl, C., Autrey, T.: Recent developments on hydrogen release from ammonia borane. Mater. Matters **2**, 6–9 (2007)
5. Riis, T., Sandrock, G.: Hydrogen storage—gaps and priorities. HIA HCG Storage paper, 11 (2005)
6. Autrey, T., Gutowska, A., Li, L., Linehan, J., Gutowski, M.: Chemical hydrogen storage in nano-structured materials, control of hydrogen release and reactivity from ammonia borane complexes. J. Am. Chem. Soc. Div. Fuel Chem. **49**(1), 150 (2004)
7. Diwan, M., Hwang, H.T., Al-Kukhun, A., Varma, A.: Hydrogen generation from noncatalytic hydrothermolysis of ammonia borane for vehicle applications. AIChE J. **57**, 259–264 (2011)
8. Gangal, A.C., Kale, P., Edla, R., Manna, J., Sharma, P.: Study of kinetics and thermal decomposition of ammonia borane in presence of silicon nanoparticles. Int. J. Hydrog. Energy (2012).
9. Dileep, K., Mahajan, S.K., Mangalvedekar, H.A.: Hydrogen storage in amine borane complexes, pp. 1–5. ICFCHT, Kuala Lumpur (2011)
10. Stowe, A.C., Shaw, W.J., Linehan, J.C., Schmid, B., Autrey, T.: In situ solid state 11B MAS-NMR studies of the thermal decomposition of ammonia borane: mechanistic studies of the hydrogen release pathways from a solid state hydrogen storage material. Phys. Chem. Chem. Phys. **9**, 1831 (2007)
11. Sit, V., Geanangel, R.A., Wendlandt, W.W.: The thermal dissociation of NH3BH3. Thermochim. Acta **113**, 379 (1987)
12. Yao, X., Li, L., Sun, C., Du, A., Cheng, L., Zhu, Z., Yu, C., Zou, J., Smith, S.C., Wang, P., Cheng, H., Frost, R.L., Lu, G.Q.: Lithium-catalyzed dehydrogenation of ammonia borane within mesoporous carbon framework for chemical hydrogen storage. Adv. Funct. Mater. **19**, 265–271 (2009)
13. Chen, P., He, T., Xiong, Z., Wu, G., Chu, H., Wu, C., Zhang, T.: Nanosized Co- and Ni-catalyzed ammonia borane for hydrogen storage. Chem. Mater. **21**, 2315–2318 (2009)
14. Burrell, A.K., Shrestha, R.P., Diyabalanage, H.V.K., Semelsberger, T.A., Ott, K.C.: Catalytic dehydrogenation of ammonia borane in non-aqueous medium. Int. J. Hydrog. Energy **34**, 2616–2621 (2009)
15. Kalidindi, S.B., Jagirdar, B.R.: Hydrogen generation from ammonia borane using nanocatalysts. J Indian Inst. Sci. **90**, 181–187 (2010)
16. Sun, S., Metin, O., Mazumder, V., Ozkar, S.: Monodisperse nickel nanoparticles and their catalysis in hydrolytic dehydrogenation of ammonia borane. J. Am. Chem. Soc. **132**, 1468–1469 (2010)
17. Manners, I., Jaska, C.A., Temple, K., Lough, A.J.: Transition metal-catalyzed formation of boron–nitrogen bonds: catalytic dehydrocoupling of amine–borane adducts to form aminoboranes and borazines. J. Am. Chem. Soc. **125**, 9424 (2003)
18. Baker, R.T., Keaton, R.J., Blacquiere, J.M.: Base metal catalyzed dehydrogenation of ammonia–borane for chemical hydrogen storage. J. Am. Chem. Soc. **129**, 1844 (2007)
19. Paolone, A., Palumbo, O., Rispoli, P., Cantelli, R., Autrey, T.: Hydrogen dynamics and characterization of the tetragonal-to-orthorhombic phase transformation in ammonia borane. J. Phy. Chem. C **113**, 5872–5878 (2009)
20. Hu, M.G., Geanangel, R.A., Wendlandt, W.: The thermal decomposition of ammonia borane. Thermochim. Acta **23**(2), 249–255 (1978)

13

Cogeneration of renewable energy from biomass (utilization of municipal solid waste as electricity production: gasification method)

Misgina Tilahun · Omprakash Sahu ·
Manohar Kotha · Hemlata Sahu

Abstract Recycling and utilization of waste is one of the key parameters of environmental issue. For this issue supercritical waste gasification has impactive impression, which has capability to convert the waste into marketable by-product. Adding catalysts or oxidants to supercritical waste gasifier can further reduce operating costs by creating self-sustaining reactions under mild conditions with even shorter residence times. The hydrogen produced by this process will be utilized for generating electricity using fuel cell technology. Besides, alkaline fuel cells appear to be an important technology in the future as they can operate at a high efficiency. Therefore, the combination of biomass gasification through supercritical water with alkaline fuel cells represents one of the most potential applications for highly efficient utilization of biomass. The main aim of the study is to recover energy from waste using alkaline fuel cell. With the different operation conditions 88.8 % of hydrogen and 45 % of carbon dioxide, maximum power density 9.24 W/cm^2 was obtained.

Keywords Biomass · Cell · Energy · Hydrogen · Reactor

Introduction

Biomass is one of the renewable and potentially sustainable energy sources and has many possible applications varying from heat generation to the production of advanced secondary energy carriers. It has almost zero or very low net CO_2 emission since carbon and energy are fixed during the biomass growth [1]. There are different types of technologies for converting biomass to electricity or to a secondary fuel such as thermal conversion, chemical conversion and bio-chemical conversion [2]. However, thermo-chemical methods such as gasification have a great potential in producing a syngas mainly composed of H_2 and CO with traces of different gases such as CH_4 in different proportions [3]. The produced fuel gas can be flexibly utilized in boilers, engines, gas turbines or fuel cells [4]. Smaller scale gasification systems with internal combustion engines can now be used for thousands of hours to give reasonably high electrical efficiencies and limited emissions [5]. However, fuel cells have the potential to operate at higher electrical efficiency and with lower emissions compared with traditional power generation techniques. Fuel cells are emerging as a leading alternative technology to the more polluting internal combustion engines in vehicle and stationary distributed energy applications. In addition, the future demand for portable electric power supplies is likely to exceed the capability of current battery technology. Hydrogen-powered fuel cells emit only water and have virtually no pollutant emissions, even nitrogen oxides, because they operate at temperatures that are much lower than internal combustion engines [6]. However, even fuel cells fuelled by hydro-carbon fuels have the potential to provide efficient, clean and quiet energy conversion, which can contribute to a significant reduction in greenhouse gases and local pollution. When heat generated in fuel cells is also utilized in combined heat and power (CHP) systems, an overall efficiency of 85 % in excess can be achieved [7]. Different types of fuel cells suitable for several energy applications at varying scales

M. Tilahun · O. Sahu (✉)
Department of Chemical Engineering, Wollo University, South Wollo, Ethiopia
e-mail: ops0121@gmail.com

M. Kotha · H. Sahu
Department of Electrical and Electronic Engineering, Wollo University, South Wollo, Ethiopia

have been developed, but all share the basic design of two electrodes (anode and cathode) separated by a solid or liquid electrolyte or a membrane. Hydrogen (or a hydrogen-containing fuel) and air are fed into the anode and cathode of the fuel cell, and the electrochemical reactions assisted by catalysts take place at the electrodes [8]. The electrolyte enables transport of ions between the electrodes while the excess electrons flow through an external circuit to provide electrical power. Fuel cells are classified according to the nature of their electrolyte, which also determines their operating temperature, the type of fuel and a range of applications [9]. The electrolyte can be acid, base, salt or a solid ceramic or polymeric membrane that conducts ions.

Gasification is a process that converts organic or fossil based carbonaceous materials into carbon monoxide, hydrogen and carbon dioxide. Hydrogen and fuel cells are often considered as a key technology for future sustainable energy supply. Renewable shares of 36 % (2025) and 69 % (2050) on the total energy demand will lead to hydrogen shares of 11 % in 2025 and 34 % in 2050 [10]. Today, hydrogen is mainly produced from natural gas via steam methane reforming, and although this process can sustain an initial foray into the hydrogen economy, it represents only a modest reduction in vehicle emissions as compared to emissions from current hybrid vehicles [11]. Biomass has been recognized as a major world renewable energy source to supplement declining fossil fuel resources [12, 13]. It will play an important role in the future global energy infrastructure for the generation of power and heat, but also for the production of chemicals and fuels. The dominant biomass conversion technology will be gasification, as the gases from biomass gasification are intermediates in the high-efficient power production or the synthesis from chemicals and fuels. Biomass gasification offers the earliest and most economical route for the production of renewable hydrogen. International Institute for Applied Systems Analysis (IIASA's) Environmentally Compatible Energy Strategies (ECS) project has developed a long-term hydrogen-based scenario (B1-H2) of the global energy system to examine the future perspectives of fuel cells [14].

The scenario illustrates the key role of hydrogen in a long-term transition towards a clean and sustainable energy future. According to this scenario, biomass gasification will become a dominant technology in the future. The main aim of the work is to utilize the sewage sludge for the production of hydrogen and uses it for electricity generation. This work concentrated on the percentage of hydrogen production with variation of temperature, pressure and residence time by gasification method. Effect of partial pressure of produced hydrogen, working temperature and electrolyte concentration on cell performance has also been studied.

Materials and methods

Material

The material was arranged from solid waste dumping area of Kombolcha town (Ethiopia).

Reactor set up

The reactor vessel was dual-shell type with an insert made of titanium, widely utilized as corrosion-resistant metal, and a pressure shell. The reactor used in this study was also equipped with auxiliaries such as a stirrer, thermocouples, nozzles, and a pressure gauge. The reaction was initiated by immersing the reactor into molten salt bath (mass ratio of salt was adjusted to K_2NO_3:$NaNO_2$:$NaNO_3$ = 6:5:1). After lapse of predefined time, the reactor was taken out of the bath and subsequently quenched to stop the reaction. HPLC high-pressure pump was used for feeding the distilled water to the reactor to adjust the reaction pressure precisely. The reaction temperature was measured by K-type thermocouple and pressure with digital pressure gauge. The reactor was loaded with deionized water and initial sewage sludge (2 wt % of deionized water) for every experiment (250 rpm and particles size 180 μm). The amount of catalyst was 20 wt % of the organic waste. Then, the air in the reactor was replaced with argon gas. The reactor was sealed and put into the sand bath heated at reaction temperature. It took about 3 min for the reactor to reach the setting reaction temperature around 700 °C. It took about 2 min for final setting of the reaction pressure and reaction time will be considered as zero. As the reaction pressure increased by about 1 MPa than the initial reaction pressure for all experiments, the reaction pressure was assumed to be the initial reaction pressure of the experiment.

Gas analysis

Produced gas was sampled from one of the sampling loop ports using a gas-tight syringe for gas analysis injection. Liquid and solid residues were collected as mixtures subsequently separated by centrifugation run at 2,500 rpm for 5 min. Moreover, the liquid phase was filtered by 0.45 μm pore size syringe filter (Millex LH, Millipore) and diluted by deionized water prior to the analysis. Separated solid (small amount) residues were dried in an oven kept at 105 °C for at least 6 h and weighed. Gas analysis was carried out with gas chromatograph (GC) GC-2014; SHMADZU equipped with Shin-carbon ST 50/80 column and thermal conductivity detector (TCD) to separate H_2, CH_4, CO, and CO_2. As for the ICP analysis of initial sewage sludge, acid decomposition by nitric acid and sulfuric acid was conducted under 210 °C using an electric hot plate [15]. Guaranteed grade of potassium hydroxide provided by Wako Pure Chemicals Industries, Ltd. was used as a catalyst for gasification.

Preparations of electrode

The anode electrode was prepared by first dispersing the required quantity of catalyst powder in a Nafion® dispersion (SE-5112) for 30 min. An ultrasonic water bath was used to prepare catalyst slurry. The Nafion® dispersions have both hydrophilic and hydrophobic features. Polytetrafluoroethylene (PTFE) is hydrophobic and when employed as a binder, it may prevent hydrophilic fuel from reaching the catalyst site. Therefore, Nafion® has been used to bind the catalyst particles on to the carbon paper. The catalyst slurry was spread on carbon paper in the form of a continuous wet film using a paint-brush technique. It was then dried in an oven for 30 min at 80 °C. Nickel meshes were used as a current-collector because of its non-corrosive nature in an alkaline medium. The catalyzed carbon paper was pressed on to the nickel mesh with application of the Teflon® dispersion. The prepared electrode was pressed at 50 kg cm^{-2} and 120 °C for 5 min to form a composite structure. The area of the working electrode was 25 cm^2. Finally, the composite was heated at 573 K for 4 h to obtain the final form of the anode electrode [16]. Similarly, we have used magnesium oxide for cathode electrode. The Teflon-coated side of the electrode was exposed to the air-side in alkaline fuel cell, and thereby prevented leakage of electrolyte to the air-side and allowed oxygen to permeate.

Experimental setup

The experiments were carried out in a 7 cm × 7 cm stainless steel plate in which a special new designed electrolyte carrier plate is fitted with bolts. The cathode (5 cm × 5 cm) and anode (5 cm × 5 cm) are placed in front and back side of electrolyte carrier. A wire connected with the anode and the cathode is used as terminals for measuring current and voltage of the alkaline fuel cell. The space between the anode and the cathode was filled with electrolyte (KOH) with the help of peristaltic pump. The electrolyte was fed 1 ml min^{-1}, such that one side of the cathode was in contact with the electrolyte and the other side was exposed to air. Oxygen present in the air acts as an oxidant. The hydrogen gas which was generated from the biomass by gasification method is used as anode feed and oxygen is taken from the air. The electrolyte in the beaker was continuously stirred by a magnetic stirrer to maintain a uniform concentration and temperature in the beaker and to reduce any concentration polarization near the electrodes. The voltage and current were measured

after a steady state is reached. The complete experimental setup for the production of hydrogen (reactor) and supply for electricity generation (fuel cell) is shown in Fig. 1.

Results and discussion

Effect of temperature, pressure and residence time

Temperature, pressure, and residence time have been noted to be the most important variables for modifying supercritical reaction conditions [17]. Optimal supercritical conditions can be experimentally derived and aided by models to induce the ideal combination of temperature, pressure, and residence time [18]. System optimization, however, involves maximizing the desired output (energy or organic destruction), while reducing reaction times to minutes or seconds versus the hours required for similar results in subcritical water [19]. The effect of temperature, pressure and residence time is shown in Fig. 2a–c.

Effect of temperature

At the chemical equilibrium state, the yields of H_2 and CO_2 increase with the increasing temperature, but the yield of CH_4 decreases sharply. The equilibrium CO yield is very small, and it is about 10^{-3} mol/kg. As temperature increases from 400 to 800 °C, the CO yield firstly increases and then drops down. The maximum CO yield is reached at about 550 °C. Hydrogen yield increases at a low speed at rather higher temperature. When the reaction temperature is above 650 °C, biomass gasification goes to completion and the equilibrium gas product consists of H_2 and CO_2 in a molar ratio equal to $(2-y + x/2)$ (x and y are the elemental molar ratios of H/C and O/C in biomass, respectively). The maximal equilibrium H_2 yield 88.623 % mol/kg of wet biomass was obtained, which is shown in Fig. 2a. From the viewpoint of thermodynamics, higher temperature is essential for hydrogen production. Temperature is considered the most sensitive variable in SCWG processes, with 600 °C serving as an often-cited, optimal target temperature due to associated high conversion [20–22].

Effect of residence time

At the chemical equilibrium state, the yields of H_2 and CO_2 increase with the increasing residence time, but the yield of CH_4 and CO decreases sharply, which is shown in Fig. 2b. Initially H_2 and CO_2 were 49 and 23 % of the yield which is slow; and after 30 min they gradually increased and at 60 min, they reached up to 88.8 and 50 %. But in the case of CH_4 and CO it decreased from 27 and 16 % to 10^{-3} mol/kg. Longer residence time can improve

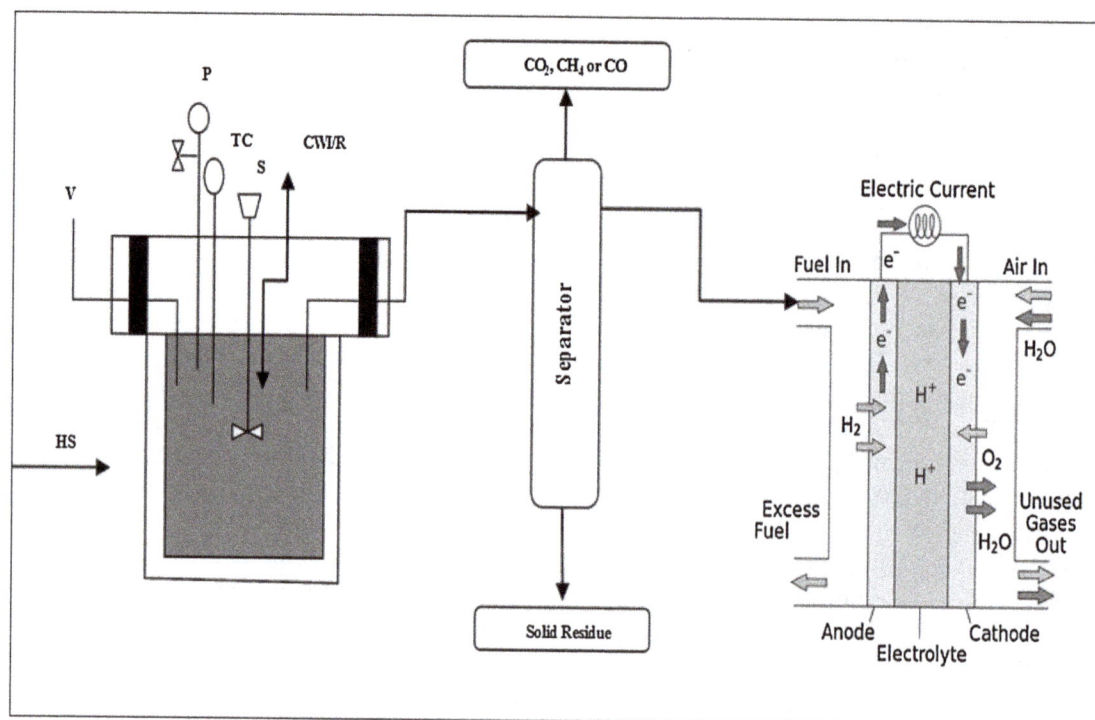

HS: Heat supply, V: Valve, P: Pressure gauge, TC: Thermocouple, S: Stirrer, CWI/R: Cooling water inlet and return

Fig. 1 Schematic diagram of experimental setup. *HS* Heat supply, *V* Valve, *P* Pressure gauge, *TC* Thermocouple, *S* Stirrer, *CWI/R* Cooling water inlet and return

gasification thoroughness, but there is also an inverse relationship between temperature and reaction completeness, dropping from a few minutes below 600 °C to a few seconds above 600 °C. The optimal temperature threshold for SCWG (i.e., 600 °C) has been shown on the lower side of the conversion range for higher concentration biomass in the absence of a catalyst [23–26]. From the viewpoint of thermodynamics, biomass can be gasified completely in SCW with a product formation of H_2 and CO_2, but adequate reaction time was required to complete the gasification process. Short residence times and high organics destruction efficiencies occur during gasification and oxidative reactions at supercritical operating conditions above 600 °C [27].

Effect of pressure

The effect of pressure on equilibrium gas yields at 600 °C at 30 min using different pressure. At 25 MPa the H_2 and CO_2 were 75 and 40 % of yeilds, with increase in pressure; little changes were found, which are shown in Fig. 2c. Pressure shows a complex effect on biomass gasification in SCW. The properties of water, such as density, static dielectric constant and ion product increase with pressure. As a result, the ion reaction rate increases and free-radical reaction is restrained with an increase of pressure.

Hydrolysis reaction plays a significant role in SCWG of biomass, which requires the presence of H^+ or OH^-. With increasing pressure, the ion product increases, therefore the hydrolysis rate also increase. Besides, high pressure favors water–gas shift reaction, but reduces decomposition reaction rate. But in the case of CH_4 and CO it was very less or negiable but with the increase in the pressure it increased slowly. The complex pressure effects can be used to fine tune the chemical composition of the solvent and control gas composition with yield [28]. Specifically, pressure has little or no influence on reaction rate, but it does affect solvent density. Density also has little effect on gasification efficiency above the critical point, but can have significant affects on gas fraction characteristics [29]. High pressures, and correspondingly higher densities, favor CH_4 production and inhibit H_2 production.

Effect of partial pressure of hydrogen in cell performance pressure

To investigate the effect of hydrogen partial pressure on the cell performance, the cell performance is studied with different hydrogen partial pressures of 0.5, 0.8, 1 and 1.2 atm, respectively. The partial pressure is adjusted by mixing argon. The cell performance with different partial pressures of hydrogen at a temperature of 65 °C is shown

(a)

(b)

(c)

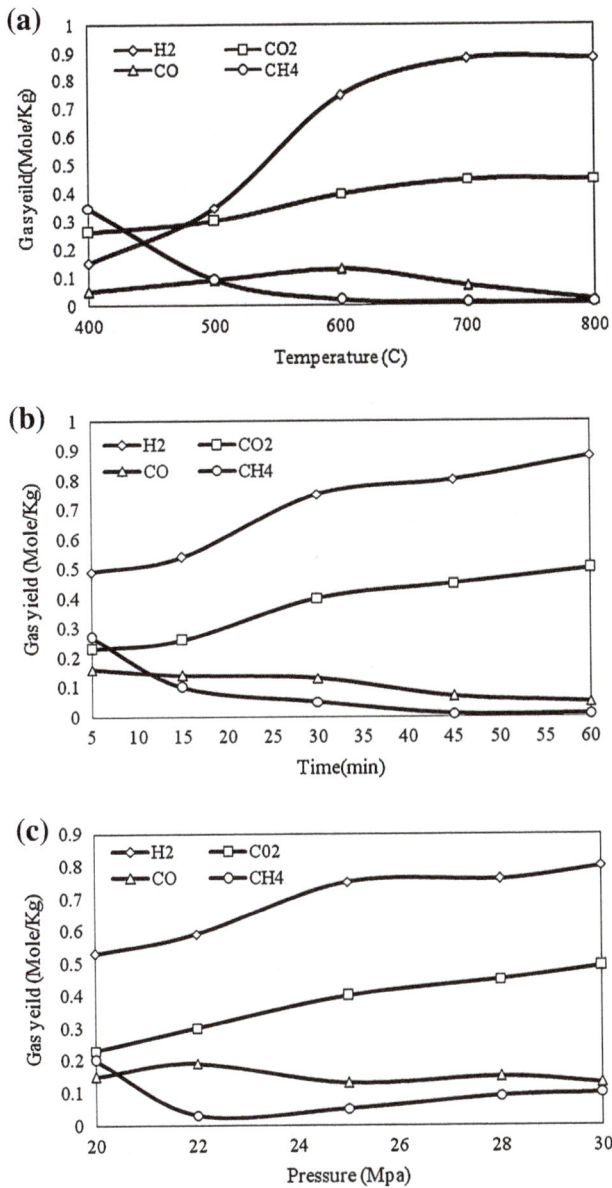

Fig. 2 a Gas yeild using different temperature at fixed 25 MPa pressure and time 30 min, **b** Gas yeild using differnet retension time at fixed 600 °C temperature and 25 Mpa pressure, **c** Gas yield using different pressure at fixed 600 °Ctemperature and time 30 min

in Fig. 3. The power density rises with the increase of the hydrogen partial pressure. The increase in cell open circuit voltage will be somewhat less because of the greater gas solubility at increasing pressure which produces higher lost currents. When the partial pressure of hydrogen is higher than 0.8 atm, the cell can keep a high output performance as that of cell using pure hydrogen. The maximum power density of 9.24 W/cm^2 was obtained using pure hydrogen. The maximum power density only decreased by about 13 % when the partial pressure of hydrogen decreased from

1 atm to 0.8 and 1.2 atm. However, the cell performance decreases dramatically as the partial pressure of hydrogen decreases to 0.5 atm. The maximum power density is only 6.5 W/cm^2 at a partial pressure of hydrogen of 0.2 atm.

The OCV can be calculated from the Nernst equation:

$$E = E^o + (RT/2F)\ln(P_{H_2}/P_{H_2O}) + (RT/2F)\ln\left(P_{O_2}^{1/2}\right) \tag{1}$$

where E^o is the open circuit voltage (OCV) at standard pressure. F is the Faraday constant and R is the gas constant. T is the absolute temperature. P_{O_2} is the partial pressure at the cathode. And P_{H_2} and P_{H_2O} are the partial pressures of hydrogen and vapour at the anode. It can be found that the OCV is dependent on cell temperature, hydrogen and water concentration in the fuel gas (anode) and oxygen in the cathode.

The above equation can be changed to:

$$E = E^o + \frac{RT}{2F}\ln\left(\frac{P_{O_2}^{1/2}}{P_{H_2O}}\right) + \frac{RT}{2F}\ln(P_{H_2}) \tag{2}$$

If the hydrogen pressure changes from P_1 to P_2 and the partial pressures of P_{H_2O} and P_{O_2} keep constant, there will be a change of voltage at 65 °C:

$$\Delta V = \frac{RT}{2F}\ln(P_2) - \frac{RT}{2F}\ln(P_1) = \frac{RT}{2F}\ln\left(\frac{P_2}{P_1}\right)$$
$$= 0.053\ln\left(\frac{P_2}{P_1}\right)V \tag{3}$$

These experimental data are in good agreement with the difference of reversible cell voltage calculated based on Nernst equation.

Effect of temperature in cell performance

In the cell, reaction process becomes faster when the electrolyte is warm rather than cold. So, the temperature plays an important role to develop the voltage across terminal. Figure 4 shows the current–voltage relationship for using different temperature at 1 atm and 2 M of electrolyte solution was fed to the alkaline fuel cell. It is seen that the cell performance increases with the increase in temperature because of decrease in the activation and concentration over-potentials [30]. In addition, mass transport limitations are reduced at higher temperatures. The overall result is an improvement in cell performance or in other words the conductivity of KOH solutions is relatively high at low temperatures. For instance an alkaline fuel cell designed to operate at 75 °C will reduce to only half power level when its operating temperature is reduced to room temperature. The maximum power density 9.36, 9.24, 6.6, and 6.12 W/cm^2 was obtained when temperature is 75, 65, 55, and

Fig. 3 Cell performance at different partial pressure at 65 °C and 2 mol electrolyte concentration

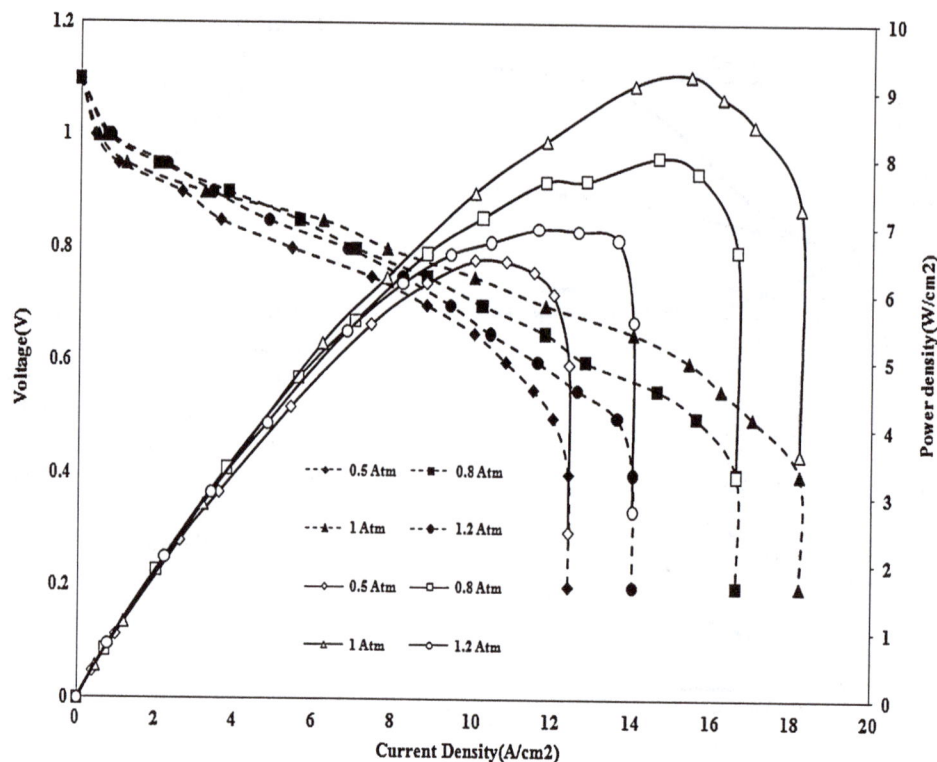

Fig. 4 Cell performance using different temperature at 1 atm. Pressure and 2 mol electrolyte concentration

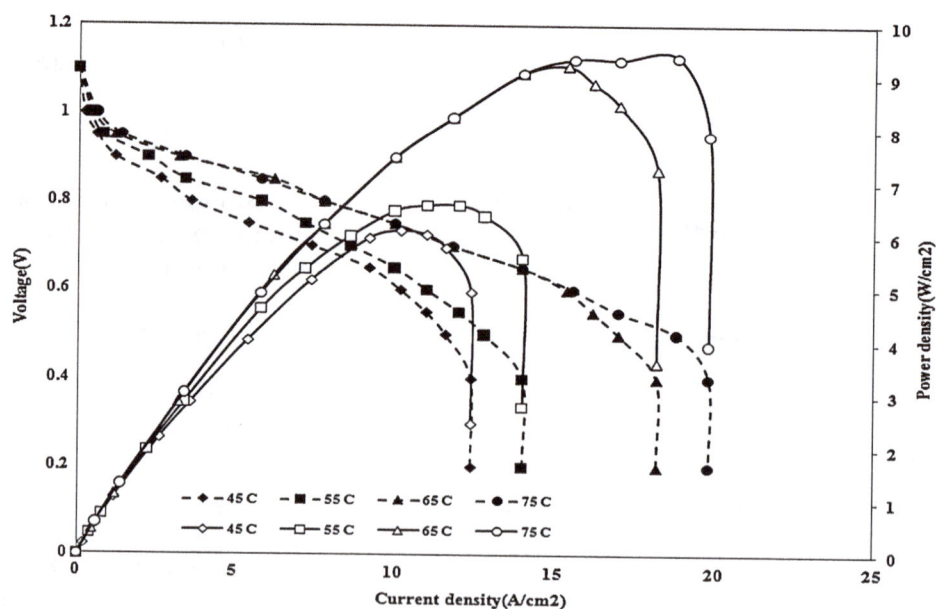

45 °C. This result was found similar to author, who used methanol and ethanol fuel at 25, 45 and 65 °C. The performance increases with the increase in temperature because of decrease in activation over potential concentration and mobility at higher temperature [31].

Effect of electrolyte concentration in cell performance

A higher current flow (amperage) through the cell means it will be passing more electrons through it at any given time. This means a faster rate of reduction at the cathode and a faster rate of oxidation at the anode. This corresponds to a greater number of moles of the product. The amount of current that passes depends on the concentration of the electrolyte; it shows different value in different concentration of electrolyte used. Figure 5 shows that the cell voltage increases with the increase in KOH concentration from 1 to 2 M for a particular load and then it decreases with further increase in KOH concentration. It is well known that the initial and final voltage losses with

Fig. 5 Cell performance using different fuel concentration at 65 °C temperature and 1 atm pressure

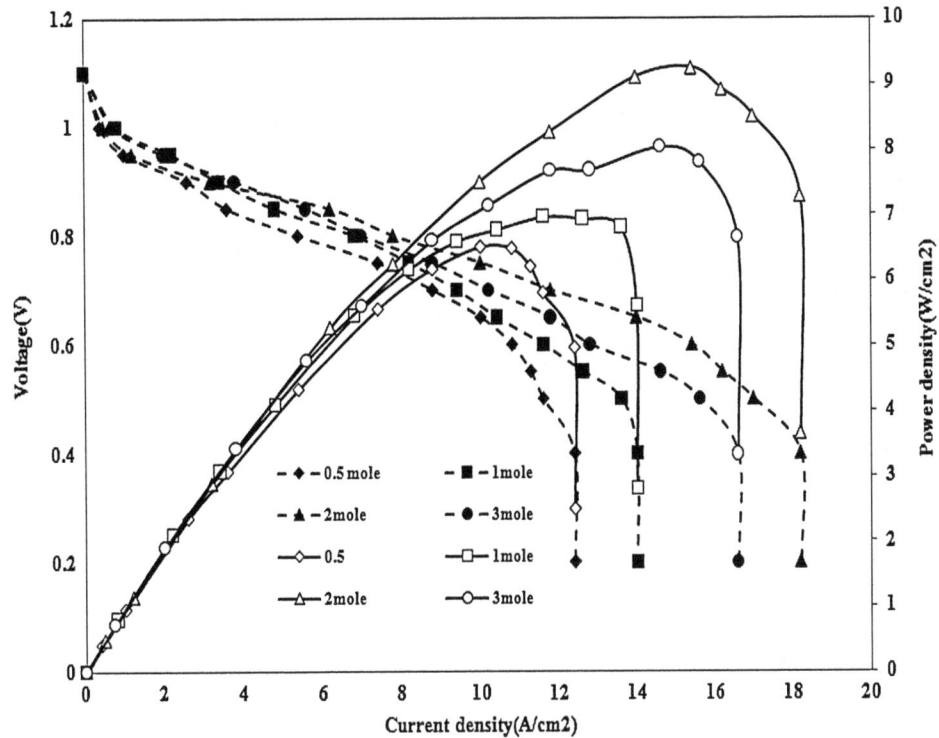

an increase in current consumption and are attributed to activation and concentration over-potentials, whereas the over-potential in the flattened portion of the curve is due to ohmic loss [32]. It is apparent from Fig. 5 that the increase in KOH concentration has minimum effect on activation over-potential while the concentration over-potential first decreases and then increases with the increase in KOH concentration. The concentration polarization increases at a higher KOH concentration because of less availability of hydrogen at the anode. On the other hand, the lowering of the KOH concentration increases the ionic conductivity of the medium or decreases the ohmic loss. The cell performance is maximum at 2 M of electrolyte concentration obtained at 9.24 W/cm^2, and lowest at 0.5 M was 6.48 W/cm^2. While further increase in electrolyte beyond 2 M, it was found the cell performance decreases.

Conclusion

It concludes that the contents of hydrogen and carbon dioxide vary with different operation conditions and obtained 88.8 % of hydrogen and approximate 45 % of carbon dioxide at temperature(600 °C), pressure(25 MPa), and residence time(60 min). Although supercritical water gasification of wet biomass seems promising for the

production of hydrogen rich gas, it should be noticed that a high concentration of biomass is necessary to reach commercial goals. From the experiment we find the maximum power density 9.24 W/cm^2 was obtained at 75 °C (Temperature) 2 M (Electrolyte concentration) and 1 atm (Pressure). The development of hydrogen and fuel-cell technologies is set to play a central role in addressing growing concerns over carbon emissions and climate change as well as the future availability and security of energy supply. Hydrogen can be generated from biomass, but this technology urgently needs further development. It is believed that in the future, biomass can become an important sustainable source of hydrogen. Due to its environmental merits, the share of hydrogen from biomass in the automotive fuel market will grow fast in the next decade. Gasification of biomass has been identified as a possible system for producing renewable hydrogen, which is beneficial to exploit biomass resources, to develop a highly efficient clean way for large-scale hydrogen production, and has less dependence on insecure fossil energy sources. Steam reforming of natural gas and gasification of biomass will become the dominant technologies by the end of the 21st century.

References

1. Dincer, I.: Technical, environmental and exergetic aspects of hydrogen energy systems. Int. J. Hydrog. Energy. **27**, 265–285 (2007)
2. Arena, U.: Process and technological aspects of municipal solid waste gasification. A review. Waste. Manag. **32**, 625–639 (2012)
3. Baratieri, M., Baggio, P., Fiori, L., Grigiante, M.: Biomass as an energy source: thermodynamic constraints on the performance of the conversion process. Biotechnol. **99**, 7063–7073 (2008)
4. Franco, A., Giannini, N.: Perspectives for the use of biomass as fuel in combined cycle power plants. Int. J. Therm. Sci. **44**, 163–168 (2005)
5. Anon, C.: Reducing greenhouse gas emissions and electrical power costs. Biocycle. **45**(10), 35–36 (2004)
6. Appleby, A.J., Foulkes, F.R.: Fuel cell handbook. Van Nostrand Reinhold, New York (1993)
7. Dutton, AG.: Hydrogen energy technology. Tyndall Working Paper TWP17, Tyndall Centre for Climate Change. papers/wp17.pdf (2008). Accessed 15 May 2008
8. Crabtree, G.W., Dresselhaus, M.S., Buchanan, M.V.: The hydrogen economy. Phy. Today. **57**, 39–44 (2004)
9. Novochinskii, I., Ma, X., Song, C., Lambert, J., Shore, L., Farrauto, R.: A ZnO-based sulfur trap for H_2S removal from reformate of hydrocarbons for fuel cell applications. Proceedings of Topical Conference on Fuel Cell Technology. AIChE Spring National Meeting, New Orleans, vol. 11–14, pp. 98–105. (2002)
10. Turner, J.A.: Sustainable hydrogen production. Inform. **305**, 971–974 (2004)
11. Unal, H., Alibas, K.: Agricultural residues as biomass energy. Energy. Sources. Part. B. **2**, 123–140 (2007)
12. Boerrigter, H., Rauch, R.: Review of applications of gases from biomass gasification. In: Knoef, HAM. (ed.) Proceedings of the handbook biomass gasification. The Netherlands: Biomass Technology Group (BTG). pp. 211–230 (2005)
13. Barreto, L., Makihira, A., Riahi, K.: Medium and long-term demand and supply prospects for fuel cells: the hydrogen economy and perspectives for the 21st century. International Institute for Applied Systems Analysis, Laxenburg (2002)
14. Sealock, L.J.J., Elliott, D.C., Baker, E.G., Fassbender, A.G., Silva, L.J.: Hydrogen production by supercritical water gasification. Ind. Eng. Chem. Res. **35**, 4111–4120 (1996)
15. Clean Renewable Fuel from the Plasma Gasification of Waste, (2011). Accessed 16 May 2011
16. Elliott, D.C., Hart, T.R., Neuenschwander, G.G.: Chemical processing in high pressure aqueous environments: improved catalysts for hydrothermal gasification. Ind. Eng. Chem. Res. **45**(11), 3776–3781 (2006)
17. Soria, J.A., McDonald, A.G., Shook, S.R.: Wood solubilization and depolymerization using supercritical methanol. Part 1: process optimization and analysis of methanol insoluble components (Bio-Char). Holzforschung **14**(4), 402–408 (2008)
18. Gloyna, E.F., Li, L.: Supercritical water oxidation: an engineering update. Waste. Manag. **14**, 379–394 (1993)
19. D'Jesus, P., Boukis, N., Kraushaar-Czarnetzki, B., Dinjus, E.: Gasification of cornand clover grass in supercritical water. Fuel **85**, 1032–1038 (2006)
20. Elliott, D.C.: Catalytic hydrothermal gasification of biomass. Bioprod. Biorefining. **2**(3), 254–265 (2008)
21. Susanti, R.F., Veriansyah, B., Kim, J.D., Kim, J., Lee, Y.W.: Continuous supercritical water gasification of isooctane: a promising reactor design. Int. J. Hydrog. Energy. **35**:51957–1970 (2010)
22. Cao, C., Guo, L., Chen, Y., Guo, S., Lu, Y.: Hydrogen production from supercritical water gasification of alkaline wheat straw pulping black liquor in continuous flow system. Int. J. Hydrog. Energy. **36**(21), 13528–13535 (2011)
23. Antal, M.J., Allen, S.G., Schulman, D., Xu, X., Divilio, R.J.: Biomass gasification in supercritical water. Ind. Eng. Chem. Res. **39**(11), 4040–4053 (2000)
24. Xu, L., Brilman, D.W.F., Withag, J.A.M., Brem, G., Kersten, S.: Assessment of a dry and a wet route for the production of biofuels from microalgae: energy balance analysis. Bioresour. Technol. **102**(8), 5113–5122 (2011)
25. Xu, X., Matsumura, Y., Stenberg, J., Antal, M.J.: Carbon-catalyzed gasification of organic feed stocks in supercritical water. Ind. Eng. Chem. Res. **35**(8), 2522–2530 (1996)
26. Savage, P.E.: A perspective on catalysis in sub- and supercritical water. J. Supercrit. Fluids. **47**(3), 407–414 (2009)
27. Du, X., Zhang, R., Gan, Z., Bi, J.: Treatment of high strength coking wastewater by supercritical water oxidation. Fuel. (2010).
28. Afif, E., Azadi, P., Farnood, R.: Catalytic hydrothermal gasification of activated sludge. Appl. Catal. B. Environ. **105**, 136–143 (2011)
29. Brunner, G.: Near and supercritical water. Part II: oxidative processes. J. Supercrit. Fluids. **47**(3):382–390 (2009)
30. Pramanik, H., Basu, S.: Modeling and experimental validation of overpotentials of a direct ethanol fuel cell. Chem. Eng. Process. **49**(7), 635–642 (2010)
31. Gaurav, D., Verma, A., Sharma, D., Basu, S.: Development direct alcohol alkaline fuel cell stack. Fuel. Cell. **10**(4), 591–596 (2010)
32. Koscher, GA., Kordesch, K.: Alkaline methanol/air power devices, in: Handbook of fuel cells—fundamentals, technology and applications. In: Vielstich, W., Gasteiger, H.A., Lamm, A. (eds.), John Wiley, **4**:1125–1129 (2003)

14

Copper nanoparticles stabilized by reduced graphene oxide for CO_2 reduction reaction

Diego C. B. Alves · Rafael Silva · Damien Voiry ·
Tewodros Asefa · Manish Chhowalla

Abstract Carbon dioxide (CO_2) is one of the main gases produced by human activity and is responsible for the green house effect. Numerous routes for CO_2 capture and reduction are currently under investigation. Another approach to mitigate the CO_2 content in the atmosphere is to convert it into useful species such as hydrocarbon molecules that can be used for fuel. In this view, copper is one of the most interesting catalyst materials for CO_2 reduction due to its remarkable ability to generate hydrocarbon fuels. However, its utilization as an effective catalyst for CO_2 reduction is hampered by its oxidation and relatively high voltages. We have fabricated hybrid materials for CO_2 reduction by combining the activity of copper and the conductivity of reduced graphene oxide (rGO). Cu nanoparticles (CuNPs) deposited on rGO have demonstrated higher current density and lower overpotential compared to other copper-based electrodes that we have tested. The CuNPs on rGO also exhibit better stability, preserving their catalytic activity without degradation for several hours.

Keywords CO_2 reduction · Synthetic photosynthesis · Reduced graphene oxide · Copper nanoparticles · Electrocatalysis

D. C. B. Alves · D. Voiry · M. Chhowalla
Materials Science and Engineering, Rutgers University, 607 Taylor Road, Piscataway, NJ 08854, USA

D. C. B. Alves (✉)
Departamento de Física, Universidade Federal de Minas Gerais, Belo Horizonte, MG 31270-901, Brazil
e-mail: diego.c.b.physicist@gmail.com;
dbarbosa@fisica.ufmg.br

R. Silva · T. Asefa
Department of Chemistry and Chemical Biology, Rutgers University, 610 Taylor Road, Piscataway, NJ 08854, USA

T. Asefa
Department of Chemical and Biochemical Engineering, Rutgers University, 98 Brett Road, Piscataway, NJ 08854, USA

Introduction

The increasing demand for renewable energy has been motivating researchers around the world to pursue the development of new technologies. Various methods and devices to generate sustainable forms of energy that are efficient and affordable have been reported [1–4]. In addition to clean energy, the high amount of CO_2 emitted in the atmosphere must also be mitigated [5, 6]. The use of CO_2 as a feedstock to generate fuels is intriguing and inspired by photosynthesis [7]. The consumption of CO_2 from the atmosphere via its conversion to fuel could provide new sources of energy. In this view, electrochemical reduction of CO_2 appears promising since it can lead to formation of methane, ethylene, methanol, formate, carbon monoxide, formic acid and other compounds that can be potentially used as fuels [8–10]. The applicability of the CO_2 reduction process is currently limited by the high overpotentials even when using noble metal catalysts as active electrodes [11].

Different materials have been investigated as catalysts for CO_2 reduction. Some candidates such as Hg, In and Cd have shown high current efficiency associated with high hydrogen overvoltage but remain poor choice for CO_2 reduction catalysts because of their low selectivity in producing the desired products [12]. On the other hand, materials such as Pt, Ni, Fe and Ti favor hydrogen evolution reaction (HER) in aqueous media and also possess high affinity to CO, leading to rapid reduction of catalytic activity [13]. Copper is the most promising catalyst for CO_2 reduction reaction because it enables the production of hydrocarbons at significant current densities, exhibits good selectivity, and has low surface affinity to CO [14]. Theoretical studies have shown that CO acts as an important precursor for hydrocarbon synthesis, which progresses in successive hydrogenation/reduction reactions [15] leading to the 16 products identified during CO_2 reduction on copper [9]. Despite its rich electrochemistry, Cu suffers from high overpotentials compared to potentials of standard catalysts for hydrocarbon production (0.17 V vs. RHE) [9, 16]. Challenges associated with high overpotential and low electrochemical stability must be overcome if copper is to be seriously considered as a practical catalyst for CO_2 reduction reaction [16]. It has been demonstrated that the oxide passivation layer on a Cu film can reduce the overpotential for the CO_2 reduction reaction as well enhance the stability [17]. Although passivated Cu films are more robust catalysts, it leads to a reduction of the current density due to the copper oxide layer. Cu nanoparticles (NPs) could be beneficial in facilitating the CO_2 reduction reaction but their instability caused by combined effects of aggregation and rapid deactivation of the catalytic surface via oxidation has limited their consideration. One strategy recently proposed to overcome the stability issues was to use Au_2Cu NPs that are less prone to oxidation instead Cu NPs for CO_2 reduction [18]. However, Au is expensive and also known to be a good catalyst for HER that can disrupt the production of hydrocarbons [19]. In the present work, we demonstrate that reduced graphene oxide decorated with copper NPs (CuNPs on rGO) is a promising catalyst system for CO_2 reduction that exhibits remarkable stability in a medium with low proton availability. CuNPs on rGO electrodes exhibit lower overpotential for CO_2 reduction reaction and show a strong increase in the current density. To assess the stability of the catalysts, we have tested CuNPs on rGO for several hours and found limited decrease in the current density and negligible change in the overpotential.

Graphene oxide (GO) has been investigated for a broad range of applications in electronics [20], biology [21], energy storage [22, 23], and more recently in catalysis [24–26]. GO consists of atomically thin sheets with lateral dimensions of a few micrometers and thickness of less than 1 nm. GO is composed of sp^2 and sp^3 carbon bonds as well as oxygen functional groups such as hydroxyls and carbonyls [27, 28]. Its chemically heterogeneous properties can be tuned to some extent by varying the oxygen content through reduction and offer a remarkable platform for

performing novel chemistry [29, 30]. The partial removal of oxygen functional groups leads to an enhancement in the π conjugation so that the semi-metallicity of graphene is partially restored and its conductivity increases by several orders of magnitude compared to as-prepared GO [31, 32]. Recently, it has been reported that reduced graphene oxide (rGO) works synergistically with catalyst particles to enhance their stability and performance [33–36].

Experimental

The synthesis was performed in two steps as shown in Scheme 1. First, an aqueous solution of GO was prepared at a concentration of 1.5 mg/mL and this solution was mixed with 9 mg of $CuSO_4$ (copper II sulfate). NaOH was then added and the mixture was vigorously stirred at 70 °C for 5 min. At this step of the synthesis, the color of the solution changed to dark grey. Then, 625 µL of hydrazine was quickly added to the mixture and stirred for additional 40 min. Upon the addition of hydrazine, the color turned dark violet. The mixture was further washed several times with EtOH by centrifugation (10 min at 30 k rpm).

Graphene oxide synthesis

Graphene oxide (GO) was prepared chemically via modified Hummer's method [37]. The graphite powder (Branwell Graphite Inc.) with size of >425 µm was chemically oxidized, exfoliated, and purified by repeated centrifugation.

Copper nanoparticles passivated with polyallylamine

Cu nanoparticles are prepared following the procedure proposed by Wang and Asefa [38], using polyallylamine as passivating agent. In the present case, 220 µL polyallylamine (Sigma-Aldrich) was completely dissolved in distilled water under vigorous stirring at 70 °C. Then, 9 mg of copper sulfate (Sigma-Aldrich) was added followed by

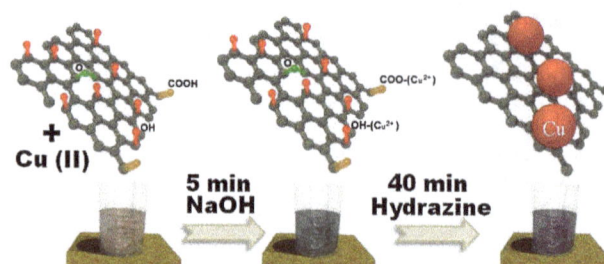

Scheme 1 Schematic representation of the synthesis method consisting of mixing GO with copper(II) and NaOH first. Hydrazine was then added and the mixture was stirred for 40 min. The final material corresponds to rGO nanosheets decorated with Cu NPs

700 µL of NaOH 0.5 M. After 5 min, ~60 µL of hydrazine was poured in the mixture. The solution color changed from an initial blue color to brown and 40 min later the solution color became red.

Copper thin film

A thin film of copper was formed by electrochemical deposition using a solution with 19 mg of Cu(I) in 10 mL acetonitrile. The copper was deposited on a glassy carbon electrode with 3 mm diameter. The deposition was carried out using a constant current of 10 µA for 10 min giving approximately 4 µg of deposited copper. Considering the density of bulk copper and the diameter of the glassy carbon, the thickness of the copper film is estimated to be around 70 nm.

Characterization

X-ray photoelectron spectroscopy (XPS) measurements were performed with a Thermo Scientific K-Alpha spectrometer. All spectra were taken using an Al Kα microfocused monochromatized source (1,486.6 eV) with a resolution of 0.6 eV. Raman spectra were obtained using a Renishaw 1,000 system operating at 514 nm. The UV–vis absorption spectra were measured with a Lambda 950 spectrophotometer (PerkinElmer). Transmission electron microscope (TEM) images of the samples were obtained with a Topcon 002B TEM that was working at an acceleration voltage of 200 kV.

Electrochemical measurements

The measurements (VersaStat 3 potentiostat from Princeton Applied Research) consist of linear sweep voltammetry (LSV) and were performed in two distinct environments (Argon and CO_2) using acetonitrile (10 mL) and tetra-n-butylammonium hexafluorophospate (NBu_4PF_6) (387 mg, Alfa Aesar) as electrolyte. The electrolyte solution was first degassed using argon for 30 min. Under this condition, no reduction reaction is expected. The same procedure was repeated under CO_2 to saturate the solution prior the measurements.

All materials were deposited by drop casting on glassy carbon electrode (CH Instruments, Inc.) and the current of the voltammograms is normalized by the geometric area ($\sim 7 \times 10^{-2}$ cm^2) of the glassy carbon. An Ag/Ag$^+$ (0.1 m) electrode in acetonitrile and a platinum wire were used as reference and counter electrode, respectively. The reference electrode was calibrated using the well-known reduction oxidation signals of ferrocene (FCN) (Figure S1). Potentials in the voltammograms are corrected to NHE.

Results and discussion

RGO decorated with CuNPs was synthesized by co-reduction of copper(II) and GO using hydrazine solution as reducing agent. Briefly, copper(II) sulfate (9 mg) and GO (3 mg) were dispersed in 10 mL of 0.5 M NaOH at 70 °C under vigorous stirring. In basic solution, copper(II) cations remain in close contact with the negatively charged oxygen functional groups of GO. 625 µL of 80 % hydrazine solution was then poured into the mixture to reduce copper(II) and the GO nanosheets (see Methods for details). The addition of hydrazine results in a significant change in the color of the solution, becoming intensely reddish due to surface plasmon resonance (SPR) effect of metallic copper nanoparticles. This effect causes strong absorption around 580–600 nm [39]. As a result of the co-reduction process, conductive rGO is obtained along with metallic copper nanoparticles formed via reduction of Cu(II) as shown in Scheme 1. It is worth noting that no additional passivating agents, such organic molecules or polymers, were used to stabilize the as-prepared copper particles. TEM images of the composite indicate that the CuNPs are uniformly distributed on the basal plane of the reduced graphene oxide and it was also observed that some nanoparticles were partially wrapped by the rGO sheets (Fig. 1a). TEM observations suggest that rGO acts as a stabilizing support for the 20–40 nm Cu nanoparticles by immobilizing and preventing their aggregation into larger particles (inset Fig. 1a).

Figure 1b presents FTIR spectra of the CuNPs on rGO and bare rGO films reduced by hydrazine. Clear differences between the two spectra can be observed. First, the rGO spectrum reveals the absence of oxygen functionalized groups, suggesting a strong reduction [40]. In contrast, the intense bands between 800 and 1,330 cm^{-1} in the CuNPs/rGO spectrum correspond to stretching vibrations (mainly from epoxy and ketone groups), which suggests that decoration with CuNPs inhibits a more severe reduction of GO. However, contributions in the 1,500–1,600 cm^{-1} range are typically not related to oxygen groups, and can be assigned to the asymmetric stretch of sp$_2$-hybridized C=C within the rGO basal plane [41]. The presence of residual hydroxyl has been demonstrated to be beneficial in catalysis through enhancement of electrochemical stability and catalytic activity [41, 42]. In addition, hydroxyl groups that remain on rGO nanosheets decrease the interface energy between rGO and copper, leading to stabilization of CuNPs.

To further investigate the structure of copper-decorated rGO, X-ray photoelectron spectroscopy (XPS) was performed. From the XPS survey spectrum of CuNPs/rGO hybrid material shown in Fig. 2, it is possible to identify signals from oxygen, carbon and copper. The upper left

(a)

(b)

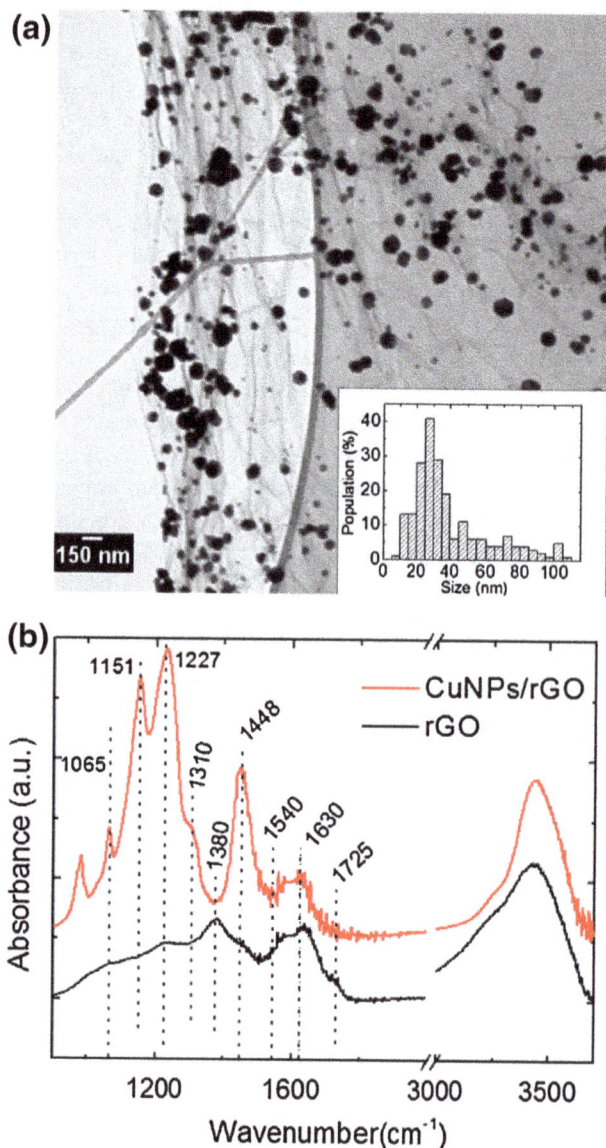

Fig. 1 TEM image and FTIR of CuNPs on rGO prepared by hydrazine reduction: **a** TEM image showing the dispersion of nanoparticles on rGO. The size distribution histogram indicates an average particle size of about 30 nm. **b** FTIR spectra of bare rGO and CuNPs on rGO. The oxygen peaks are virtually absent in bare rGO while they are present in Cu NP decorated rGO and are presumably helping the stabilization of the Cu NPs

inset shows the high-resolution signal of the C1s region from rGO with C–C/C=C, C–OH/C–O, C=O and C=O–OH peaks at 284.4, 286, 287.3, 289 and the satellite peak at 290.6 eV. After hydrazine treatment, the carbon/oxygen ratio in the CuNPs on rGO sample increased from 2.1 up to 7.3. This value is in good agreement with those previously reported for mildly reduced GO [43, 44]. On the other hand, peaks from Cu $2p_{3/2}$ and $2p_{1/2}$ can be identified at 932.5 and 952.4 eV, respectively. The fitting of the Cu peaks reveals a main signal at 932.5 eV that can be attributed to the zero-valence states of Cu and smaller peaks

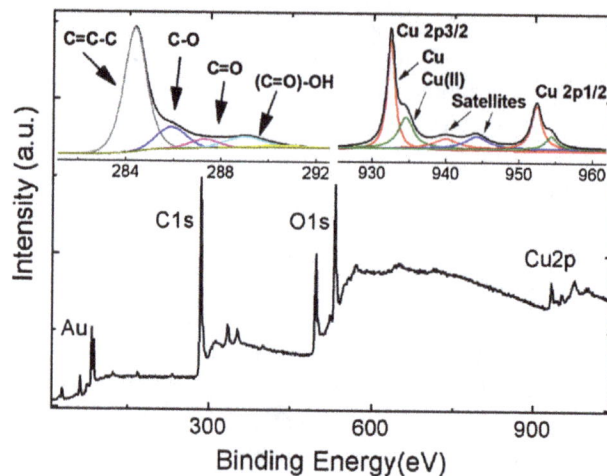

Fig. 2 XPS survey spectrum of CuNPs on rGO. High-resolution spectra from the C1s and Cu2p regions can be deconvoluted with different components from various chemical bonds, as indicated by the fits in the *upper portion* of the figure

at 934.4 eV that are due to the presence of Cu(II) [17]. The latter indicates partial oxidation of the Cu nanoparticle surface in the form of CuO. We believe that the CuO shell is formed during exposure to air after the synthesis since the presence of hydrazine during nanoparticle synthesis is likely to prevent copper from being oxidized. It is also possible to verify the presence of two Cu satellites bands at 940 and 943.9 eV. Moreover, the presence of copper in zero-valence state is also confirmed by our UV–Vis spectroscopy results (see Supplementary Information, Figure S2) that show an intense absorption band at around 600 nm attributed to the surface plasmon resonance effect (SPR) in metallic copper NPs as previously reported [45–47]. This peak is not observed in the UV–Vis spectrum of the bare rGO, Figure S2. The SPR band was also observed at 560 nm for CuNPs prepared using polyallylamine as capping agent. The shift in position of copper SPR band between CuNPs prepared with polyallylamine and CuNPs on rGO can be attributed to differences in the size of the nanoparticles. That is, our TEM image analyses show that the Cu NPs prepared using polyallylamine are significantly smaller in size than the CuNPs on rGO (Figure S3).

CO_2 reduction reaction

To gain insight into the catalytic activity of copper nanoparticles on rGO, cyclic voltammetry was performed in argon (Ar)- and CO_2-saturated electrolyte solutions (Fig. 3). It has been recently reported that the selectivity toward the chemical species is highly dependent on the morphology of polycrystalline copper [48]. To limit the number of products from the CO_2 reduction, all the measurements reported here have been performed in the

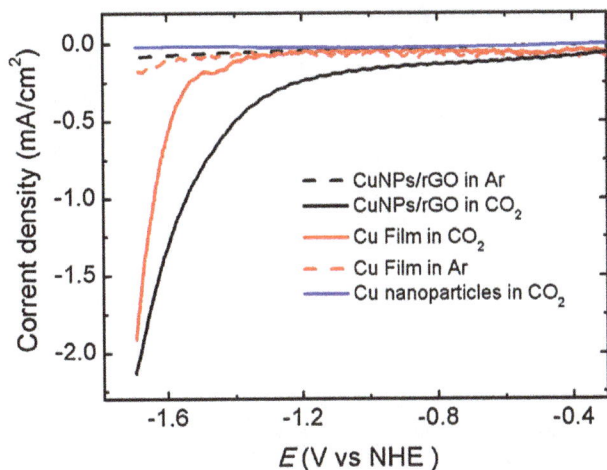

Fig. 3 Linear sweep voltammetry (LSV) of CuNPs/rGO, Cu thin film and polymer passivated CuNPs under Argon- and CO$_2$-saturated solutions. The Cu NPs on rGO electrodes exhibit lower overpotential and higher current density compared to other forms of copper tested in this study

Fig. 4 **a** Linear sweep voltammetry (LSV) curves for CuNPs on rGO as a function of time (0–3 h). The electrode was held at −1.4 V vs. NHE during testing. The corresponding chronoamperometric curve for CuNPs/rGO is shown in inset. **b** Turnover frequency (TOF) as a function of the overpotential calculated for CuNPs/rGO (*black*) and Cu thin film (*red*). Inset is the evolution of the turnover from CuNPs on rGO measured over 24 h with an overpotential of 1.4 V. The *black arrow* indicates the change of the electrolyte solution. Calculations are based considering the total amount of copper and a 2-electron process (CO, CO$_3^{2-}$ and C$_2$O$_4^{2-}$ products)

absence of water. Specifically, the experiments were performed in acetonitrile using Tetra-n-butylammonium hexafluorophosphate (NBu$_4$PF$_6$, 0.1 M) electrolyte. Acetonitrile was chosen because it can dissolve approximately eightfold more CO$_2$ than water [49], and also because it is a low hydrogen availability medium so that the contribution from HER on the cathodic current can be largely disregarded. In addition, the mechanism for CO$_2$ reduction in low hydrogen availability medium is well known. Only three species can be obtained: carbon monoxide (CO), carbonate (CO$_3^{2-}$) or oxalate (C$_2$O$_4^{2-}$) involve 2 electrons each (Equation I and II). Oxalate formation requires C–C coupling which is known to occur at higher overpotentials (>0.9 V vs. RHE) [9, 10]. Experiments in DMF with low proton content similar to acetonitrile have demonstrated that oxalate and CO are the two main products of the reaction and only traces of HCO$_2^-$ have been detected [50].

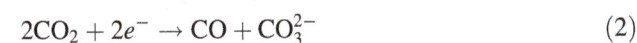

$$2CO_2 + 2e^- \rightarrow C_2O_4^{2-} \qquad (1)$$

$$2CO_2 + 2e^- \rightarrow CO + CO_3^{2-} \qquad (2)$$

The activity of the electrodes can be easily tested by comparing the cathodic current in solution saturated with CO$_2$ with the cathodic current in solution saturated with Ar. It is worth noting that the potentials to reduce Cu (II) to Cu (I) and Cu(I) to Cu metal in organic medium are approximately −0.3 and −0.6 V vs. NHE, respectively [51]. Thus, at the range of negative potentials used here for CO$_2$ reduction, the reduction of the copper oxide shell is thermodynamically favored and the nanoparticles are most likely being reduced to copper metal [48]. From Fig. 3, the CO$_2$ reduction is clearly demonstrated by the high cathodic

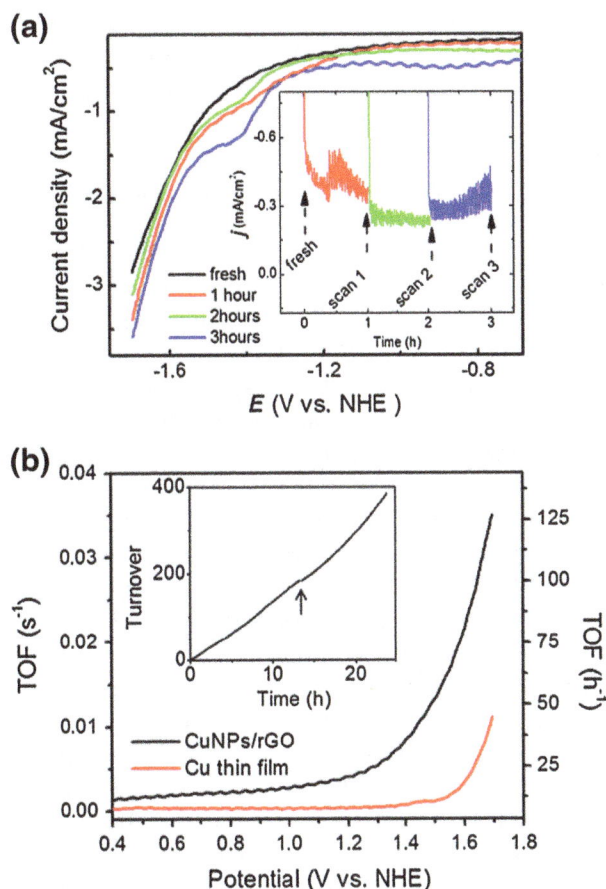

currents obtained from CuNPs on rGO electrodes immersed in CO$_2$-saturated solution compared to Ar-saturated solution, whereas rGO alone shows a limited catalytic activity (Figure S4). Furthermore, catalytic activity of CuNPs on rGO electrodes is significantly higher for ∼70 nm-thick Cu film or CuNPs stabilized by polyallylamine (see Methods for details). Interestingly, a significant cathodic current was not observed using an electrode with passivated CuNPs with polyallylamine. Therefore, it can be stated that the presence of the polymer on the surface of CuNPs inhibits the copper activity. These results demonstrated the clear superior activity of the CuNPs supported on rGO nanosheets. Comparing the electrocatalytic activity of CuNPs on rGO and Cu thin film, it can be observed that higher current density can be achieved with lower overpotentials with rGO decorated with CuNPs. For instance, at -1.2 V vs. NHE the current density observed in the CuNPs/rGO electrode is -0.24 mA cm^{-2}, while the same current density is reached at -1.54 V for Cu thin film. The current density obtained in the CuNPs/rGO at -1.54 V vs. NHE is -0.97 mA cm^{-2}, fourfold better than the Cu thin film.

Stability is an additional parameter in catalysis. To demonstrate the potential use of CuNPs on rGO for CO_2 reduction reaction, chronoamperometric measurements were carried out to investigate the stability (Fig. 4a). A constant overpotential of 1.4 V vs. NHE was applied on the CuNPs/rGO electrode for 3 h. Linear sweep voltammetry (LSV) was also performed with intervals of 1 h during the chronoamperometric measurements to check any overpotential shift. CuNPs on rGO exhibit stable current during the CO_2 reduction reaction (inset Fig. 4a). The improvement of the overpotential in the first hour under negative potential is likely due to the partial reduction of the oxide passivation layer on the copper nanoparticles as demonstrated by XPS and supported by Li et al. [17]. To gain insight into the reaction kinetics of the reduction process, we have calculated the turnover frequency (TOF) of the CuNP on rGO and Cu thin-film catalysts. The results of which are shown in Fig. 4b. It can be seen that the TOF obtained considering two-electron process (CO, CO_3^{2-} and $C_2O_4^{2-}$ products) is larger for CuNPs on rGO than the Cu thin films, suggesting that the CO_2 reduction is more favorable for CuNPs on rGO. It can also be seen from Fig. 4b that the number of turnovers increases linearly, indicating no substantial degradation of the activity of the Cu nanoparticles when deposited on rGO.

Conclusions

Our results suggest that rGO is a promising support for nanosized catalysts. The stabilization of Cu NPs on rGO not only improves the overall electrochemical stability but also enhances the catalytic activity for CO_2 reduction, in contrast with Cu NPs with polymer capping layer. Specifically, Cu nanoparticles on rGO have demonstrated lower overpotential and higher current density in comparison to pure Cu thin films and polymer capped Cu NPs. The enhanced stability may be attributed to partial covering of the NPs by a shell of rGO as observed in the TEM. Recent work proposes rough copper surface like nanoparticles as a way to enhance the selectivity towards hydrocarbons through uncoordinated copper sites [48]. These results suggest that rGO decorated with CuNPs may be a promising support for reducing CO_2.

Acknowledgments TA gratefully acknowledges the partial financial assistance provided by the US National Science Foundation (NSF) (Grant Nos: CAREER CHE-1004218, NSF DMR-0968937, NSF NanoEHS-1134289, NSF-ACIF for 2010, and NSF special creativity grant in 2011). DCBA and RS acknowledge the CAPES (Coordenação de Aperfeiçoamento de Pessoal de Nível Superior, Brazil) and the Fulbright Agency, USA for their fellowships and financial support for his graduate study.

References

1. Turner, J.A.: A realizable renewable energy future. Science **285**, 687–689 (1999)
2. Somorjai, G.A., Frei, H., Park, J.Y.: Advancing the frontiers in nanocatalysis, biointerfaces, and renewable energy conversion by innovations of surface techniques. J. Am. Chem. Soc. **131**, 16589–16605 (2009)
3. Silva, R., Asefa, T.: Noble metal-free oxidative electrocatalysts: polyaniline and Co(II)-polyaniline nanostructures hosted in nanoporous silica. Adv. Mater. **24**, 1878–1883 (2012).
4. Silva, R., Al-Sharab, J., Asefa, T.: Edge-Plane-Rich Nitrogen-Doped Carbon Nanoneedles and Efficient Metal-Free Electrocatalysts. Angew. Chemie. **124**, 7283–7287 (2012)
5. Gattrell, M., Gupta, N., Co, A.: A review of the aqueous electrochemical reduction of CO_2 to hydrocarbons at copper. J. Electroanal. Chem. **594**, 1–19 (2006)
6. Varghese, O.K., Paulose, M., Latempa, T.J., Grimes, C.A.: High-rate solar photocatalytic conversion of CO2 and water vapor to hydrocarbon fuels. Nano. Lett. **9**, 731–737 (2009)
7. Olah, G.A., Goeppert, A., Prakash, G.K.S.: Chemical recycling of carbon dioxide to methanol and dimethyl ether: from greenhouse gas to renewable, environmentally carbon neutral fuels and synthetic hydrocarbons. J. Org. Chem. **74**, 487–498 (2009)
8. Whipple, D.T., Kenis, P.J.A.: Prospects of CO_2 utilization via direct heterogeneous electrochemical reduction. J. Phys. Chem. Lett. **1**, 3451–3458 (2010)
9. Kuhl, K.P., Cave, E.R., Abram, D.N., Jaramillo, T.F.: New insights into the electrochemical reduction of carbon dioxide on metallic copper surfaces. Energy. Environ. Sci. **5**, 7050 (2012)
10. Centi, G., Perathoner, S.: Opportunities and prospects in the chemical recycling of carbon dioxide to fuels. Catal. Today **148**, 191–205 (2009)
11. Dewulf, D.W., Jin, T., Bard, A.J.: Electrochemical and surface studies of carbon-dioxide reduction to methane and ethylene at copper electrodes in aqueous-solutions. J. Electrochem. Soc. **136**, 1686–1691 (1989)
12. Hori, Y., Wakebe, H., Tsukamoto, T., Koga, O.: Electrocatalytic process of CO selectivity in electrochemical reduction of CO_2 at metal electrodes in aqueous media. Electrochim. Acta **39**, 1833–1839 (1994)
13. Azuma, M., Hashimoto, K.: Electrochemical reduction of carbon dioxide on various metal electrodes in low-temperature aqueous $KHCO_3$ media. J. Electrochem. Soc. **137**, 1772–1778 (1990)
14. Kaneco, S., Ueno, Y., Katsumata, H., Suzukib, T., Ohta, K.: Electrochemical reduction of CO_2 in copper particle-suspended methanol. Chem. Eng. J. **119**, 107–112 (2006)
15. Peterson, A.A., Abild-Pedersen, F., Studt, F., Rossmeisl, J., Norskov, J.L.: How copper catalyzes the electroreduction of carbon dioxide into hydrocarbon fuels. Energy. Environ. Sci. **3**, 1311 (2010)

16. Hori, Y., Murata, A., Takahashi, R.: Formation of hydrocarbons in the electrochemical reduction of carbon dioxide at a copper electrode in aqueous solution. J. Chem. Soc. Faraday. Trans. 1(85), 2309 (1989)

17. Li, C.W., Kanan, M.W.: CO_2 reduction at low overpotential on Cu electrodes resulting from the reduction of thick Cu_2O films. J. Am. Chem. Soc. 134, 7231–7234 (2012)

18. Xu, Z., Lai, E., Shao-Horn, Y., Hamad-Schifferli, K.: Compositional dependence of the stability of AuCu alloy nanoparticles. Chem. Commun. (Camb.) 48, 5626–5628 (2012)

19. Perez, J., Gonzalez, E.R., Villullas, H.M.: Hydrogen evolution reaction on gold single-crystal electrodes in acid solutions. J. Phys. Chem. B. 102, 10931–10935 (1998)

20. Eda, G., Chhowalla, M.: Chemically derived graphene oxide: towards large-area thin-film electronics and optoelectronics. Adv. Mater. 22, 2392–2415 (2010)

21. Tang, L., Chang, H., Liu, Y., Li, J.: Duplex DNA/graphene oxide biointerface: from fundamental understanding to specific enzymatic effects. Adv. Funct. Mater. 22, 3083–3088 (2012)

22. Liu, F., Song, S., Xue, D., Zhang, H.: Folded structured graphene paper for high performance electrode materials. Adv. Mater. 24, 1089–1094 (2012)

23. Wu, Z.-S., Winter, A., Chen, L., Sun, Y., Turchanin, A., Feng, X., Müllen, K.: Three-dimensional nitrogen and boron co-doped graphene for high-performance all-solid-state supercapacitors. Adv. Mater. 24, 5130–5135 (2012)

24. Wang, J., Zhang, X.-B., Wang, Z.-L., Wang, L.-M., Zhang, Y.: Rhodium–nickel nanoparticles grown on graphene as highly efficient catalyst for complete decomposition of hydrous hydrazine at room temperature for chemical hydrogen storage. Energy. Environ. Sci. 5, 6885 (2012)

25. Li, Y., Wang, H., Xie, L., Liang, Y., Hong, G., Dai, H.: MoS2 nanoparticles grown on graphene: an advanced catalyst for the hydrogen evolution reaction. J. Am. Chem. Soc. 133, 7296–7299 (2011)

26. Wu, Z., Yang, S., Sun, Y., Parvez, K., Feng, X., Müllen, K.: 3D nitrogen-doped graphene aerogel-supported Fe_3O_4 nanoparticles as efficient electrocatalysts for the oxygen reduction reaction. J. Am. Chem. Soc. 134, 9082–9085 (2012)

27. Bagri, A., Mattevi, C., Acik, M., Chabal, Y.J., Chhowalla, M., Shenoy, V.B., Vivek, B.: Structural evolution during the reduction of chemically derived graphene oxide. Nat. Chem. 2, 581–587 (2010)

28. Bagri, A., Grantab, R., Medhekar, N.V., Shenoy, V.B.: Stability and formation mechanisms of carbonyl- and hydroxyl-decorated holes in graphene oxide. J. Phys. Chem. C 114, 12053–12061 (2010)

29. Dreyer, D.R., Park, S., Bielawski, C.W., Ruoff, R.S.: The chemistry of graphene oxide. Chem. Soc. Rev. 39, 228–240 (2010)

30. Loh, K.P., Bao, Q., Eda, G., Chhowalla, M.: Graphene oxide as a chemically tunable platform for optical applications. Nat. Chem. 2, 1015–1024 (2010)

31. Eda, G., Fanchini, G., Chhowalla, M.: Large-area ultrathin films of reduced graphene oxide as a transparent and flexible electronic material. Nat. Nanotechnol. 3, 270–274 (2008)

32. Gómez-Navarro, C., Weitz, R.T., Bittner, A.M., Scolari, M., Mews, A., Burghard, M., Kern, K.: Electronic transport properties of individual chemically reduced graphene oxide sheets. Nano. Lett. 7, 3499–3503 (2007)

33. Sahoo, N.G., Pan, Y., Li, L., Chan, S.H.: Graphene-based materials for energy conversion. Adv. Mater. 24, 4203–4210 (2012)

34. Wen, Z., Cui, S., Pu, H., Mao, S., Yu, K.H., Feng, X.L., Chen, J.H.: Metal nitride/graphene nanohybrids: general synthesis and multifunctional titanium nitride/graphene electrocatalyst. Adv. Mater. 23, 5445–5450 (2011)

35. Yu, X., Kuai, L., Geng, B.: CeO2/rGO/Pt sandwich nanostructure: rGO-enhanced electron transmission between metal oxide and metal nanoparticles for anodic methanol oxidation of direct methanol fuel cells. Nanoscale 4, 5738–5743 (2012)

36. Li, Y., Zhou, W., Wang, H., Xie, L., Liang, Y., Wei, F., Idrobo, J.-C., Pennycook, S.J., Dai, H.: An oxygen reduction electrocatalyst based on carbon nanotube–graphene complexes. Nat. Nanotechnol. 7, 394–400 (2012)

37. Hirata, M., Gotou, T., Horiuchi, S., Fujiwara, M., Ohba, M.: Thin-film particles of graphite oxide 1. Carbon. N. Y. 42, 2929–2937 (2004)

38. Wang, Y., Asefa, T.: Poly(allylamine)-stabilized colloidal copper nanoparticles: synthesis, morphology, and their surface-enhanced Raman scattering properties. Langmuir 26, 7469–7474 (2010)

39. Pedersen, D.B., Wang, S.: Surface plasmon resonance spectra of 2.8 ± 0.5 nm diameter copper nanoparticles in both near and far fields. J. Phys. Chem. C 111, 17493–17499 (2007)

40. Goncalves, G., Marques, P.A.A.P., Granadeiro, C.M., Nogueira, H.I.S., Singh, M.K., Gracio, J.: Surface modification of graphene nanosheets with gold nanoparticles: the role of oxygen moieties at graphene surface on gold nucleation and growth. Chem. Mater. 21, 4796–4802 (2009)

41. Acik, M., Lee, G., Mattevi, C., Pirkle, A., Wallace, R.M., Chhowalla, M., Cho, K., Chabal, Y.: The role of oxygen during thermal reduction of graphene oxide studied by infrared absorption spectroscopy. J. Phys. Chem. C 115, 19761–19781 (2011)

42. Silva, R., Kunita, M.H., Girotto, E.M., Radovanovic, E., Muniz, E.C., Carvalho, G.M., Rubira, A.F.: Synthesis of Ag-PVA and Ag-PVA/PET-s20 composites by supercritical CO 2 method and study of silver nanoparticle. Growth. 19, 1224–1229 (2008)

43. Kou, R., Shao, Y., Mei, D., Nie, Z., Wang, D., Wang, C., Viswanathan, V.V., Park, S., Aksay, I.A., Lin, Y., Wang, Y., Liu, J.: Stabilization of electrocatalytic metal nanoparticles at metal–metal oxide–graphene triple junction points. J. Am. Chem. Soc. 133(8), 2541–2547 (2011)

44. Compton, O.C., Jain, B., Dikin, D.A., Abouimrane, A., Amine, K., Nguyen, S.T.: Chemically active reduced graphene oxide with tunable C/O ratios. ACS. Nano. 5, 4380–4391 (2011)

45. Mattevi, C., Eda, G., Agnoli, S., Miller, S., Mkhoyan, K.A., Celik, O., Mastrogiovanni, D., Granozzi, G., Garfunkel, E., Chhowalla, M.: Evolution of electrical, chemical, and structural properties of transparent and conducting chemically derived graphene thin films. Adv. Funct. Mater. 19, 2577–2583 (2009)

46. Hu, J., Liu, P., Chen, L.: Comparison of surface plasmon resonance responses to dry/wet air for Ag, Cu, and Au/SiO2. Appl. Opt. 51, 1357–1360 (2012)

47. Blosi, M., Albonetti, S., Dondi, M., Martelli, C., Bald, G.: Microwave-assisted polyol synthesis of Cu nanoparticles. J. Nanoparticle. Res. 13, 127–138 (2010)

48. Tang, W., Peterson, A.A., Varela, A.S., Jovanov, Z., Bech, L., Durand, W.J., Dahl, S., Norskov, J.K., Chorkendorff, I.: The importance of surface morphology in controlling the selectivity of polycrystalline copper for CO_2 electroreduction. Phys. Chem. Chem. Phys. 14, 76–81 (2012)

49. Tomita, Y., Teruya, S., Koga, O., Hori, Y.: Electrochemical reduction of carbon dioxide at a platinum electrode in acetonitrile–water mixtures. J. Electrochem. Soc. 147, 4164 (2000)

50. Amatore, C., Saveant, J.M.: Mechanism and kinetic characteristics of the electrochemical reduction of carbon dioxide in media of low proton availability. J. Am. Chem. Soc. 103, 5021–5023 (1981)

51. Calderón, C.A., Ojeda, C., Macagno, V.A., Paredes-Olivera, P., Patrito, E.M.: Interaction of oxidized copper surfaces with alkanethiols in organic and aqueous solvents. the mechanism of Cu_2O reduction. J. Phys. Chem. C 114, 3945–3957 (2010)

15

Electrochemical stability of the polymer-derived nitrogen-doped carbon: an elusive goal?

Kun Cong · Mariusz Radtke · Steffi Stumpf ·
Bernd Schröter · Duncan G. G. McMillan ·
Markus Rettenmayr · Anna Ignaszak

Abstract Nitrogen-doped carbon is a promising metal-free catalyst for oxygen reduction reaction in fuel cells and metal-air batteries. However, its practical application necessitates a significant cost reduction, which can be achieved in part by using new synthetic methods and improvement of catalytic activity by increasing the density of redox active centers. This can be modulated by using polymer as the carbon and nitrogen sources. Although, superior catalytic activity of such N-doped C has been investigated in details, the electrochemical long-term stability of polymer-derived doped-carbon is still unclear. Herein, in this study we generated N-doped carbon from the most recommended polymer that is comparable to the state-of-the-art materials with porosity as high as 2,086 m^2 g^{-1} and a nitrogen doping level of 3–4 at.%, of which 56 % is pyrrolic N, 36.1 % pyridinic and ∼8 % graphitic. The electrochemical characterization shows that N-doped carbon is catalytic toward oxygen reduction in an alkaline electrolyte via a favorable four-electron process, however, not stable under long-term potential scanning. The irreversible electrochemical oxidation of this material is associated with the presence of a significant content or pyrrolic and pyridinic N close to the edge of the carbon network originating from the polypyrrole precursor. These structures are less stable under operating electrochemical potential. The role of polypyrrole as the precursor of N-doped carbons has to be carefully revised since it supplies sufficient number of catalytic sites, but also generates unstable functionalities on the carbon surface.

Keywords Stability · Electrocatalyst · Nitrogen-doped carbon · Surface functionality

K. Cong · M. Radtke · A. Ignaszak (✉)
Institute of Organic and Macromolecular Chemistry, Friedrich-Schiller University, Lessingstrasse 12, 07743 Jena, Germany
e-mail: anna.ignaszak@uni-jena.de

S. Stumpf
Jena Center of Soft Mater (JCSM), Friedrich-Schiller University, Philosophenweg 7, 07743 Jena, Germany

B. Schröter
Institute of Solid State Physics, Friedrich-Schiller University, Philosophenweg 7, 07743 Jena, Germany

D. G. G. McMillan
University Hospital Jena, Friedrich-Schiller University, Bachstrasse 18, 07743 Jena, Germany

M. Rettenmayr
Institute of Metallic Materials, Friedrich-Schiller University, Löbdergaben 32, 07743 Jena, Germany

Introduction

Porous doped carbons are prosperous materials due to valuable features such as relatively low production cost, high electrical conductivity, chemical and thermal stabilities, to name a few of the reasons for their becoming attractive candidates in a wide range of applications [1]. Specifically, nanostructured doped carbons have been extensively used as electrodes, electrocatalysts or catalyst supports in energy storage and conversion devices [2–4]. Among these potential applications, a great deal of research has been devoted to the use of porous carbons as catalysts in fuel cell technology [5]. Cheaper and easily available electrocatalysts for oxygen reduction reaction (ORR) based on nitrogen-doped carbon (NDC) are great alternatives to the noble-metal Pt/C system with already proven higher activity and stronger resistance against poisoning effects when comparing to the platinum [6]. This remarkable property of nitrogen-doped carbon stems from its unique electronic structure compared to bare carbon. Since nitrogen is considered as n-type carbon dopant, it introduces an electronically disordered carbon structure by donating electrons to carbon, consequently. The local centers with different electron density on the catalyst surface attract and stabilize the absorbed O_2, resulting in superior catalytic activity toward electrochemical oxygen reduction [7].

Generally, there are two main approaches to synthesize N-doped carbon. First is a post-synthesis doping of carbon in the presence of nitrogen source, with the most frequent a high-temperature annealing of carbon in the flow of gaseous ammonia [8–10]. Although this method is efficient with respect to the doping level, relatively harsh conditions used during ammonia treatment generate a serious safety risk, requiring special corrosion-resistant equipment and a storage-controlling system. For this reason, extensive studies are prompted to search for alternative methods. One of the most promising is to adapt nitrogen-rich carbon precursors, e.g. organic compounds [11], such as polypyrrole (PPy) [12], polyaniline (PANI) [2] and polyacrylonitrile [13]. In summary, this method provides a high surface area product with homogenous morphology due to low conversion (polymer-to-carbon) temperature, allowing to control better the final properties, such as the surface concentration of nitrogen, and the way it is incorporated in the carbon structure (e.g., as pyrrolic, graphitic, pyridinic or quaternary N–C states) [14]. Although the effect of nitrogen doping level (i.e., an established optimum at 2–15 at.% of N) on the catalytic activity toward oxygen reduction has been well described, there are discrepancies with regards to the long-term electrochemical stability of these compounds [15, 16]. In general, we observe that all N-doped carbons consisting of graphitized carbon including carbon nanotubes [17] or graphene [18] show superior electrochemical stability under extensive potential cyclic in alkaline media, owing to the presence of graphitic basal plane structure that is resistant to irreversible chemical or electrochemical oxidation [19]. The greater extent of graphitization stabilizes the carbon against oxidation in comparison to the amorphous phase. In contrary, there is a lack of information with regard to the electrochemical durability of the polymer-derived N-doped carbons with a polymer as the only sources of carbon and nitrogen. Current literature supports the claim that an ultra-high surface area is achievable, resulting in excellent catalytic activity [20, 21], however the electrochemical stability of this type of metal-free catalyst over a commercially useful potential range is unclear. For example, some groups observed

instability of the polymer-derived N-doped carbon and associated it with chemical incompatibility of a binder used to prepare the electrode [20]. Another group claims the electrochemical stability, but the degradation conditions applied in their work are not relevant to the operating potential window expected for the ORR catalysts [21]. This motivates us to critically assess the long-term electrochemical durability of N-doped carbon made from the polymer as only carbon and nitrogen precursor. In this work we generate a model compound, whose properties are comparable to the current state-of-the-art ultra-porous N-doped carbon with the N-doping level in the range (2–15 at.%). By employing polypyrrole, the most frequently used and promising precursor for N-doped C, we ensure optimum catalytic activity toward electrochemical oxygen reduction. Here we adapt a template-assisted synthesis of the polymer precursor and a robust post chemical and thermal annealing in order to achieve the representative product. This will be analyzed in detail with regard to the chemical structure, electrochemical stability, and electrochemical performance. The stability of this model system will be studied using voltammetry and impedance spectroscopy correlated with the surface elemental composition and the structure of the both the fresh and degraded compound.

Experimental

Materials

All reagents and solvents were purchased from Sigma-Aldrich and used without further purification unless described. Pyrrole (Py) was distilled and stored under argon. $FeCl_3$ was used as an oxidizing agent for the polymerization. Nafion monomer was purchased as 5 wt% solution in mixture of alcohols. Silica spheres of 0.15–0.55 μm diameter were received from Polysciences GmbH as 5.0 wt% aqueous suspensions. Nitric acids (HNO_3) and potassium hydroxide (KOH) were used for washing and to prepare electrolyte, respectively. All solutions were prepared from deionized water of resistivity 18 MΩ and degassed prior measurement with high purity argon (99.999 %).

Methods

The product identification was performed using a solid state FTIR spectrophotometer for the chemical composition (IR Affinity-1 Shimadzu, a pure KBr tablet was used as a blank for the background subtraction). Thermal analysis (STA 449F1 Jupiter, Netzsch) was carried out in Ar with heating/cooling step of 5 °C min^{-1} to determine the optimum temperature of the polymer-to-carbon conversion

(Fig. S1). The surface area, pore size, and pore volume were determined with a Quantachrome Autosorb analyzer (Quantachrome Corporation). Adsorption–desorption isotherms were recorded at 77 K using N_2 as the adsorbate. The adsorption data was analyzed using the Autosorb software version 1.16 based on the BET isotherm and the BJH method allowing estimating the pore size distribution and a cumulative pore volume. X-ray photoelectron spectroscopy (XPS) measurements were performed on Leybold Max 200 spectrometer using Mg KR radiation (1,253.6 eV) operated at 200 W (10 kV, 20 mA). All data were calibrated with respect to the C 1s excitation (284.6 eV) to account for charging effects by the electron beam. Prior to individual elemental scans, survey scans were acquired with a 1 eV step to estimate the surface (depth <10 nm) composition. Narrow scans for C 1s and N 1s were carried out with 0.2 eV step size, revealing information of the chemical state of each sample. The spectra were fitted and evaluated by XPSPEAK 4.1 software: a mixed Gaussian–Lorentzian function (G:L 85:15) was used for the peak shapes, while the Shirley function was applied to subtract the background. X-ray diffraction (XRD) was performed using Inel 120 CPS instrument with Cu-Kα radiation in the range of 2θ from 10° to 55°. Scanning electron microscopy (SEM) observations were performed using a LEO-1450 VP microscope at 8.0 kV.

Electrochemical experiments were carried out using a Princeton Applied Research VersaSTAT MC potentiostat and an electrochemical cell containing platinum counter electrode, Ag/AgCl (in saturated KCl solution) as a reference electrode and N-doped carbon ink casted onto glassy carbon as a working electrode. All potentials are quoted versus Ag/AgCl (+197 mV). All electrochemical measurements were carried out in 0.1 M KOH purged with N_2 or O_2 for 30 min prior to experiments. Before casting the glassy carbon disk of 0.5 cm diameter (PINE Instrument Company, USA) was mechanically polished with 0.05 μm Al_2O_3 slurry (Cypress Systems, Inc., USA), then rinsed in deionized water, cleaned ultrasonically in ultrapure water for 5 min, followed by rinsing in ultrapure isopropanol and acetone. A standard ink containing 5 mg of N-doped carbon, 1,080 μL ethanol and 180 μL 5 wt% Nafion was sonicated for 30 min, and a 10 μL of suspension was cast on glassy carbon disk, and dried in ambient conditions. Various concentrations of Nafion binder were used in order to investigate the mechanical stability of the N-doped carbon electrodes. Cyclic voltammetry (CV) in O_2- and N_2-saturated electrolyte was carried out in the potential range from −0.8 to 0.6 V versus Ag/AgCl, and a scan rate of 50 mV s^{-1}. The catalytic activity for oxygen reduction was investigated using linear scan voltammetry (LSV) applied to the rotating disk electrode in the potential range from 0.4 to −1.0 V with the scan rate of 10 mV s^{-1} and the rotation

speed from 0 to 2,500 rpm in O_2-saturated electrolyte. A long-term stability test was performed by scanning potential at 200 mV s^{-1} using electrolyte with an O_2 blanket (typically 2,000 scans without rotation of the electrode) using CV and LSV voltammograms for the comparison with relevant scans for the fresh electrode.

Results and discussion

Synthesis of porous N-doped carbon

In a typical synthesis procedure, 2.3 g oxidant $FeCl_3$ with the ratio of 2.2:1 (oxidant to monomer) was dissolved in 30 ml H_2O followed by the addition of 3.2 g of a 10 % silica aqueous suspension. The $FeCl_3$ and SiO_2 template mixture was sonicated for 10 min and then combined with 0.42 mL of a freshly distilled pyrrole under vigorous stirring. The reaction was carried out for 8 h until the color of the suspension changed from light yellow to the characteristic dark green associated with polypyrrole formation. The mixture was then filtered, washed with ethanol and water until the filtrate became colorless and dried at 80 °C for 12 h. The yield of the silica enclosed polypyrrole precursor was around 90 %. The chemical activation of polypyrrole prior to thermal treatment was performed by mixing (SiO_2-PPy) with 3 M KOH at a weight ratio of 1 to 3. After the solvent evaporation, the mixture was annealed under flow of argon at 675 °C for 2 h. SiO_2 was leached out in 10 M KOH solution under a reflux at 110 °C for 48 h. Finally, the product was pH neutralized with 1 M HCl, washed with distilled water and dried at 120 °C overnight.

Structure and chemical composition

For phase identification, an FTIR spectrum was compared with intermediate compound, SiO_2-reinforced polypyrrole, and the final product after the carbonization and cleaning (Fig. S2). The characteristic peak of polypyrrole at 751 cm^{-1} corresponds to the out of plane C–H and N–H deformations together with a C=C stretching vibration at 1,600 cm^{-1} and C–N stretching vibration at 1,340 cm^{-1} (Fig. S2, red line). A significant difference in FTIR identities between the precursor and the final product occurs within the region from 500 to 1,500 cm^{-1} wavenumber, where a symmetric and anti-symmetric ring-stretching vibrations of C=C bond at 1,540 and 1,454 cm^{-1}, 1,149 cm^{-1} vibration of pyrrole ring (disappear for N-doped carbon), 1,033 cm^{-1} in plane C–H and N–H deformations are present. Also the peak near 1,000 cm^{-1} attributes to C–N stretching is weaker for carbonized material [22].

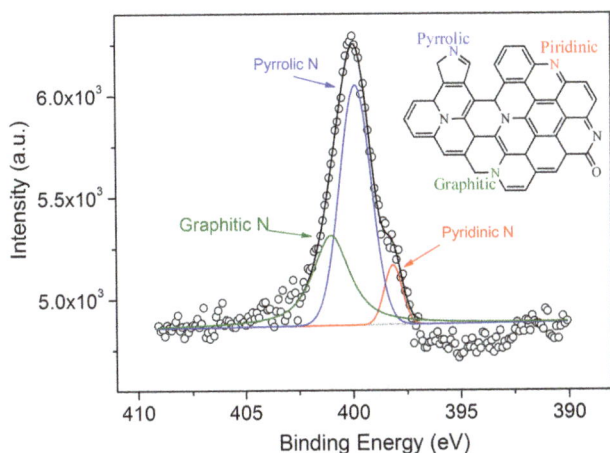

Fig. 1 XPS N1 s narrow scan of N-doped carbon

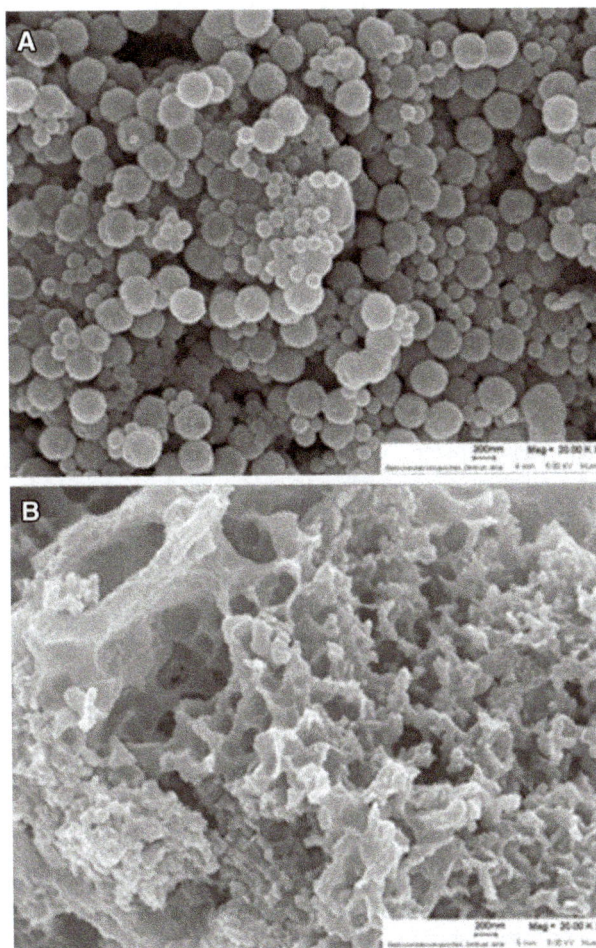

Fig. 2 SEM image of silica reinforced polypyrrole (**a**) and nitrogen-doped carbon after template removal and washing (**b**)

The nitrogen functional groups are usually in the following molecular structures (chemical states) in N-doped carbons (Fig. 1, insert): pyridinic N refers to nitrogen atoms at the edge of graphite planes, each of which is bonded to two carbon atoms and donates one π-electron to the aromatic π system; pyrrolic N refers to nitrogen atoms that are bonded to two carbon atoms and contribute to the π system with two π-electrons; quaternary nitrogen is also called ''graphitic nitrogen'' or ''substituted nitrogen'', in which nitrogen atoms are incorporated into the carbon network. The nitrogen state and content is of great significance with regard to the electrochemical activity, however, the influence of nitrogen chemical states on ORR is still yet to be revealed. Some groups claim that the pyridinic N plays the most important role in ORR activity [23, 24], while others consider graphitic nitrogen as the most active centers [25, 26], however it is generally acknowledged that both may equally contribute to the catalytic activity. For the N-doped carbon generated in this study, the total amount of surface nitrogen was 3.05 at.% with respect to C and O content, and is sufficient for the catalytic effect toward ORR [27]. The XPS narrow scan of the N 1s signal presented in Fig. 1 identifies three types of nitrogen states: graphitic N (398.1 eV), pyrrolic N (399.9 eV) and pyridinic N (401.0 eV) with corresponding fractions of 8.07, 56.81 and 35.12 %, respectively. The composition is governed by the structure of polypyrrole precursor that is mostly composed of pyrrolic N.

Morphology of N-doped carbon

The morphology of N-doped carbon is shown in Fig. 2. SEM micrographs of SiO_2-reinforced polypyrrole precursor (Fig. 2a) reveal a uniform, thin-shell polymer coating over oxide templates of various sizes, and the absence of silica-free polypyrrole phase. The following synthesis step ensures a regular porosity of the final product, composed of three-dimensional framework with micron-scale voids separated by thin carbon walls (Fig. 2b). The specific surface area and pore structure were measured by N_2 adsorption–desorption method at 77 K. The hysteresis loop (Fig. S3) corresponds to type-IV according to IUPAC classification, demonstrating mesoporous range. The multipoint BET surface is very large, 2,086 m^2 g^{-1}, compared to various, currently competing N-doped carbon materials (e.g., 2,370 m^2 g^{-1} [28]). Regarding the pore size, an average pore diameter calculated by BHJ method was 3 nm, the total pore volume for pores with radius <42 nm was 1.1 cm^3 g^{-1}, and for the pore size larger than 100 nm (based on mathematical model for meso- and macro-pores) the total volume was about 0.096 cm^3 g^{-1}. The original microstructure makes this carbon a competitive alternative for future energy storage and conversion systems.

Validation of electrocatalytic activity for oxygen reduction reaction

The cyclic voltammetry (CV) and rotating disk electrode (RDE) voltammetry were carried out to assess the electrocatalytic activity of N-doped carbon towards oxygen reduction. Figure S4a represents cyclic voltammograms recorded under nitrogen (black line) and in a O_2-saturated electrolyte (red line); from this comparison we identify the onset potential corresponding to the oxygen reduction; in our study it is −0.15 V. The large double-layer current is related to the capacitive behavior of the high surface area carbon materials, which makes N-doped carbon also a promising candidate for the capacitor electrodes. Furthermore, Fig. S4b shows linear sweep voltammograms of oxygen reduction process at a different rotation speed of the electrode. A well-pronounced kinetically controlled region up to −0.42 V is observed along with a mixed kinetic and mass transfer-controlled step, where current is a function on rotation rate, and in this case, at the potential more negative than −0.42 V. The corresponding Levich plots [current density as function of square root of rotation speed, $J = f(\omega^{-0.5})$; Fig. S4c] at different potentials are in next step used to calculate the number of electrons exchanged in the process (n), according to Eqs. (1) and (2) [29–31]:

$$J^{-1} = J_L^{-1} + J_K^{-1} = (B\omega^{1/2})^{-1} + J_K^{-1} \quad (1)$$

$$B = 0.62nFC_o(D_o)^{2/3}v^{-1/6} \quad (2)$$

where J is the measured current density, J_k and J_L are kinetic and diffusion-limiting current densities, ω is the angular speed of the rotating electrode, n is the overall number of electrons transferred in oxygen reduction process, F is the Faraday constant ($F = 96,485$ C mol^{-1}), C_0 is the volume concentration of O_2 ($C_0 = 1.2 \times 10^{-6}$ mol cm^{-3}), v is the kinematic viscosity of the electrolyte ($v = 0.01$ cm^2 s^{-1}), D_0 is diffusion coefficient of O_2 in 0.1 M KOH (1.9×10^{-5} cm^2 s^{-1}). The calculated number of electron exchange in ORR according to above models is 3.7, close to theoretically predicated for direct 4-e^- process. This also indicates some influence of an indirect (2-e^- path) [29–31]. Mixed four- and two-electron mechanisms are apparently influenced by the well-known electrochemical oxygen evolution via 2e^- exchange, while H_2O_2 is evolved as an intermediate product on pure carbon [29, 30] (in our sample the carbon is predominant phase with only 3.04 at.% of the surface N). The correlation of ORR mechanism with respect to various nitrogen states on N-doped carbon was discussed in detail by other groups [32, 33].

Electrochemical stability

The electrochemical stability of the N-doped carbon catalyst is still unclear with some groups claiming the N-doped C is stable thermally and chemically [34], electrochemically in both acidic [35] and alkaline medium [36],

and the loss of electrochemical performance under extensive potential cycling is only related to the delamination of the catalyst layer (mechanical instability) [37]. After careful observation, we concluded that most studies dealing with the electrochemical durability of N-doped carbons are carried out for the N-doped graphitized structures, such as N-CNs [17] or N-doped graphene [38], and correlate the superior stability with the presence of substantial amount of quaternary and graphitic N–C structures (Fig. 1 insert). The electrochemical stability for the polypyrrole-derived N-doped carbon was recently reported [21] with emphasis on the type of carbon precursor (notice: an amorphous carbon black and graphitized tubes were used as carbon precursor, not like in our case, where PPy was the only N and C source). This study revealed that the electrochemical instability of PPy-derived N-doped C is correlated with structural changes of amorphous phase—for comparison, the sample derived from the commercial graphitized structures remains stable. Thus, our goal was to correlate the effect of polymer precursor adapted for the synthesis of N-doped carbon with a long-term durability over a commercially pertinent potential window. The electrochemical degradation was conducted in O_2-saturated 0.1 M KOH by scanning potential (at least 2,000 CV cycles) and

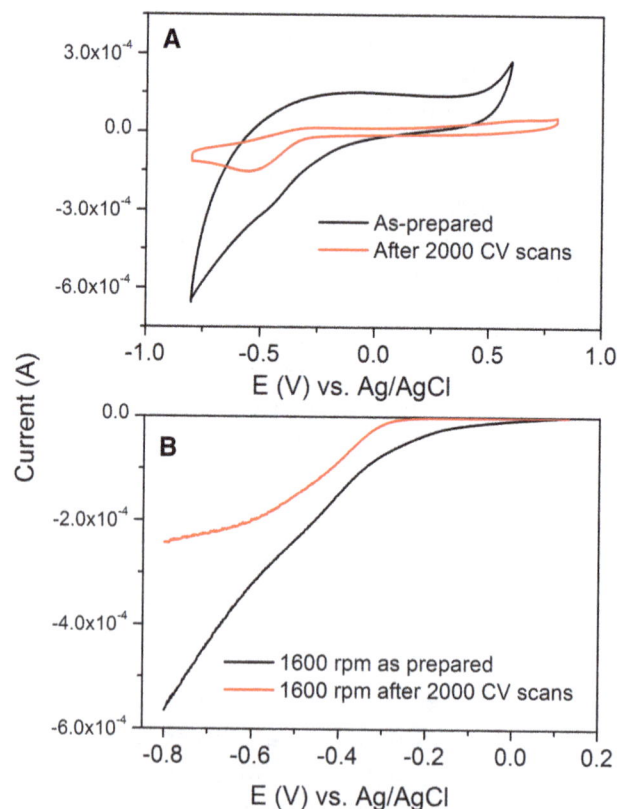

Fig. 3 CV (a) and LSV at 1,600 rpm (b) of fresh-prepared (black) and after degradation (red) in O_2-saturated 0.1 M KOH on glassy carbon disk electrode

comparing CV and LSV of the ORR process (Fig. 3) for the freshly prepared electrode and after accelerated durability test. Figure 3 demonstrates that the sample synthesized in this work is not stable under cathode conditions. The electrochemical features, including redox activity and double-layer capacitance of porous carbon are much weaker after degradation test, regardless the amount of Nafion binder used for the electrode stability (Fig. 4a, b correspond to the catalyst layer with doubled Nafion content performed with the goal of making mechanically stronger electrode). From the LSV (Fig. 3b), the current at −0.45 V for the fresh electrode drops for about 38 % after 2,000 cycles, which is significant loss when compared to previous reports [36]. Since there is a lack of standardized protocol of the durability test for the alkaline ORR catalyst as well as a lack of reliable information with regards to the stability of polypyrrole-derived N-doped carbon we can conclude that the rapid electrochemical degradation of our materials originates from the low content of graphitized carbon and from the minor amount of graphitic N-C structures. With respect to the degradation test experimental conditions, we observe that the sample was unstable regardless the speed of potential scanning (data not shown) and the choice of the potential range in our

work was the most relevant to operating conductions of the ORR cathode in alkali. The N-doped C layer casted on a porous Ni mesh also revealed significant instability (Fig. 4c, d), which indicates that the roughness of the electrode substrate and a potential risk of the catalyst layer delamination is not an issue here (as suggested by some groups [37]). With respect to the degradation test of N-doped carbon on Ni, we did not account the possible Ni activity toward ORR since there is no evidence of ORR catalysis on metallic Ni (please note we did not observe any resposne from Ni and the Ni substrate was fully covered by N-doped C and Nafion binder—that minimized the contact of Ni metal with electrolyte). The test of electrochemical durability on Ni was carried out in order to exclude the effect of electrode substrate roughness on the detachment (physical destruction) of the N-doped C layer.

Effect of carbon graphitization on electrochemical stability

XRD of different specimens was recorded in order to compare the crystalline structure of the N-doped carbon [commercial control materials: Vulcan XC-72R carbon black and single walled carbon nanotubes (SWCNs) are

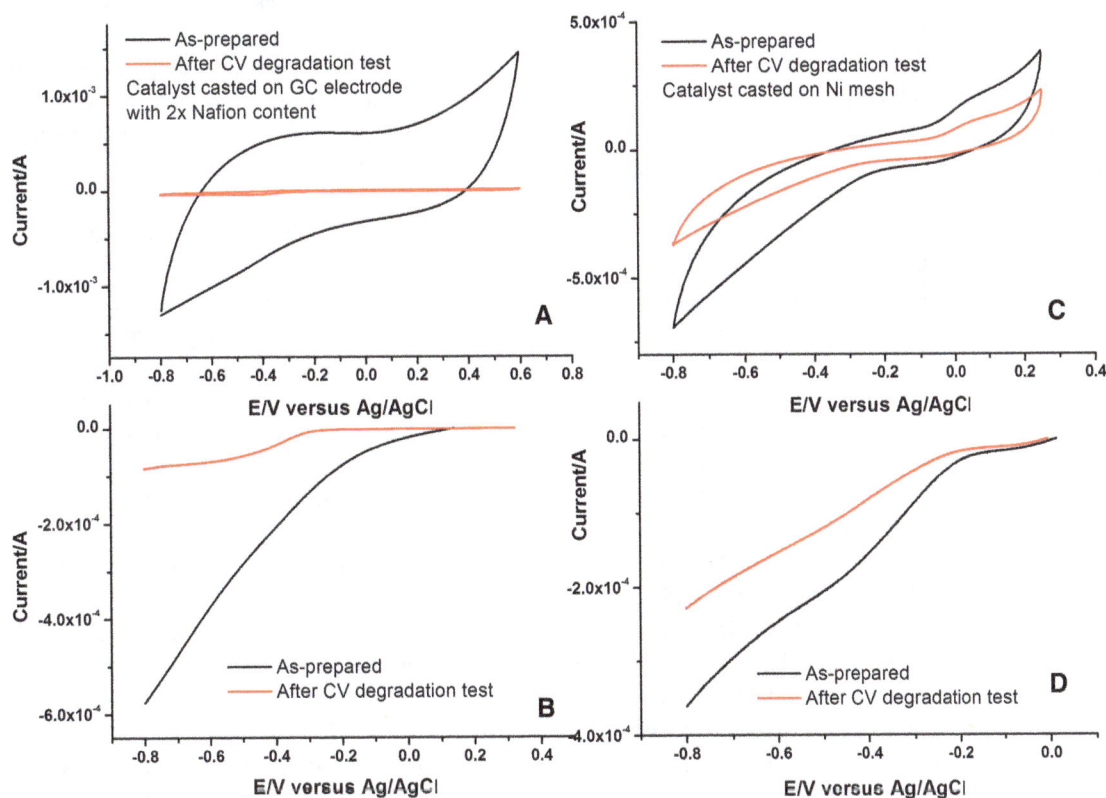

Fig. 4 CV (a) and LSV at 1,600 rpm (b) of fresh-prepared (black) and after degradation (red) in O_2-saturated 0.1 M KOH on glassy carbon disk electrode for the catalyst layer with doubled amount of Nafion binder and CV (c) and LSV (d) on catalyst casted on rough Ni mesh with standard Nafion content

used as references of amorphous and crystalline structures]. This structural investigation reveals the level of crystallization, since there is a potential risk that the amorphous carbon (that is less resistant to the corrosion) collapses during the electrochemical cycling [39]. The diffraction profiles of all carbons (Fig. S5) show the presence of an asymmetric band around ∼25.5° also designated to the (002) plane and is associated with the spacing of aromatic ring layer of the graphite-like structures (crystalline carbon), and a weaker at ∼42.3° of the (101) plane also denoted as a "10 peak" corresponding to the two-dimensional orientation of graphene planes [40, 42]). The major difference is observed between the symmetry of (002) and the intensity of "10 peak" for the N-doped C and a commercial carbon black (Vulcan XC-72R) when compared to SWCNs, indicating that first two samples have intermediate structures between graphite and amorphous state called turbostratic structure or random layer lattice structure. The "10 peak" at 43.5° of the SWCNs is stronger in comparison to other carbons, signifying a higher degree of crystallinity. Also, in the case of SWCNs asymmetric (002) band at ∼25.5° showed an additional peak shifted toward lower angle (∼20°) called γ band and is attributed to the presence of saturated structures such aliphatic side chains that are attached to the edge of crystallites [41]. This γ shoulder reflects the packing distance of saturated structures (however is not a direct indication on higher crystalline phase of the SWCNs). XRD analysis is further correlated with the XPS studies (Fig. 5) of the degraded N-doped carbon and the proposed mechanism of degradation of C–C and N–C bonds.

The C 1s core level peak (Fig. 5a) can be resolved into four components with maxima at 284.9, 286.2, 288.4 and 289.7 eV, representing sp^2C–sp^2C graphite bonds, N-sp^2C (N substitution in graphite-like configuration), C–O and C=O functionalities and N-sp^3C (pyrrolic), respectively [43–45]. Also, a small peak at 292.5 eV of the C-F_2 bonding is localized and associated with Nafion binder [46] and the intensity of C–F peaks for the fresh and degraded samples depends on the effectiveness of the washing step and do not affect the C–C, C–O and the C–N signals. Furthermore, the C 1s narrow scan of the fresh and degraded sample revealed strikingly different distribution of carbon chemical states (Fig. 5c). The graphitic carbon disappeared completely for both C–C and N–C contributions, the signal of C–O, C=O and indicating opening of aromatic ring [47] increased dramatically and the presence of pyrrolic N-sp^3C bonds is no longer obvious. This clearly indicates on irreversible degradation of the whole carbon network, including breaking of the strongest graphitic C–C and C–N bonds. This is further complemented by the analysis of N 1s core level spectrum (Fig. 5b, d). The narrow scan of N 1s of the fresh N-doped carbon (Fig. 5b)

containing all N–C components (graphitic, pyrrolic, pyridinic, and quaternary—as discussed in detail in conjunction with Fig. 1), changed significantly after the degradation (Fig. 5d). All signals assigned to the N–C functionalities of N-doped C disappeared, similar to the C–N bonds in Fig. 5a. Taken together, this clearly demonstrates that the chemical structure of material changed during the accelerated degradation test. The signal assigned to the open form of N–C pyrrolic ring at 401.7 eV [47], and new signals at 402.7 eV corresponding to –C=N–OH, at 403.8 eV assigned to –C–N=O group can be identified [43, 48, 49]. The analysis of both C and N 1s core signals before and after degradation confirmed irreversible changes in the carbon structure.

Effect of chemical composition of N-doped carbon: role of polypyrrole precursor: proposed mechanisms for the degradation of N-doped carbon

Scheme 1 represents possible reaction paths of the polypyrrole-derived N-doped carbon. The steps 1–5 correspond to the ORR process on N–C active centers that proceeds through reversible ring opening-cleavage route [7], and secondary degradation mechanism following steps 6–16. Regardless the electrochemical oxidation of carbon due to the massive amount of amorphous phase, especially for the most outer carbon planes (Fig. 5c; C–O functionalities are predominant structures), more complex degradation mechanism related to the N-graphitic active sites should be taken into account. In the first step (1) the appropriate double bonds rearrangements are the driving force for the primary attack on the triplet oxygen atom. The high activation energy of kinetically unfavorable structure (3) is overcome at higher operating temperature, which lowers the barrier of Gibb's free energy required for this reaction to proceed. The electrochemically driven homolytic cleavage of the peroxy bond (3), which spontaneously forms an aldehyde group (5) can be considered. Furthermore, an unpaired electron in the pyridinic ring together with the hydroxyl radical anion will form a radical 6-exo-trig ring closure (classified according to Baldwin rules), generating the molecular pattern (6–7). Under an oxidizing potential, the free electron pair of nitrogen attacks an adjacent carbon atom, leading to the peroxy group (7 and 10). The unfavorable molecule (11) can further react with hydroxyl radical anion generated in the reaction mixture through the electrochemically driven homolytic cleavage and create structure (12). The latest may react with hydrogen peroxide (by-product of 2-electron ORR process [29–31]) and form N-oxide (13–16). Furthermore, we speculate that the structure (5) instead of recovering to step (4), is modified under attack of a strong oxidizing agent, H_2O_2 [29–31], generating products (8–9) in Scheme 1. The

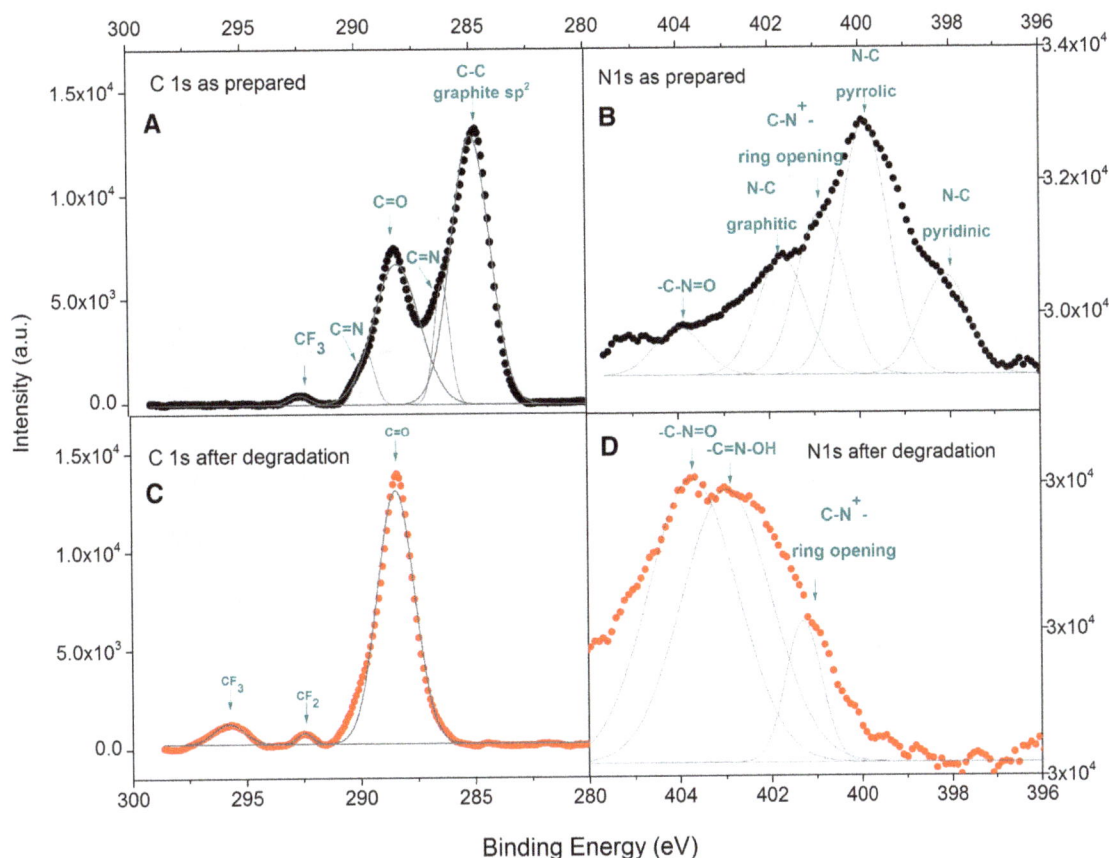

Fig. 5 XPS narrow scans of the C 1s (**a, c**) and N 1s (**b, d**) signals before (*black*) and after electrochemical durability test (*red*)

structure (9) we consider as a very active for oxygen reduction based on similarity the hydroquinone structure, thus may react in the same pattern as anthraquinone in the industrial production of hydrogen peroxide [50]. Under reducing conditions the step 5 in Scheme 1 will recover to the initial graphitic state of 1, which is a well-known path for the ORR on N–C graphitic sites [7]. There is also possibility of a side reaction (another degradation path) called Dakin reaction, which takes place on the aldehyde's formyl group (Scheme 2). The attack of the free electron pair of hydrogen peroxide will be favored under basic conditions, which forms a bicyclic molecule, pyrido[1,2-b] [1, 2] oxazine (Scheme 2, B4). Since XPS analysis identified nitroso and oxime functionalities in the degraded sample, we speculate that they are generated via steps C1–C3 in Scheme 3 (C1 is formed by the reaction of B1 or 5 with hydrogen peroxide). Meanwhile the –C=N–OH and –C–N=O groups produced during degradation of the pyrrolic nitrogen (predominant in our sample) might have also significant contribution (degradation mechanism not included). In summary, we observe that the degradation of N-doped carbon is related to an excess of pyrrolic and pyridinic nitrogen in our product. The oxidation of pyridinic nitrogen is correlated with the weakening or breaking

C–C bonds presumably from the less crystalline phase. Based on our observations, we propose that the electrochemically stable N-doped carbons should possess quaternary (or graphitic) nitrogen as the main structure in crystalline carbon network. In addition, the polymer precursor should deliver stable, preferably aromatic carbon structures, for this reason other precursors (e.g., polyaniline), should generate more electrochemically stable N-doped carbons, and will be the subject of our future study.

Conclusion

In this work we validate the electrochemical stability of polypyrrole-derived N-doped carbon, by generating a model compound that is comparable to the current state-of-the-art doped carbon with porous structure and a nitrogen doping level of 3–4 at.%, consisting of 56 % pyrrolic N, 36.1 % pyridinic N and only ∼8 % graphitic N centers. Although, the electrochemical studies show that N-doped carbon can catalyze ORR in alkaline electrolyte via a favorable four-electron pathway, the long-term electrochemical stability of this material is very poor comparing

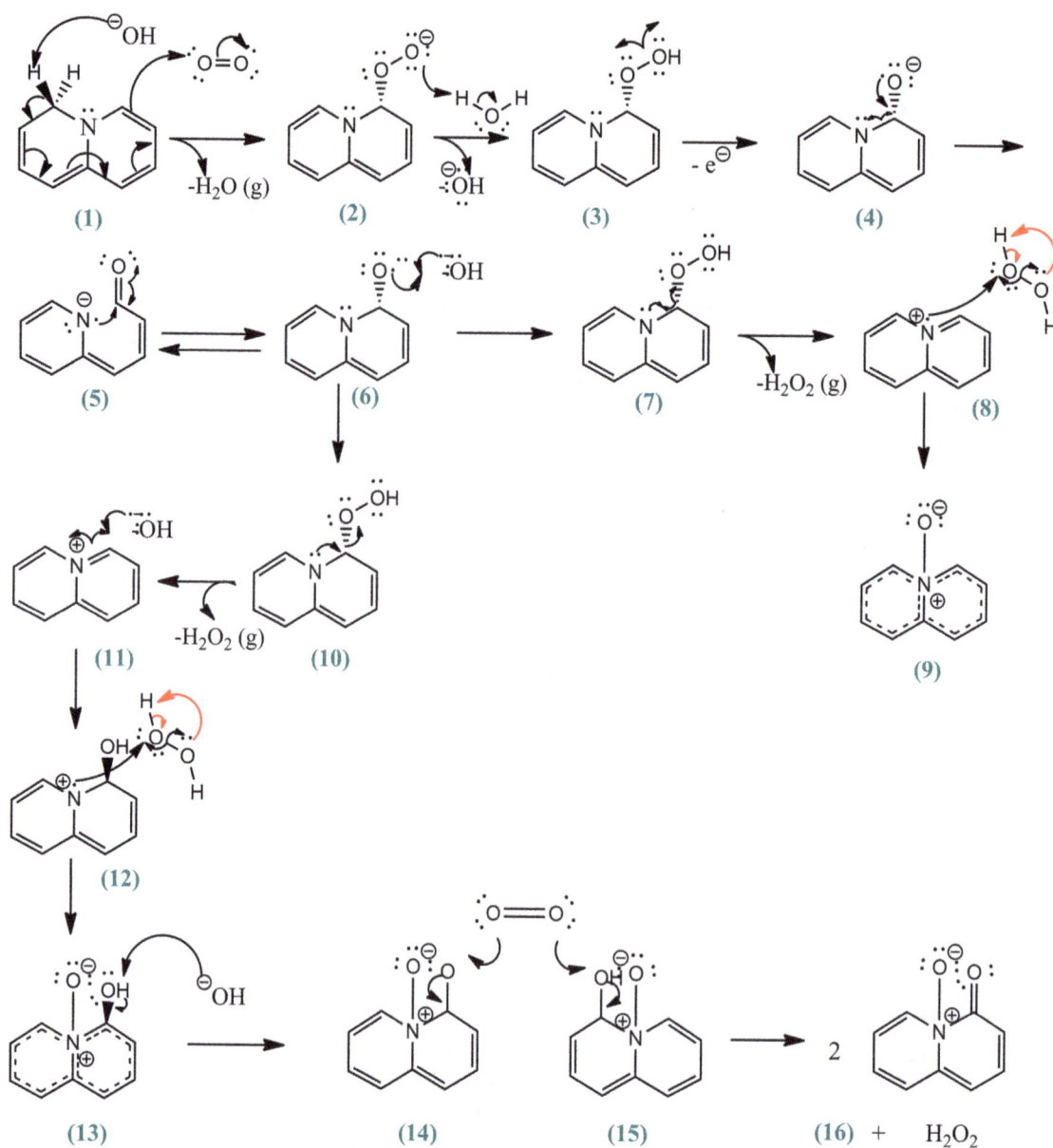

Scheme 1 Oxygen evaluation on N–C active center during ORR through ring opening-to-cleavage (steps 1–5) and irreversible degradation mechanism of N-doped carbon (steps 6–16)

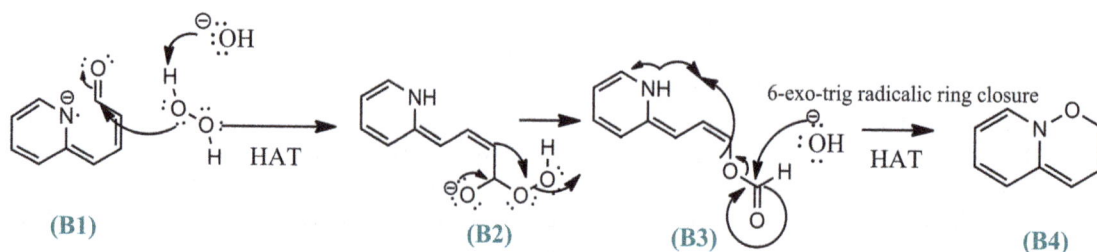

Scheme 2 Dakin–West mechanism with hydrogen atom transfer (HAT) from the solvent and hydrogen peroxide

to N-doped CNs or N-doped graphene. There are two reasons for the electrochemical instability, first is a very low content of graphitized carbon in comparison to commercial carbon black or SWCNs. Second, is a low content of graphitic nitrogen (only 8 at.%), which is the most electronically and chemically stable structure in this

Scheme 3 The formation of nitroso and oxime groups during electrochemical degradation of N-doped C

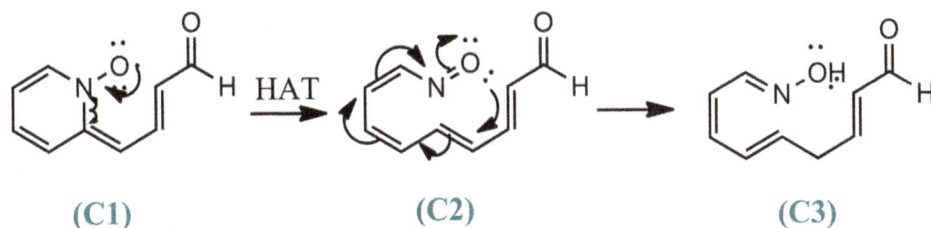

(C1)   (C2)   (C3)

type of materials. Furthermore, the XPS analysis of degraded samples revealed that both C–C and N–C bonds are broken and massive amount of oxygen containing functionalities are generated on carbon and nitrogen centers after long-term electrochemical durability test, based on the surface analysis we proposed several degradation paths. This irreversible oxidation is correlated with the presence of significant content or pyrrolic and pyridinic N close to the edge of the carbon network. These are generated by the polypyrrole precursor. Although N as n-type dopant stabilizes aromatic rings, giving high electrochemical stability for the graphitic and quaternary nitrogen centers, these are minor structures in our material, consequently, the polypyrrole-derived carbon is less stable under oxidizing potential. As a solution, a careful selection of the polymer precursor has to be carried out in order to create highly graphitic carbon with nitrogen in quaternary (or graphitic) states. These studies critically revised the influence of polymer precursor on the electrochemical stability of the resulting N-doped carbon materials, closing the knowledge gap with respect to their long-term durability under oxidizing potential.

Acknowledgments We thank you Carl-Zeiss Foundation for financial support.

References

1. Wang, H., Maiyalagan, T., Wang, X.: Review on recent progress in nitrogen-doped graphene: synthesis, characterization, and its potential applications. ACS Catal. **2**(5), 781–794 (2012)
2. Han, J., Xu, G., Ding, B., Pan, J., Dou, H., MacFarlane, D.R.: Porous nitrogen-doped hollow carbon spheres derived from polyaniline for high performance supercapacitors. J. Mater. Chem. A **2**, 5352–5357 (2014)
3. Li, X.H., Antonietti, M.: Metal nanoparticles at mesoporous N-doped carbons and carbon nitrides: functional Mott–Schottky heterojunctions for catalysis. Chem. Soc. Rev. **42**, 6593–6604 (2013)
4. Zhu, P., Song, J., Lu, D., Wang, D., Jaye, C., Fischer, D.A., Wu, T., Chen, Y.: Mechanism of enhanced carbon cathode performance by nitrogen doping in lithium–sulphur battery: an X-ray absorption spectroscopic study. J. Phys. Chem. C **118**, 7765–7771 (2014)
5. Sevilla, M., Yu, L., Fellinger, T.P., Fuertesa, A.B., Titiricic, M.M.: Polypyrrole-derived mesoporous nitrogen-doped carbons with intrinsic catalytic activity in the oxygen reduction reaction. RSC Adv. **3**, 9904–9910 (2013)
6. Gong, K., Du, F., Xia, Z., Durstock, M., Dai, L.: Nitrogen-doped carbon nanotube arrays with high electrocatalytic activity for oxygen reduction. Science **323**, 760–764 (2009)
7. Kim, H., Lee, K., Woo, S.I., Jung, Y.: On the mechanism of enhanced oxygen reduction reaction in nitrogen-doped graphene nanoribbons. Phys. Chem. Chem. Phys. **13**, 17505–17510 (2011)
8. Liu, S., Tian, J., Wang, L., Zhang, Y., Luo, Y., Asiri, A.M., Al-Youbi, A.O., Sun, X.: A novel acid-driven, microwave-assisted, one-pot strategy toward rapid production of graphitic N-doped carbon nanoparticles-decorated carbon flakes from *N, N*-dimethylformamide and their application in removal of dye from water. RSC Adv. **2**, 4632–4635 (2012)
9. Geng, D., Chen, Y., Chen, Y., Li, Y., Li, R., Sun, X., Ye, S., Knights, S.: High oxygen-reduction activity and durability of nitrogen-doped graphene. Energy Environ. Sci. **4**, 760–764 (2011)
10. Zhang, Y., Fugane, K., Mori, T., Niu, L., Ye, J.: Wet chemical synthesis of nitrogen-doped graphene towards oxygen reduction electrocatalysts without high-temperature pyrolysis. J. Mater. Chem. **22**, 6575–6580 (2012)
11. Wang, Y., Shao, Y., Matson, D.W., Li, J., Lin, Y.: Nitrogen-doped graphene and its application in electrochemical biosensing. ACS Nano. **4**, 1790–1798 (2010)
12. Liu, Z., Zhang, X., Poyraz, S., Surwade, S.P., Manohar, S.K.: Oxidative template for conducting polymer nanoclips. J. Am. Chem. Soc. **132**, 13158–13159 (2010)
13. Cao, C., Zhuang, X., Su, Y., Zhang, Y., Zhang, F., Wu, D., Feng, X.: 2D Polyacrylonitrile brush derived nitrogen-doped carbon nanosheets for high-performance electrocatalysts in oxygen reduction reaction. Polym. Chem. **5**, 2057–2064 (2014)
14. Geng, D., Liu, H., Chena, Y., Li, R., Sun, X., Ye, S., Knights, S.: Non-noble metal oxygen reduction electrocatalysts based on carbon nanotubes with controlled nitrogen contents. J. Power Sour. **196**, 1795–1801 (2011)
15. Liu, D., Zhang, X., Sun, Z., You, T.: Free-standing nitrogen-doped carbon nanofiber films as highly efficient electrocatalysts for oxygen reduction. Nanoscale **5**, 9528–9531 (2013)
16. Liu, J., Sasaki, K., Lyth, S.M.: Electrochemical oxygen reduction on metal-free nitrogen-doped graphene foam in acidic media. ECS Trans. **58**(1), 1529–1540 (2013)
17. Li, H., Liu, H., Jong, Z., Qu, W., Geng, D., Sun, X., Wang, H.: Nitrogen-doped carbon nanotubes with high activity for oxygen reduction in alkaline media. Int. J. Hydrogen Energy **36**, 2258–2265 (2011)
18. Qu, L., Liu, Y., Baek, J.B., Dai, L.: Nitrogen-doped graphene as efficient metal-free electrocatalyst for oxygen reduction in fuel cells. ACS Nano. **4**, 1321–1326 (2010)
19. Stevens, D.A., Hicks, M.T., Haugen, G.M., Dahn, J.R.: Ex situ and in situ stability studies of PEMFC catalysts: effect of carbon

type and humidification on degradation of the carbon. J. Electrochem. Soc. **152**, A2309–A2315 (2005)

20. Li, Y.S., Zhao, T.S., Liang, Z.X.: Effect of polymer binders in anode catalyst layer on performance of alkaline direct ethanol fuel cells. J. Power Sour. **190**, 223–229 (2009)

21. Zhao, A., Masa, J., Muhler, M., Schuhmann, W., Xia, W.: N-doped carbon synthesized from N-containing polymers as metal-free catalysts for the oxygen reduction under alkaline conditions. Electrochim. Acta **98**, 139–145 (2013)

22. Chougulea, M.A., Pawara, S.G., Godse, P.R., Mulik, R.N., Sen, S., Patila, V.B.: Synthesis and characterization of polypyrrole (PPy) thin films. Soft Nanosci. Lett. **1**, 6–10 (2011)

23. Lee, K.R., Lee, K.U., Lee, J.W., Ahn, B.T., Woo, S.I.: Electrochemical oxygen reduction on nitrogen doped graphene sheets in acid media. Electrochem. Commun. **12**, 1052–1055 (2010)

24. Subramanian, N.P., Li, X., Nallathambi, V., Kumaraguru, S.P., Colon-Mercado, H., Wu, G., Lee, J.W., Popov, B.N.: Nitrogen-modified carbon-based catalysts for oxygen reduction reaction in polymer electrolyte membrane fuel cells. J. Power Sour. **188**, 38–44 (2009)

25. Luo, Z., Lim, S., Tian, Z., Shang, J., Lai, L., MacDonald, B., Fu, C., Shen, Z., Yu, T., Lin, J.: Pyridinic N doped graphene: synthesis, electronic structure, and electrocatalytic property. J. Mater. Chem. **21**, 8038–8044 (2011)

26. Niwa, H., Horiba, K., Harada, Y., Oshima, M., Ikeda, T., Terakura, K., Ozaki, J., Miyata, S.: X-ray absorption analysis of nitrogen contribution to oxygen reduction reaction in carbon alloy cathode catalysts for polymer electrolyte fuel cells. J. Power Sources **187**, 93–97 (2009)

27. Brun, N., Wohlgemuth, S.A., Osiceanub, P., Titiricic, M.M.: Original design of nitrogen-doped carbon aerogels from sustainable precursors: application as metal-free oxygen reduction catalysts. Green Chem. **15**, 2514–2524 (2013)

28. Liu, H., Cao, Y., Wang, F., Huang, Y.: Nitrogen-doped hierarchical lamellar porous carbon synthesized from the fish scale as support material for platinum nanoparticle electrocatalyst toward the oxygen reduction reaction. ACS Appl. Mater. Interfaces **6**(2), 819–825 (2014)

29. Jin, Z., Nie, H., Yang, Z., Zhang, J., Liu, Z., Xu, X., Huang, S.: Metal-free selenium doped carbon nanotube/graphene networks as a synergistically improved cathode catalyst for oxygen reduction reaction. Nanoscale **4**, 6455–6460 (2012)

30. Zhang J (ed.) PEM Fuel Cell Electrocatalysts and Catalyst Layers: Fundamentals and Applications. Springer, New York, pp. 102 (2008)

31. Chen, Z., Higgins, D., Chen, Z.: Nitrogen doped carbon nanotubes and their impact on the oxygen reduction reaction in fuel cells. Carbon **48**, 3057–3065 (2010)

32. Sidik, R.A., Anderson, A.B., Subramanian, N.P., Kumaraguru, S.P., Popov, B.N.: O_2 reduction on graphite and nitrogen-doped graphite: experiment and theory. J. Phys. Chem. B **110**, 1787–1793 (2006)

33. Shao, Y., Zhang, S., Engelhard, M.H., Li, G., Shao, G., Wang, Y., Liu, J., Aksay, I.A., Lin, Y.: Nitrogen-doped graphene and its electrochemical applications. J. Mater. Chem. **20**, 7491–7496 (2010)

34. Liu, H., Zhang, Y., Li, R., Sun, X., Abou-Rachid, H.: Thermal and chemical durability of nitrogen-doped carbon nanotubes. J. Nanopart. Res. **14**, 1016 (2–8) (2012)

35. Gang, W., More, K.L., Johnston, C.M., Zelenay, P.: High-performance electrocatalysts for oxygen reduction derived from polyaniline, iron, and cobalt. Science **332**, 443–447 (2011)

36. Feng, L., Yang, L., Huang, Z., Luo, J., Li, M., Wang, D.: Enhancing electrocatalytic oxygen reduction on nitrogen-doped graphene by active sites implantation. Sci. Rep. **3-3306**, 1–8 (2013)

37. Zhao, Y., Nakamura, R., Kamiya, K., Nakanishi, S., Hashimoto, K.: Nitrogen-doped carbon nanomaterials as non-metal electrocatalysts for water oxidation. Nat. Commun. **4**, 1–4 (2013)

38. Peng, H., Mo, Z., Liao, S., Liang, H., Yang, L., Luo, F., Song, H., Zhong, Y., Zhang, B.: High performance Fe- and N- doped carbon catalyst with graphene structure for oxygen reduction. Sci. Rep. **3–1765**, 1–7 (2013)

39. Ignaszak, A., Song, C., Zhu, W., Wang, Y., Zhang, J., Bauer, A., Baker, R., Neburchilov, V., Ye, S., Campbell, S.: Carbon–$Nb_{0.07}Ti_{0.93}O_2$ composite supported Pt–Pd electrocatalysts for PEM fuel cell oxygen reduction reaction. Electrochim. Acta **75**, 220–228 (2012)

40. Saikia, B.K., Boruah, R.K., Gogoi, P.K.: A X-ray diffraction analysis on graphene layers of Assam coal. J. Chem. Sci. **121**, 103–106 (2009)

41. Manoj, B., Kunjomana, A.G.: Study of stacking structure of amorphous carbon by X-ray diffraction technique. Int. J. Electrochem. Sci. **7**, 3127–3134 (2012)

42. Hussain, R., Quadeer, R., Ahmad, M., Salem, M.: X-ray diffraction study of heat-treated graphitized and ungraphitized carbon. Turk. J. Chem. **24**, 177–183 (2000)

43. Lim, S.H., Elim, H.I., Gao, X.Y., Wee, A., Ji, W., Lee, J.Y., Lin, J.: Electronic and optical properties of nitrogen-doped multi-walled carbon nanotubes. Phys. Rev. B **73**, 045402-1 (2006)

44. Chao, S., Lu, Z., Bai, Z., Cui, Q., Qiao, J., Yang, Z., Yang, L.: Tuning synthesis of highly active nitrogen-doped graphite and determining the optimal structure from first-principles calculations. Int. J. Electrochem. Sci. **8**, 8786–8799 (2013)

45. Nolan, H., Mendoza-Sanchez, B., Kumar, N.A., McEvoy, N., O'Brien, S., Nicolosi, V., Duesberg, G.S.: Nitrogen-doped reduced graphene oxide electrodes for electrochemical supercapacitors. Phys. Chem. Chem. Phys. **16**, 2280–2284 (2014)

46. XPS database: http://xpssimplified.com

47. Allouche, J., Le Beulze, A., Dupin, J.C., Ledeuil, J.B., Blanc, S., Gonbeau, D.: Hybrid spiropyran-silica nanoparticles with a core-shell structure: sol–gel synthesis and photochromic properties. J. Mater. Chem. **20**, 9370–9378 (2010)

48. Wilken, R., Holländer, A., Behnisch, J.: Surface radical analysis on plasma-treated polymers. Surf. Coat. Technol. **116–119**, 991–995 (1999)

49. Wilken, R., Holländer, A., Behnisch, J.: Nitric oxide radical trapping analysis on vacuum-ultraviolet treated polymers. Macromolecules **31**, 7613–7617 (1998)

50. Goor, G., Glenneberg, J., Jacobi, S.: "Hydrogen Peroxide". Ullmann's Encyclopedia of Industrial Chemistry. Wiley, Weinheim (2007)

16

Microencapsulation of a PCM through membrane emulsification and nanocompression-based determination of microcapsule strength

Asif Rahman · Michelle E. Dickinson ·
Mohammed M. Farid

Abstract Microencapsulating a phase-change material (PCM) has become a prominent method of creating a stable environment in which the PCM can undergo its phase change without affecting the environment in which it is used. The method of encapsulation used in this study takes advantage of a new technology known as membrane emulsification and suspension polymerization. This study investigates the encapsulation of the paraffin wax RT21® in a poly(methyl methacrylate) shell, which could be used to increase the thermal mass of a building. The objectives of the study are: (1) to encapsulate RT21® through the use of membrane emulsification and (2) to test the mechanical properties of the microcapsules under nanocompression. Membrane emulsification was carried out using Shirasu porous glass hydrophilic membranes of pore sizes 10, 10.2, and 20 μm. Polymerization was conducted in a batch reactor with methyl methacrylate as the monomer in the temperature range 70–90 °C. The thermal properties (the latent heat of melting and melting temperature) of the microcapsules were tested using a differential scanning calorimeter. Particle size analysis was conducted to determine the average size distribution of the microcapsules produced. Membranes with pore sizes of 10, 10.2, and 20 μm produced microcapsules with average diameters of 22.40 ± 1.47, 25.38 ± 0.80, and 37.50 ± 1.69 μm, and average latent heats of 113.91 ± 12, 116.69 ± 1.40, and 109.89 ± 8.69 J/g, respectively. In order to determine the mechanical properties of these microcapsules, a modified nanoindentation compression technique was used to test the bursting force for individual microcapsules.

Keywords Microcapsules · Phase-change materials (PCM) · Microencapsulation of PCM (MPCM) · Thermal energy storage · Nanoindentation

Present Address:
A. Rahman (✉)
Department of Biological Engineering, Utah State University,
4105 Old Main Hill, Logan, UT 84322, USA
e-mail: arah045@aucklanduni.ac.nz;
asif.rahman@aggiemail.usu.edu

A. Rahman · M. E. Dickinson · M. M. Farid
Department of Chemical and Materials Engineering, University
of Auckland, Private Bag 92019, Auckland, New Zealand
e-mail: m.dickinson@auckland.ac.nz

M. M. Farid
e-mail: m.farid@auckland.ac.nz

Introduction

Increasing the thermal inertia of buildings through the use of phase-change materials (PCMs) can be an effective method of reducing energy consumption [1–3]. An extensive review of PCMs can be found in [1, 4, 5].

The PCM RT21® has a melting temperature range of 20–22 °C, which is close to the human comfort temperature [6]. Due to the nature of PCMs, the bulk usage of PCM in buildings can become problematic. Microencapsulated PCMs can be advantageous for the thermal insulation of buildings [7].

The microencapsulation of PCMs (MPCM) is a technique that has received considerable attention recently. Microcapsules comprise of an internal core material and an external encapsulating shell material. MPCM allows the core material to undergo phase (and volume) changes without affecting the bulk structure or integrity of a building [7]. Comprehensive reviews of the microencapsulation of PCMs can be found in Tyagi et al. [7] and Zhao et al. [8].

There are two major ways in which microcapsule shells can be formed: via an out-inside process or an inside-out process. Examples of each method are provided in [9] and [10].

Membrane emulsification can be used as an alternative to traditional emulsification to create microcapsules. While traditional methods of emulsification via mechanical means are sometimes faster than membrane emulsification, a broad range of particle sizes are produced. Furthermore, as mentioned by Charcosset et al., the process of membrane emulsification utilizes less energy than traditional mechanical methods. The principle of membrane emulsification, much like traditional emulsification, is that a dispersed phase is created in a continuous phase. A key characteristic of membrane emulsification is that the droplet size is controlled primarily by the size of the pores of the membrane itself. This allows specific size droplets to be generated [11, 12]. Membrane emulsification was used in this study to create a colloidal mixture using Shirasu porous glass membranes (SPG) developed by Nakashima et al. [13].

Previous work by Omi et al. [14] proposed the production of poly(methyl methacrylate) microcapsules (PMMA) through the use of SPG membrane emulsification.

Elucidating the mechanical properties of microcapsules provides crucial information about the extent to which microcapsules can be implemented for different applications. MPCM are commonly mixed with with building materials, and are vulnerable to rupture during mixing or compaction. One requirement of the microcapsules is to maintain structural integrity in order to prevent the PCM from leaking out, which would drastically affect the efficiency of the PCM and lead to potential interactions of the PCM with other materials in the surrounding environment [1].

While some initial studies have been performed to determine the mechanical properties of microcapsules [15–17] and other materials of a similar size [18], a standard method of testing microcapsules is yet to be fully developed, although Rahman et al. [19] proposed a new standard method of measuring the mechanical properties of individual commercially available Micronal® DS5008 microcapsules.

The aim of this study was to encapsulate RT21® (core) in poly(methyl methacrylate) (PMMA, shell) by suspension polymerization, using hydrophilic SPG membranes with pore sizes of 10, 10.2, and 20 μm in the emulsification step. While membrane emulsification has been used in prior studies, to the author's knowledge, the encapsulation of RT21® in a PMMA shell via membrane emulsification has not been mentioned elsewhere. This study also builds upon studies of the rupture of individual microcapsules using a nanomechanical setup. It is surprising that even commercially produced PCM microcapsules have not been widely tested for mechanical strength. An additional study of the mass loss of microcapsules under heating is also discussed in this paper.

Materials and methods

Two separate solutions were prepared: the aqueous phase and the organic phase. Emulsification was carried out first, followed by polymerization, washing steps, and drying. The chemicals used in this study (unless otherwise stated) were obtained from Sigma–Aldrich (St. Louis, MO, USA).

Aqueous phase

Polyvinylalcohol (PVA; 5.70 g) and sodium nitrite (polymerization inhibitor; 0.21 g) were dissolved in 150 ml of distilled water.

Organic phase

The organic phase consisted of 7.20 g methyl methacrylate (MMA), 2.87 g ethylene glycol dimethacrylate (EDMA, crosslinker), 0.19 g benzoyl peroxide (BPO, initiator), and 24.95 g of RT21® (PCM, Rubitherm GmbH, Berlin, Germany). BPO was dissolved in EDMA first, and then MMA was added, followed by the RT21®.

Emulsification

The SPG hydrophilic membranes are tubular, with the continuous phase passing through the interior of the tube and the dispersed phase entering the interior of the tube through the membrane pores from the shell side. This study used membranes with pore sizes of 10, 10.2, and 20 μm (part numbers: SPG: PJN03D25, PJN08C03, and PJN03K20, respectively), which were pre-soaked in the aqueous phase (continuous phase) for 10 min prior to emulsification [12].

Due to the complex nature of the membrane emulsification step, a diagram that better explains the process is provided (see Fig. 1). The organic phase (dispersed phase) was loaded into vessel D and sealed. A 500 ml beaker under the outlet from the membrane section contained the aqueous phase. Dispersion of the organic phase through the pores of the membrane and into the aqueous phase allowed the formation of an emulsion. Vessel B was subjected to 10 kPa of N_2 to ensure the organic phase was completely dispersed in the aqueous phase.

Polymerization

The emulsification product was transferred to a 400 ml glass reactor (6.5 cm diameter, 12.05 cm height) in a prewarmed water bath (80 °C initial temperature), and agitation was set to 240 rpm. The temperature of the water bath was maintained at 80 °C for 2 h, increased to 90 °C for 4 h, and then set to ambient room temperature. The process

Fig. 1 Schematic of a SPG
Technology (Miyazaki, Japan)
fast mini-kit (adapted from the
SPG Technology manual).
1 Needle valve, *2* vessel
vent valve, *3* pressure valve,
4 module vent valve, *5* valve
controlling flow into SPG
module, *6* valve controlling flow
exiting SPG module, *A* digital
pressure gauge, *B* pressure
vessel, *C* vessel vent, *D* SPG
module, *E* pump, *F* beaker,
G module vent

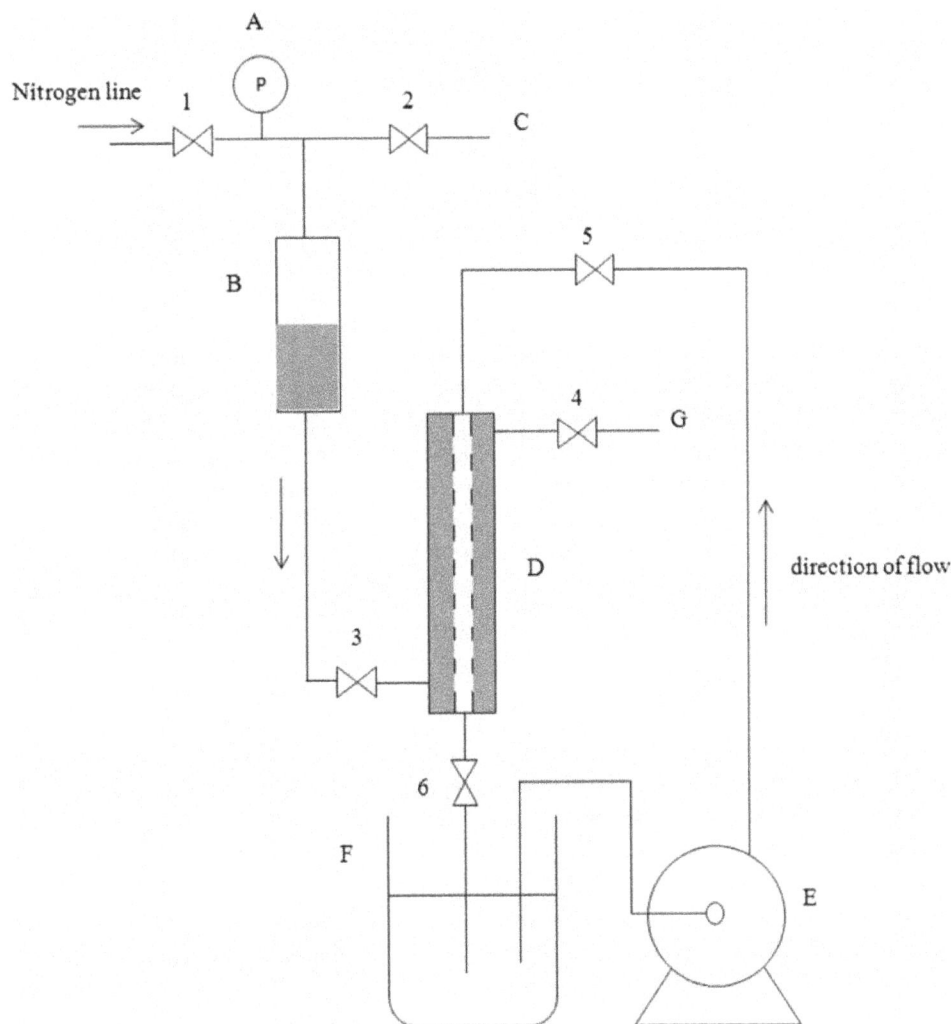

that was used to produce PMMA is known as free radical
vinyl polymerization. This method utilizes BPO as the
radical initiator and MMA as the monomer. The advantage
of BPO is that it is a relatively nonpolar molecule and thus
dissolves easily when mixed with other nonpolar molecules
such as paraffin waxes.

Extraction of microcapsules

After polymerization, the microcapsules were washed three
times with deionized water to remove impurities. The
purified microcapsules were dried at room temperature for
24 h.

Differential scanning calorimetry (DSC)

DSC (DSC-60, Shimadzu, Kyoto, Japan) was used to
determine the melting range and latent heat of the microcapsules. DSC was carried out within the temperature
range of −20 to 60 °C at a heating rate of 3 °C/min or less.

At 60 °C, the temperature was held constant for 10 min.
Air was used to purge the system at 100 ml/min and liquid
nitrogen was used as the refrigerant. Calibration of the
DSC was conducted with *n*-octadecane (GC grade 99 %,
Merck, Darmstadt, Germany).

Each 4–6 mg sample of microcapsules was weighed in
an aluminum DSC pan. An aluminum reference pan was
used in the DSC. Each sample of microcapsules was
examined twice, and the resulting melting temperatures
and latent heats were averaged. The latent heat was calculated from a plot of power versus temperature (°C), and
integration was carried out using the DSC software package. The amount of RT21® encapsulated was calculated
using:

$$\% \text{ Paraffin in microcapsule by mass}$$
$$= \Delta H_{microcapsule} / \Delta H_{pure\ paraffin} \times 100\ \%, \qquad (1)$$

where $\Delta H_{microcapsule}$ is the latent heat of the microcapsule
and $\Delta H_{pure\ paraffin}$ is the latent heat of pure paraffin wax
[10].

Particle size analysis (PSA)

A particle size analyzer (Mastersizer 2000, Malvern Instruments, Malvern, UK) was used to determine the sizes of the microcapsules. The microcapsules were dispersed in distilled water, and the refractive index of MMA was used (1.412 at 25 °C) [20].

Mass loss

To simulate real-life applications of MPCM in buildings, a simple mass loss analysis was conducted. Samples of microcapsules (0.5 g each sample) were weighed out and spread into labeled aluminum pans (53.5 mm in diameter). Each pan was fully covered with microcapsules to give a constant surface area. The pans were loaded onto a tray in a drying oven pre-set to 50 °C. Every 24 h, for a period of one week, the samples were removed from the drying oven and weighed.

Nanoindentation

Nanoindentation of individual microcapsules was carried out as described in Rahman et al. [19]. The microcapsules were fixed onto the surface of an epoxy substrate and indentation testing was carried out using an MTS nano-indenter system (MTS, TN, USA) equipped with a 10 µm, 60° conospherical diamond tip, as well as an optical microscope with a 40× objective.

Results and discussion

Microencapsulation experiments were carried out in triplicate. The results for the microcapsules generated using 10 and 20 µm pore membranes are discussed in detail below as examples of the analysis conducted (see the section "Microencapsulation experiments involving 10 and 20 µm pore size membranes" below), although the section "Summary of the microencapsulation experiments performed using 10, 10.2, and 20 µm pore size membranes" summarizes the results for all of the samples carried out in this study. The "Nanoindentation" section discusses the use of nanocompression.

Pure RT21®

Prior to the encapsulation of RT21®, the properties of a pure sample of RT21® were verified in duplicate. Approximately 5–10 mg of RT21® were used in each DSC run. A sample of the DSC curve generated by the software is shown in Fig. 2. Integration was also performed using the DSC software package, which gave an average latent

heat value of 132.75 J/g, as compared to the value of 134 J/g reported by the manufacturer, and an average peak melting temperature of 23.12 °C. The average temperature corresponding to the onset of melting in the two runs was 17.09 °C, and the average endset temperature was 27.26 °C.

Microencapsulation experiments involving 10 and 20 µm pore size membranes

Microscopic images of the two samples are shown in Fig. 3. The microscopic images show that the microcapsules produced using the 10 µm membrane are smaller than those produced with the 20 µm membrane. This is confirmed by particle size analysis (Fig. 4): microcapsules produced with the 10 µm membrane had an average diameter of 25.65 µm, while microcapsules produced with the 20 µm membrane had an average diameter of 43.52 µm. Comparing the size distributions for the two curves in Fig. 3, the volume percentages of the samples are seen to be quite similar (approximately 9.5 and 9.0 % at their respective average microcapsule diameters). This is an indication that membranes with different pore sizes produce microcapsules with similar size distributions when the membrane emulsification method is used.

It is important to have a narrow size distribution, and also important that the thermal properties of the microcapsules produced are acceptable for use in buildings. The DSC thermographs of microcapsules produced using the 10 and 20 µm pore size membranes showed peak melting temperatures of 21.24 and 21.16 °C, and their latent heats of melting were 112.27 and 117.19 J/g, respectively (Figs. 5, 6). These latent heat values indicate that the thermal properties of the microcapsules are appropriate for use in buildings.

Figure 7 shows the mass loss from microcapsules over a period of one week when placed in an oven at 50 °C. Microcapsules produced using the 10 and 20 µm pore size membranes lost 0.71 and 4.2 % of their initial masses, respectively. The larger microcapsules lost more mass than the smaller microcapsules over the same time period. This is due to the fact that, even though both sets of microcapsules had the same temperature profile for polymerization, the larger droplets of organic phase material had less polymer deposited on the surface of the droplet and thus did not form a sufficiently thick shell. This is also apparent from the DSC result, as microcapsules produced from the 20 µm pore size membrane had a higher latent heat of melting than microcapsules produced from the 10 µm pore size membrane. Using Eq. 1, the microcapsules produced using the 10 and 20 µm pore size membranes were calculated to have core:shell ratio of 0.84:0.16 and 0.88:0.12, respectively.

Fig. 2 Two DSC thermographs for pure RT21® (average latent heat of melting: 132.75 J/g; average peak melting temperature: 23.12 °C; average temperature corresponding to the onset of melting: 17.09 °C; and average endset temperature: 27.26 °C). *Blue lines* show the results of peak integration

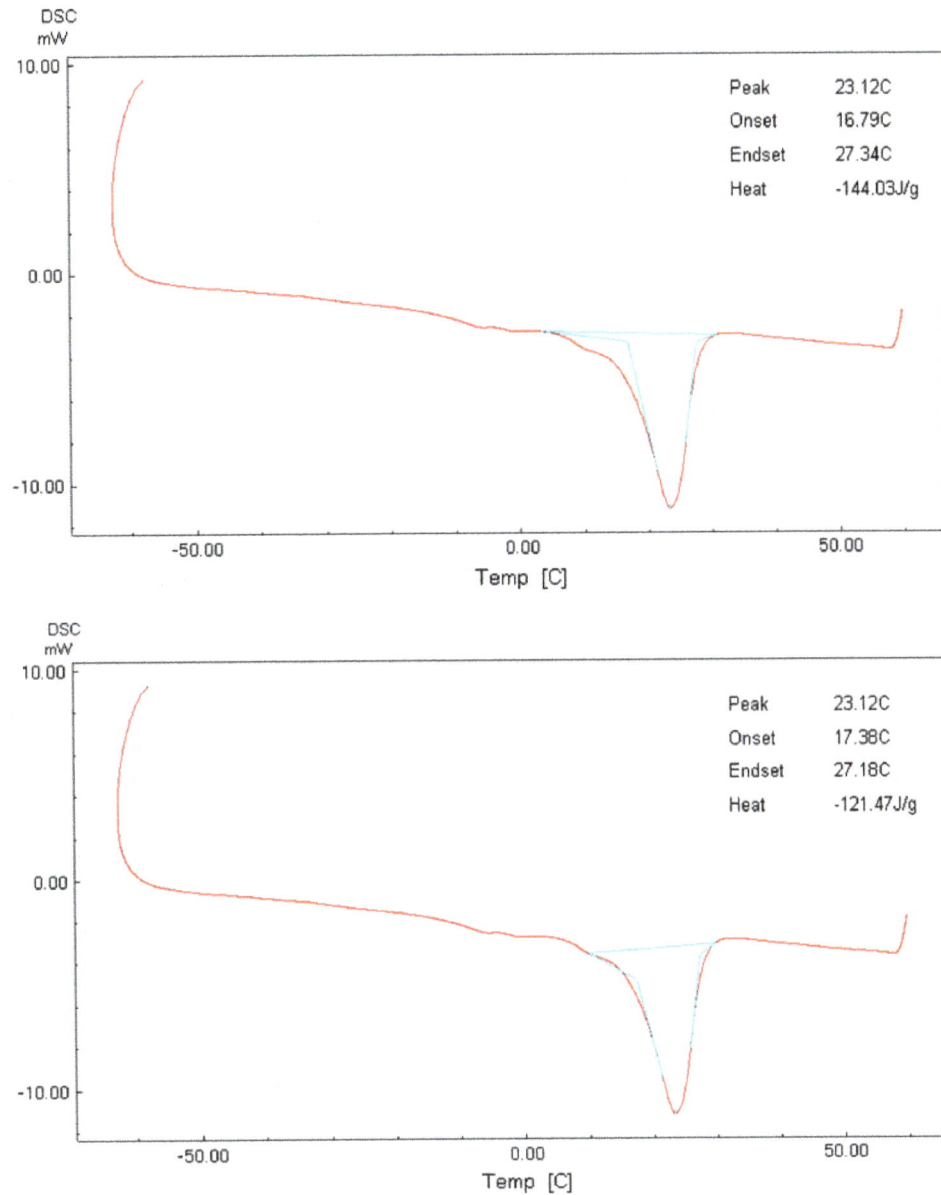

DSC
mW

Peak	23.12C
Onset	16.79C
Endset	27.34C
Heat	-144.03J/g

DSC
mW

Peak	23.12C
Onset	17.38C
Endset	27.18C
Heat	-121.47J/g

Fig. 3 Sample microscopic photographs of microcapsules produced using SPG membranes with 10 μm (*left*) and 20 μm (*right*) pores, respectively. *Scale* represents 30 μm

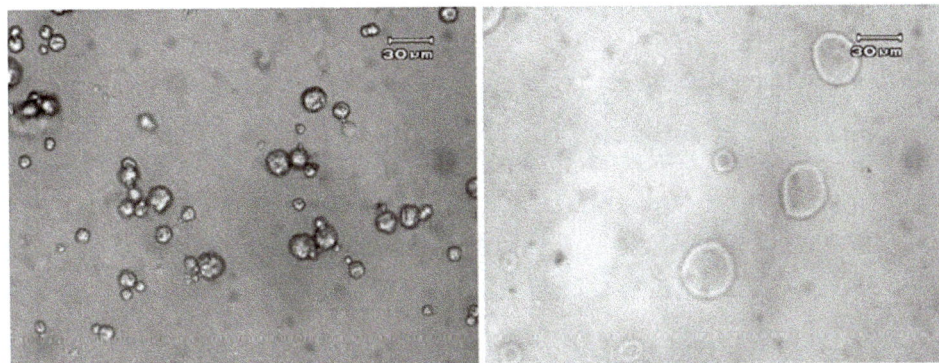

Summary of the microencapsulation experiments performed using the 10, 10.2, and 20 μm pore size membranes

The previous section described an example of the successful encapsulation of RT21® using membrane emulsification with 10 and 20 μm pore size hydrophilic membranes. To demonstrate the results reported in that section were reproducible, triplicate studies were carried out on microcapsules produced using the 10, 10.2, and 20 μm pore size hydrophilic membranes. These results are summarized in Table 1.

According to the results presented in Table 1, the thermal properties of RT21® are conserved upon microencapsulation

in membranes with different pore sizes. The average peak melting temperatures of the RT21® encapsulated in 10, 10.2, and 20 μm pore size membranes were 20.73 ± 0.13, 21.10 ± 0.14, and 20.72 ± 0.20 °C, respectively, which are close to that seen for unencapsulated RT21®. The lower melting temperatures of the microencapsulated RT21® samples can be attributed to the presence of impurities in the final product.

The latent heats of melting for the microcapsules produced using the 10, 10.2, and 20 μm pore size membranes were 113.91 ± 12.0, 116.69 ± 1.40, and 109.89 ± 8.69 J/g, respectively. These high values for the latent heat of each sample demonstrate that microcapsules can be produced using this method without any significant loss of the thermal properties of the PCM RT21®.

The level of encapsulation of the PCM RT21® was above 82 % for all of the samples prepared using the 10, 10.2, and 20 μm membranes. This is significant, as it means that the majority of the PCM is encapsulated, showing that the polymerization and crosslinking are successful. Regardless of the size of the microcapsules created, at least 82 % of the PCM RT21® will be encapsulated. Minimizing the percentage of PCM lost during the encapsulation processes will be vital when scaling up this process.

The diameters of the microcapsules produced using the 10, 10.2, and 20 μm pore size membranes were 22.40 ± 1.47, 25.38 ± 0.80, and 37.50 ± 1.69 μm, respectively. The size of the microcapsules produced was linearly dependent on the pore size of the membrane used during emulsification—the larger the pore size, the larger the microcapsules produced. This is advantageous, as different microcapsule sizes may be required for different MPCM applications.

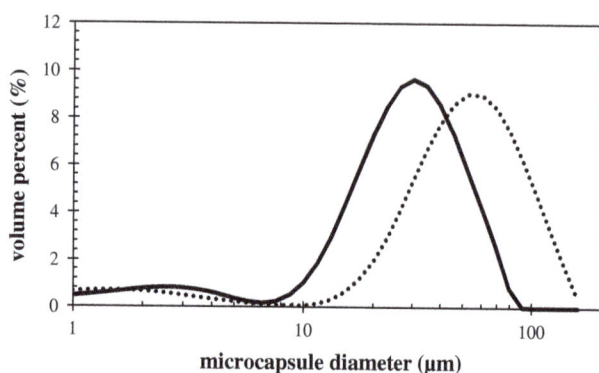

Fig. 4 Sample particle size distributions for microcapsules produced using SPG membranes with 10 and 20 μm pores, respectively. The average diameters of the microcapsules produced using the 10 and 20 μm pore size membranes were 25.65 and 43.52 μm, respectively. The volume percentages were 9.0 and 9.5 % at these average diameters. *Solid* and *dashed lines* indicate 10 and 20 μm pore size membranes, respectively

Fig. 5 Sample DSC thermograph for microcapsules produced using the 10 μm pore size SPG membrane (latent heat of melting: 112.27 J/g; peak melting temperature: 21.24 °C). The onset of melting occurs at 15.35 °C and the endset of melting at 23.61 °C. *Blue lines* show the results of melting peak integration

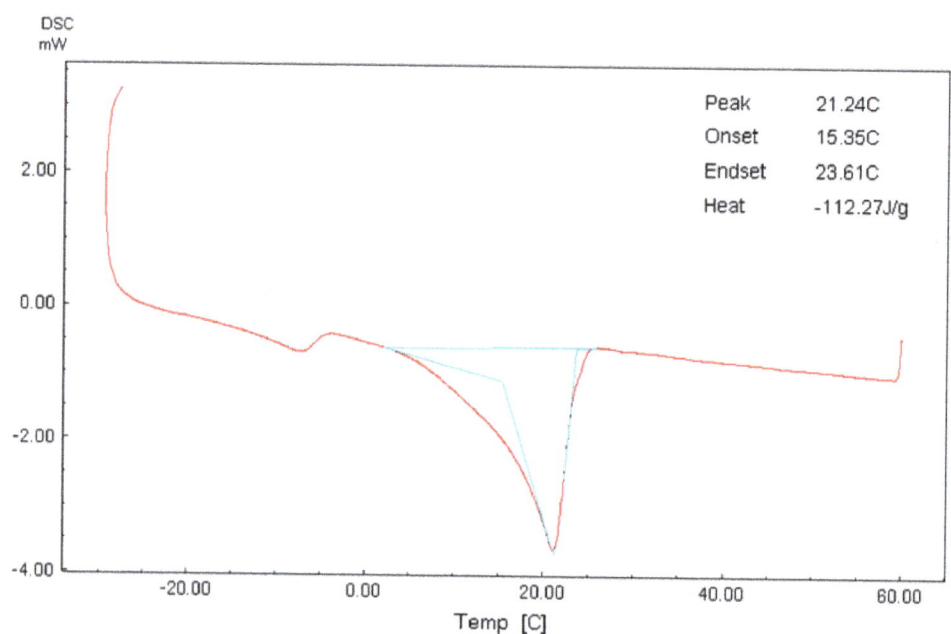

Peak	21.24C
Onset	15.35C
Endset	23.61C
Heat	-112.27J/g

Fig. 6 Sample DSC thermograph for microcapsules produced using the 20 μm pore size SPG membrane (latent heat of melting: 117.19 J/g; peak melting temperature: 21.16 °C). The onset of melting occurs at 14.35 °C and the endset of melting at 23.45 °C. *Blue lines* show the results of melting peak integration

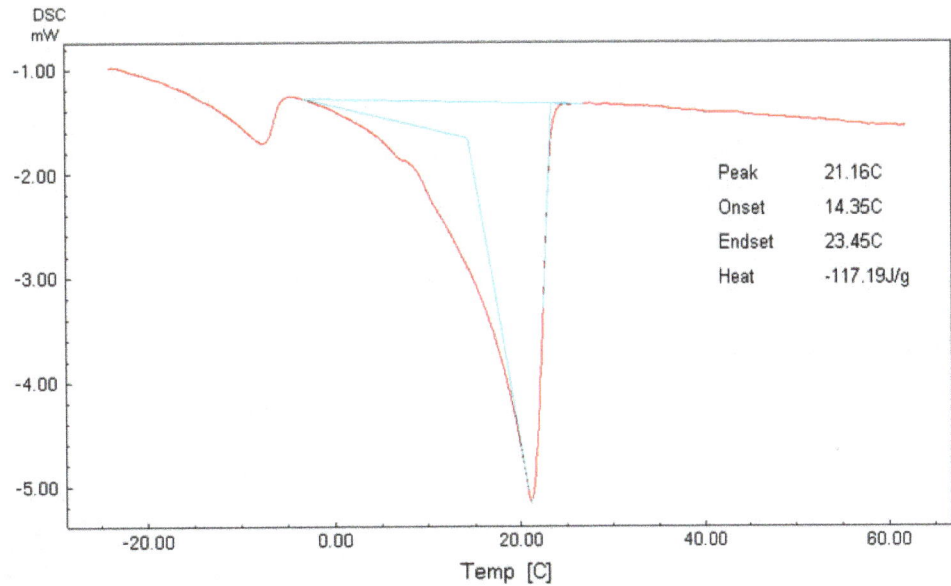

Fig. 7 The *cross symbols* indicate 10 μm and the *open circles* indicate 20 μm pore size membrane

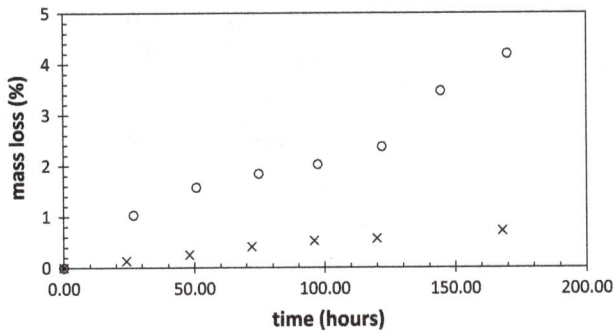

In addition, this study demonstrates that membrane emulsification using SPG membranes can be used not only for Thermal energy storage in PCMs but also for other applications requiring microencapsulation.

Nanoindentation

In order to validate this test method, initial nanocompression tests were conducted on commercially available samples (Microtek MPCM-18D and Microtek MPCM-24D; Microtek Laboratories, Inc., Dayton, OH, USA), and then on the microcapsules created in this study.

The average size of the tested microcapsules from MPCM-18D was measured optically as 24.64 μm. Care was taken to select microcapsules with diameters close to 20 μm, as it was assumed that they would have similar shapes and sizes, facilitating comparisons between the tests.

Microscopic images were taken before and after the nanocompression of a microcapsule (Fig. 8). The images showed that the microcapsule ruptured during testing. The microcapsule shown in the images had an initial diameter of 18.75 μm, and the ruptured shell diameter was measured as 28.11 μm. In addition to optical images, load–displacement curves from the compression test were used to calculate the force required to rupture the microcapsule. To ensure that the whole microcapsule had been compressed, a force larger than that required to collapse the microcapsule was used, producing a loading curve with two slopes. The initial slope corresponded to microcapsule compression. The second slope corresponded to indentation of the underlying substrate. The

Table 1 Average parameter values for all of the different microcapsule samples produced in this study based on membrane type

Membrane pore size (μm)	SPG membrane type	Peak melting temperature (°C)	Latent heat of melting (J/g)	Core:shell ratio (core %)	Size of microcapsule produced (μm)	Mass loss after 1 week (%)
10	PJN03D25	20.73 ± 0.13	113.91 ± 12.0	85 + 9	22.40 ± 1.47	4.42 ± 2.40
10.2	PJN08C03	21.10 ± 0.14	116.69 ± 1.40	87 ± 1	25.38 ± 0.80	3.36 ± 1.28
20	PJN03K20	20.72 ± 0.20	109.89 ± 8.69	82 ± 6	37.50 ± 1.69	5.79 ± 5.19

$n = 3$ samples were evaluated for each membrane pore size investigated

point at which the initial slope changed into the second slope is indicated by an arrow labeled "A" in Fig. 9. Indentation tests were carried out on blank substrate samples as a reference, and these loading curves were used to confirm that the microcapsule had been fully compressed and the substrate was being indented (when the secondary loading slope is the same as the substrate loading slope, it can be assumed that the secondary slope is due to substrate indentation, as shown in the upper left plot of Fig. 9).

The microcapsules made in this study using RT21® as the core material and MMA shells were dried in the laboratory. However, some samples agglomerated, which

made it difficult to test individual microcapsules. Therefore, we decided only to test samples that did not have agglomerated microcapsules (sample numbers 17, 18, 20, and 28) to ensure that the test protocols were being followed precisely.

Table 2 displays the results from nanocompressing all of the microcapsule samples tested in this study, where each run number corresponds to a different microcapsule in the same sample set. According to these preliminary experiments, the microcapsules produced in this study via membrane emulsification are comparable to commercially available microcapsules (MPCM-18D and MPCM24D).

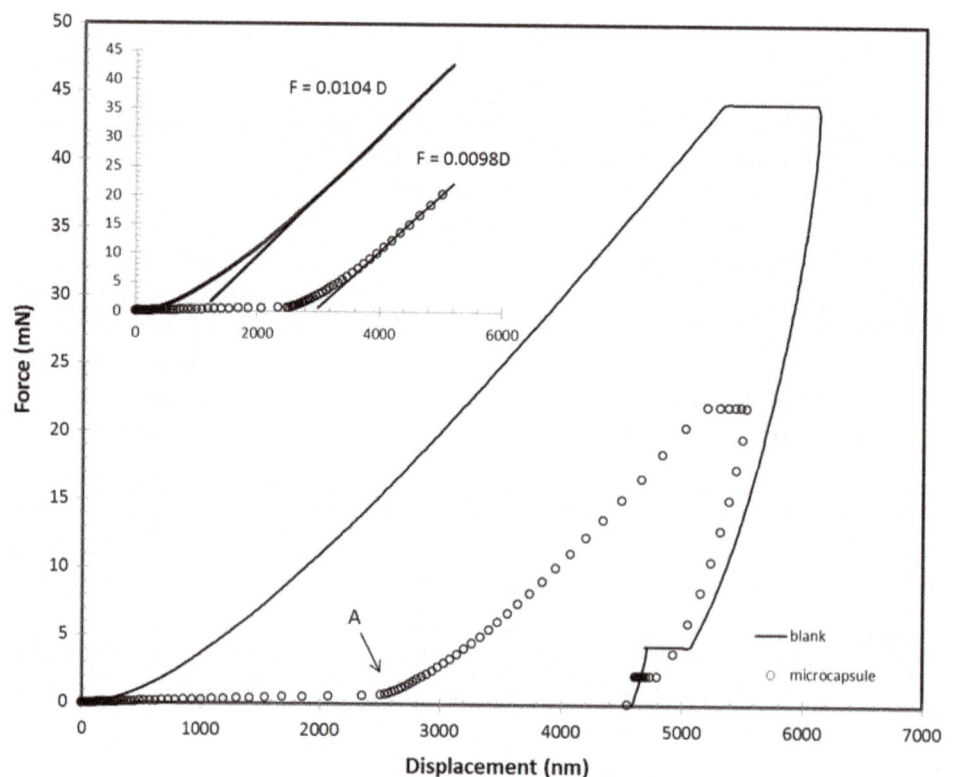

Fig. 8 Microscopic images. **a** Before nanoindentation, showing the spherical microcapsule, and **b** after nanoindentation, showing the ruptured sample

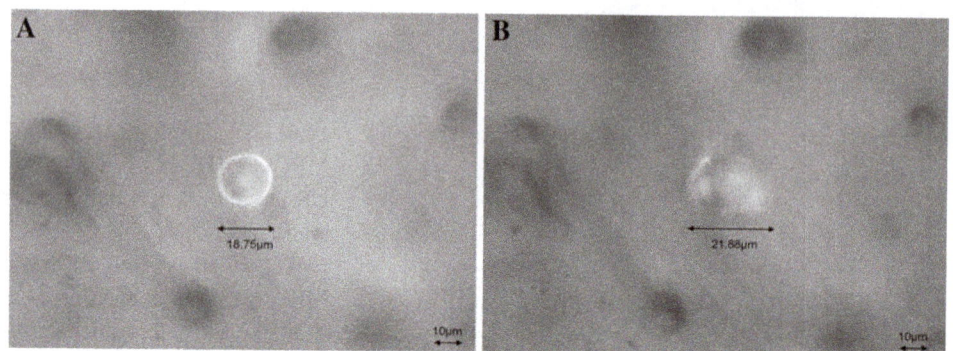

Fig. 9 Force versus displacement curve for the nanoindentation of MPCM-18D run 20 and blank substrate samples. Point A indicates the microcapsule rupture point at a displacement (D) of 2,503 nm and a force (F) of 0.67 mN. The *upper left curve* is an enhanced view of the regions of graph between 0 and 6,000 nm, demonstrating that both the blank and microcapsule curves have similar slopes

Table 2 The force and displacement required to rupture different-sized microcapsules (nanocompression test)

Sample name	Microcapsule diameter (μm)	Rupture force (mN)	Displacement (nm)
MPCM-18D run 19	20.25	0.33	1,925
MPCM-18D run 20	18.75	0.67	2,503
MPCM-18D run 24	28.13	0.29	1,449
MPCM-24D run 1	31.05	1.24	2,830
MPCM-24D run 2	21.25	2.05	2,174
MPCM-24D run 4	19.85	0.37	2,339
MPCM-24D run 7	27.25	0.75	1,704
Sample 17 run 4	21.25	0.54	940
Sample 17 run 6	22.15	0.81	1,039
Sample 17 run 11	32.75	1.08	1,386
Sample 17 run 12	19.85	0.36	472
Sample 18 run 2	28.25	0.79	921
Sample 18 run 4	30.50	0.93	1,034
Sample 18 run 7	29.75	0.27	495
Sample 18 run 17	21.25	0.89	1,591
Sample 20 run 1	32.15	0.25	324
Sample 20 run 2	33.25	0.46	467
Sample 20 run 3	22.65	0.89	1,421
Sample 20 run 6	27.35	0.51	502
Sample 28 run 6	19.75	1.71	1,818

MPCM-18D and MPCM-24D are commercial microcapsules from Microtek Laboratories, Inc. (Dayton, OH, USA). Microencapsulated RT21® samples 17, 18, and 20 were produced with a 10 μm pore size SPG membrane. Sample 28 was produced with a 20 μm pore size SPG membrane

Conclusions

This study demonstrated a new process for microencapsulating a paraffin wax, RT21®, with poly(methyl methacrylate) via membrane emulsification and suspension polymerization. Membrane emulsification was carried out with hydrophilic SPG membranes with pore sizes of 10, 10.2, and 20 μm, producing microcapsules with average diameters of 22.40 ± 1.47, 25.38 ± 0.80, and 37.50 ± 1.69 μm and average latent heats of 113.91 ± 12, 116.69 ± 1.40, and 109.89 ± 8.69 J/g, respectively. The thermal characteristics of the microcapsules produced in this study indicate that they can be used to increase the thermal capacity of buildings.

Nanocompression of microcapsules produced in this study, in addition to commercially available microcapsules, was also conducted. Based on the preliminary results, it is clear that there is a further need to investigate the nanocompression of individual microcapsules and to develop a correlation between microcapsule size and the force required for rupture.

References

1. Khudhair, A.M., Farid, M.M.: A review on energy conservation in building applications with thermal storage by latent heat using phase change materials. Energy Convers. Manag. **45**, 263–275 (2004)
2. Cabeza, L.F., Castell, A., Barreneche, C., De Gracia, A., Fernández, A.I.: Materials used as PCM in thermal energy storage in buildings: a review. Renew. Sustain. Energy Rev. **15**, 1675–1695 (2011)
3. Zhou, D., Zhao, C.Y., Tian, Y.: Review on thermal energy storage with phase change materials (PCMs) in building applications. Appl. Energy **92**, 593–605 (2012)
4. Zalba, B., Maran, J.M., Cabeza, L.F., Mehling, H.: Review on thermal energy storage with phase change: materials, heat transfer analysis and applications. Appl. Thermal Eng. **23**, 251–283 (2003)
5. Farid, M.M., Khudhair, A.M., Razack, S.A.K., Al-Hallaj, S.: A review on phase change energy storage: materials and applications. Energy Convers. Manag. **45**, 1597–1615 (2004)
6. Behzadi, S., Farid, M.M.: Experimental and numerical investigations on the effect of using phase change materials for energy conservation in residential buildings. HVAC R Res. **17**, 366–376 (2011)
7. Tyagi, V.V., Kaushik, S.C., Tyagi, S.K., Akiyama, T.: Development of phase change materials based microencapsulated technology for buildings: a review. Renew. Sustain. Energy Rev. **15**, 1373–1391 (2011)
8. Zhao, C.Y., Zhang, G.H.: Review on microencapsulated phase change materials (MEPCMs): fabrication, characterization and applications. Renew. Sustain. Energy Rev. **15**, 3813–3832 (2011)

9. Bayés-García, L., Ventolà, L., Cordobilla, R., Benages, R., Calvet, T., Cuevas-Diarte, M.A.: Phase change materials (PCM) microcapsules with different shell compositions: preparation, characterization and thermal stability. Solar Energy Mater. Solar Cells **94**, 1235–1240 (2010)

10. Sánchez-Silva, L., Rodríguez, J.F., Romero, A., Borreguero, A.M., Carmona, M., Sánchez, P.: Microencapsulation of PCMs with a styrene-methyl methacrylate copolymer shell by suspension-like polymerisation. Chem. Eng. J. **157**, 216–222 (2010)

11. Charcosset, C., Limayem, I., Fessi, H.: The membrane emulsification process—a review. J. Chem. Technol. Biotechnol. **79**, 209–218 (2004)

12. Joscelyne, S.M., Trägårdh, G.: Membrane emulsification—a literature review. J. Membr. Sci. **169**, 107–117 (2000)

13. Nakashima, T., Shimizu, M., Kukizaki, M.: Membrane emulsification by microporous glass. Key Eng. Mater. **61–62**, 513–516 (1991)

14. Omi, S., Katami, K., Taguchi, T., Kaneko, K., Iso, M.: Synthesis of uniform PMMA microspheres employing modified SPG (Shirasu porous glass) emulsification technique. J. Appl. Polym. Sci. **57**, 1013–1024 (1995)

15. Sun, G., Zhang, Z.: Mechanical strength of microcapsules made of different wall materials. Int. J. Pharm. **242**, 307–311 (2002)

16. Zhang, Z., Sun, G.: Mechanical properties of melamine-formaldehyde microcapsules. J. Microencapsul. **18**, 593–602 (2001)

17. Su, J., Ren, L., Wang, L.: Preparation and mechanical properties of thermal energy storage microcapsules. Colloid Polym. Sci. **284**, 224–228 (2005)

18. Arfsten, J., Bradtmoller, C., Kampen, I., Kwade, A.: Compressive testing of single yeast cells in liquid environment using a nanoindentation system. J. Mater. Res. **23**, 3153–3160 (2008)

19. Rahman, A., Dickinson, M., Farid, M.: Microindentation of microencapsulated phase change materials. Adv. Mater. Res. **275**, 85–88 (2011)

20. Yaws, C.L. (ed.): Yaws' handbook of thermodynamic and physical properties of chemical compounds. Knovel, Beaumont; 2003

17

Study of hydrogen absorption kinetics of Mg$_2$Ni-based powders produced by high-injected shock power mechanical alloying and subsequent annealing

Moomen Marzouki · Ouassim Ghodbane · Mohieddine Abdellaoui

Abstract Mg$_2$Ni-based compounds were prepared using a high-energy milling technique, with a planetary ball mill, and subsequent annealing at 350 °C. X-ray diffraction analyses revealed that Mg$_2$Ni phase was obtained after 10 milling hours, in addition to Ni phase. Increasing the ball-milling duration from 10 to 17 h causes a decrease in the crystal size of particles from 93 to 76 nm. However, the subsequent annealing of all Mg$_2$Ni-based materials highly increases their crystallite sizes with calculated values in the range of 261–235 nm. In the same way, X-ray diffraction patterns of annealed compounds show the presence of highly crystallized Mg$_2$Ni phases. The particle size of Mg$_2$Ni/Ni particles is estimated at 25 μm after 10 milling hours and drops to 5 μm at longer times. The scanning electron microscopy images of Mg$_2$Ni/Ni particles demonstrated a drastic increase of their sizes up to 100 μm upon annealing. The hydrogenation reactivity and kinetic of Mg$_2$Ni/Ni were both characterized by solid–gas reactions. The hydrogen absorption capacity value was about 3.5 H/f.u for milled and annealed Mg$_2$Ni/Ni compound. The highest hydrogen absorption kinetic was obtained during the first 5 h of absorption time, where 95 % of the maximum absorption capacity was reached.

Keywords Mg$_2$Ni-based compounds · Mechanical alloying · X-ray diffraction · Hydrogen absorption kinetic · Annealing treatment

M. Marzouki · O. Ghodbane (✉) · M. Abdellaoui
Laboratoire des Matériaux Utiles, Pôle Technologique
de Sidi Thabet, Institut National de Recherche et d'Analyse
Physico-chimique, 2020 Sidi Thabet, Tunisia
e-mail: ouassim.ghodbane@inrap.rnrt.tn

Introduction

Ni-metal hydrides (Ni-MH) are extensively studied as secondary batteries. Their applications include portable equipments and hydrogen fuel cell transportations. Among the materials investigated as possible Ni-MH negative electrode, Mg-based alloys exhibit promising performances [1–4]. Furthermore, the intermetallic Mg$_2$Ni compound received a great attention for the reversible hydrogen storage [5]. Mg$_2$Ni combines with hydrogen to form Mg$_2$NiH$_4$ hydride and highly improves the hydrogenation kinetic of magnesium [6]. Concerning synthesis processes of materials, melting is the conventional technique to prepare Mg$_2$Ni [7]. However, the large difference in melting points and vapor pressures between Mg and Ni make difficult the formation of high quality Mg$_2$Ni [7]. These drawbacks could be avoided by using the mechanical alloying (MA) process since the reaction between Mg and Ni occurs easily and reliably in the solid state [8–10]. Moreover, the MA process leads to the formation of surface defects and favors the formation of nanocrystalline materials. Commonly, amorphous phases are obtained from the MA of Mg and Ni, and may act as precursors for the formation of the crystalline Mg$_2$Ni phase. Such properties are required for the improvement of the hydrogenation kinetic and the absorption capacity [11]. Combining ball-milling with a heat-treatment step generally increases the yield of the synthesis reaction [12–14]. Spassov et al. [15, 16] showed that Mg$_2$Ni-based alloys subjected to an annealing step after the ball-milling procedure consist in homogenously nanocrystalline particles. On the other hand, Rojas et al. [17] succeeded in reducing the long milling time (14 h), needed for the transformation of Ni and Mg into Mg$_2$Ni, to a shorter time (5 h) by a subsequent annealing at 673 K for 1 h. We have previously reported that the heat-treatment of ball-milled Mg_2Ni materials leads to a high absorption capacity of 3.5 wt% [8].

In this work, Mg_2Ni-based powders were prepared by mechanical alloying and subsequent annealing at 350 °C for 24 h. This temperature induces relevant microstructural modifications in the milled phase and enhances the hydrogen absorption/desorption properties. The control of the synthesis parameters plays an important role in increasing the equilibrium pressure and reducing the activation time of the hydrogen absorption reaction. The present paper presents an investigation of the hydrogenation kinetics of numerous Mg_2Ni-based powders obtained by varying the cumulated energy of milling. Series of Mg_2Ni phases were prepared and characterized by X-ray diffraction (XRD), scanning electron microscopy (SEM) and thermogravimetric analyses before being tested for the hydrogen storage.

Experimental section

A mixture of elemental Mg (VWR, 99.8 %) and Ni (VWR, 99.9 %), with an atomic ratio of 2:1, was sealed into a stainless steel vial (50 cm^3 in volume) with 5 stainless steel balls (15 mm in diameter and 13.6 g in mass) in a glove box filled with purified argon gas. The ball-to-powder weight ratio was equal to 68:1. The MA experiments were performed at room temperature using a Retsh PM400 planetary ball miller. The disc rotation speed and the vial rotation speed were equal to 250 and 500 rpm, respectively. These milling conditions correspond to kinetic shock energy of 0.63 J/hit, shock frequency of 45.6 Hz, and injected shock power of 5.75 W/g.

The crystallographic characterization of synthesized powders was carried out by XRD using a (θ–2θ) Panalytical XPERT PRO MPD diffractometer operating with Cu Kα radiation ($\lambda = 0.15406$ nm). The powder morphology of the samples was characterized with a FEI Quanta 200 environmental scanning electron microscope. Thermogravimetric measurements were based on the differential scanning calorimetry (DSC) technique using a Setaram 131 instrument. The analyses were realized at a heating rate of 20 °C/min, under nitrogen gas atmosphere. Prior to the hydrogen absorption measurements, the sample was pulverized mechanically, by a metallographic hammer and an agate mortar, into a powder of 63 μm in size. Hydrogenations of synthesized compounds were performed using solid–gas reactions. The hydrogen absorption capacity was measured with a home-made Sievert's apparatus at a pressure of 11 bars and a temperature of 280 °C.

Results and discussion

Structural characterization

XRD patterns of mechanically alloyed powders are shown as a function of the alloying time in Fig. 1. After 10 h of milling, diffraction peaks relative to Mg_2Ni phase (JCPDS 01-75-1249) are observed at 2θ (hkl) = 20° (003), 23° (102), 37° (112), 40° (200), 45° (203), 72° (220) and 86° (226). At this stage of milling, the pattern indicates the coexistence of elemental Ni and Mg_2Ni phases. The presence of Ni is evidenced by diffraction peaks located at 2θ (hkl) = 44° (111), 52° (200) and 78° (220). Surprisingly, peaks relative to elemental Mg are not observed in any pattern. The absence of Mg peaks contrasts with previous studies, where elemental Mg and Ni still present in composites milled during 10 h [18–20], this behavior could be explained by the distinct shock powers injected during the mechanical alloying. In fact, in the same kinetic conditions (Ω_{disc} is the disc rotation speed, ω_{vial} is the vial rotation speed, R_{disc} is the disc radii, r_{vial} is the vial radii and r_{ball} is the ball radii), the injected shock power increases by increasing the ball number and weight, or by reducing the material weight. For these reasons, the injected shock power increases by increasing the ball-to-powder weight ratio (BPR). The injected shock power in this study corresponds to a BPR of 68:1. Such a value is more important than the ones considered elsewhere by Gennari et al. [18] (BPR = 42:1) and Ebrahimi-Purkani et al. [19] (BPR = 20:1). In our previous works [21, 22], we reported that the structure of the stationary state was only a function of the injected shock power. Nevertheless, we reported that the amount of intermediary phases depends on the cumulated energy of milling (E_{cum}), defined by the following equation [5, 6]:

$$E_{cum} = P_{inj} \times \Delta t \qquad (1)$$

where E_{cum} is expressed in [Wh/g], P_{inj} is the injected shock power expressed in [W/g] and Δt is the alloying duration expressed in [h]. Consequently, the increase of the injected shock power allows the formation of the same intermediary states at lesser alloying durations.

In the present work, the absence of Mg diffraction peaks and the presence of remaining Ni suggest that amorphous Mg is confined in the Mg_2Ni/Ni composite [18]. From 10 to 15 h of milling, i.e. cumulated energy from 57 to 86 Wh/g, the increase in the peak intensity of Mg_2Ni (203) simultaneously occur with a decrease in the peak intensity of Ni (111) (Fig. 1b). After 17 h of milling, the cumulated energy is about 98 Wh/g. At this stage, the diffraction line

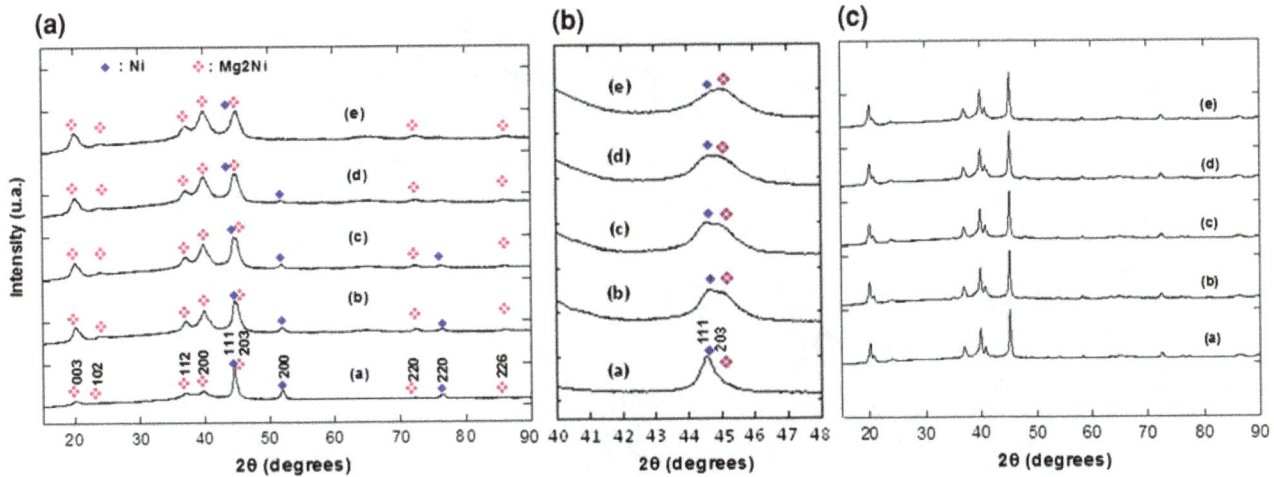

Fig. 1 XRD patterns of mechanically alloyed (**A**, **B**) and annealed (**C**) Mg_2Ni powders. Milling durations are 10 h (*a*), 12 h (*b*), 13 h (*c*), 15 h (*d*) and 17 h (*e*)

of Ni (200) disappears, while the intensity ratio of Ni (111)/ Mg_2Ni (203) diminishes. These observations suggest a progressive formation of Mg_2Ni phase accompanied by a loss of residual Ni in the composite. On the other hand, the increase of milling time results in broadening of peaks. Two factors may explain such a behavior: (i) the decrease of the crystallite size and, in a lesser extent, (ii) the increase of the lattice strains [23]. The average crystallite size of Mg_2Ni powders were calculated from the corresponding (003) peak [8], using the following Debye-Sherrer equation [24]. The corresponding values are presented in Table 1 as a function of the milling time:

$$D_{003} = 0.89\lambda/\beta\cos(\theta) \qquad (2)$$

where D_{003} is the crystallite size of Mg_2Ni, λ is the wavelength of the X-radiation, β is the full-width at half maximum (FWHM) of the peak and θ is the diffraction angle corresponding to (003) peak. Instrumental broadening and lattice distortion contributions were subtracted to values of peak broadening. Table 1 shows a continuous decrease of the crystallite size from 93 to 76 nm when the milling time varies from 10 to 17 h. The crystallite sizes estimated for as-prepared powder are higher than values reported in the literature [15, 18, 19, 25].

The heat-treatment of Mg_2Ni powders was carried out at 350 °C for 24 h under vacuum. The resulting powders were characterized by XRD as shown in Fig. 1c. Independently on the milling duration, annealed powders show similar patterns reflecting a similar structural rearrangement upon annealing. The heat-treatment leads to sharper peaks, the disappearance of Ni diffraction peaks and a pure and highly crystalline Mg_2Ni phase. The evaluation of the crystallite size for annealed powders indicates a steep increase of all D_{003} values upon annealing (Table 1) and demonstrates that heat-treating particles enhance their agglomeration.

SEM characterization

SEM micrographs of Mg_2Ni powders mechanically alloyed during 10 and 12 h are shown in Fig. 2. The powder obtained after 10 h of MA shows an irregular shape of porous particles with an average size of 25 μm (Fig. 2a). As milling progresses, powder particles are fractured and their size becomes 5–6 μm, as displayed in Fig. 2b and Table 1. During the mechanical alloying, the elemental powder is blended, cold worked, welded and fragmented repeatedly. After a long period of milling, fracturing

Table 1 Average crystallite size (based on XRD patterns) and particle size (measured by SEM technique) of Mg_2Ni-based powders

Milling time (h)	10	12	13	15	17
Cumulated energy of milling (E_{cum}, Wh/g)	57	70	75	86	98
Crystallite size of MA powders (nm)	93	87	86	79	76
Crystallite size of annealed powders (nm)	261	248	236	214	235
Particle size of as MA powders (μm)	25	5	6	6	5
Particle size of annealed powders (μm)	100	100	100	100	100

Annealing temperature: 350 °C, annealing duration: 24 h

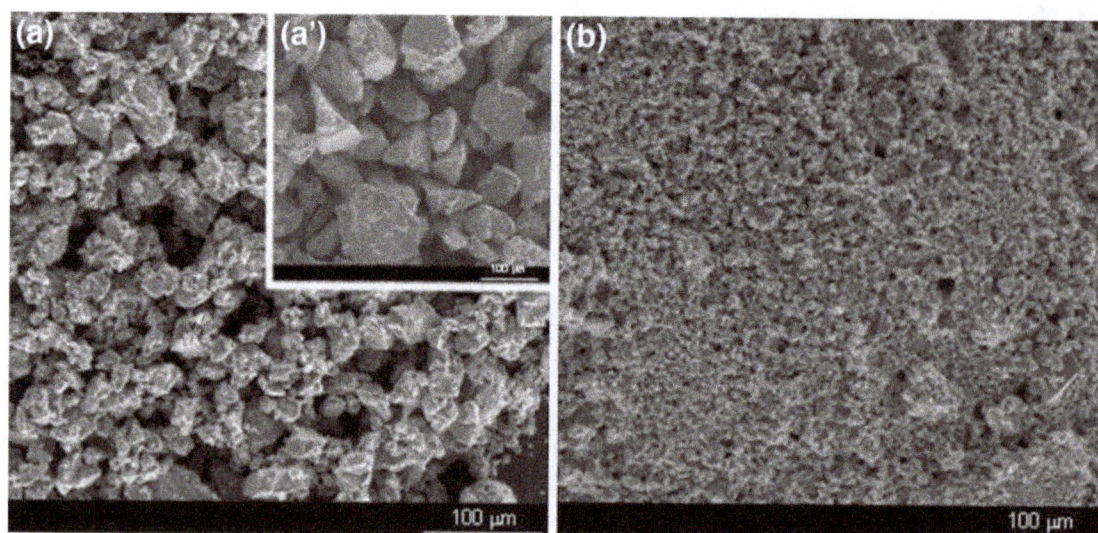

Fig. 2 SEM micrographs (secondaries electrons mode) of Mg_2Ni powders prepared by mechanical alloying during **a** 10 h, **b** 12 h and **a'** 10 h with subsequent annealing at 350 °C for 24 h

becomes the predominant event because of the brittle nature of Mg_2Ni materials [19]. Further mechanical alloying from 12 to 17 h slightly affects the powder morphology (Table 1). On the other hand, Fig. 2a' shows that subsequent annealing of Mg_2Ni-based powders leads to a substantial increase of their particle size up to 100 μm. The same value was obtained for the whole series of milled and annealed Mg_2Ni powders (Table 1). These observations are in a good agreement with the XRD results. The heat-treatment promotes the diffusion and the coalescence of Mg_2Ni particles during the temperature increase. Thermodynamically, the particle size extends and converges to a value corresponding to the lowest free energy of Mg_2Ni composite. During the growth of particles, the volume energy decreases, while the surface energy increases [26].

Thermogravimetric investigation

Figure 3 shows the DSC curves of mechanically alloyed and annealed products. During the temperature increase, the composite formed upon 10 h of milling time exhibits two exothermic peaks located at 132 and 220 °C (Fig. 3a). Both peaks were assigned elsewhere to the crystallization of amorphous Mg_2Ni [17, 20]. In the present case, the first peak is assigned to the conversion of amorphous Mg_2Ni to crystalline Mg_2Ni, while the second one is attributed to the formation of highly crystallized Mg_2Ni from residual Ni and amorphous Mg [17, 20]. The same phenomena are considered for the composite milled during 12 h since two thermal events are observed at 132 and 200 °C (Fig. 3b). It should be noticed that the second peak shifts to lower temperatures. When the milling time is increased up to 13 h (Fig. 3d), the DSC curve shows the presence of three exothermic peaks. Thermal events located at 132 and

200 °C are similar to those observed for compounds milled during 10 and 12 h and, therefore, correspond to the same reactions. However, the new peak appearing at 280 °C is associated to a further formation of Mg_2Ni compound [27, 28]. This is thermodynamically favorable when the overall energy is increased by either increasing the milling time or through a thermal activation. For milling durations of 15 and 17 h, the peak located at 200 °C diminishes and only peaks located at 132 and 280 °C still present (Fig. 3f, g). XRD data demonstrated that increasing the milling time leads to Mg_2Ni-rich phase (Fig. 1). The fraction of residual Ni that is transformed during the DSC scan becomes lower since it was already been transformed into Mg_2Ni during the milling process. For this reason, the absence of the

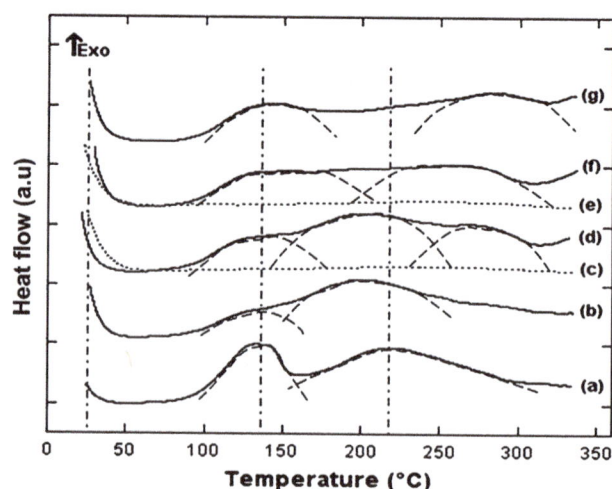

Fig. 3 DSC curves of Mg_2Ni prepared by mechanical alloying during 10 h (**a**), 12 h (**b**), 13 h (**c, d**), 15 h (**e, f**) and 17 h (**g**), and subsequent annealing at 350 °C for 24 h (**c, e**)

second exothermic peak is associated with residual precursors being mechanically transformed into Mg_2Ni.

DSC curves of powders milled during 13 and 15 h and heat-treated at 350 °C for 24 h are shown in Fig. 3c, e. They are characterized by the absence of any thermal events. Curves of all annealed powders present a similar shape (data not shown). This is due to the formation of a highly crystalline Mg_2Ni phase following the heat-treatment (Fig. 1c) and elucidates the role of the amorphous precursors in the Mg–Ni system.

Hydrogen storage properties

Figure 4 shows the evolution of the capacity during the hydrogen absorption for different ball-milled and annealed Mg_2Ni-based materials. Independently on the milling

Fig. 4 Evolution of the hydrogen absorption capacity during time for ball-milled Mg_2Ni powders before (*full lozenges*) and after (*open squares*) annealing at 350 °C for 24 h. Milling durations are 10 h (**a**), 12 h (**b**), 13 h (**c**), 15 h (**d**) and 17 h (**e**). The hydrogen absorption was performed at a pressure of 11 bars H_2 and a temperature of 280 °C

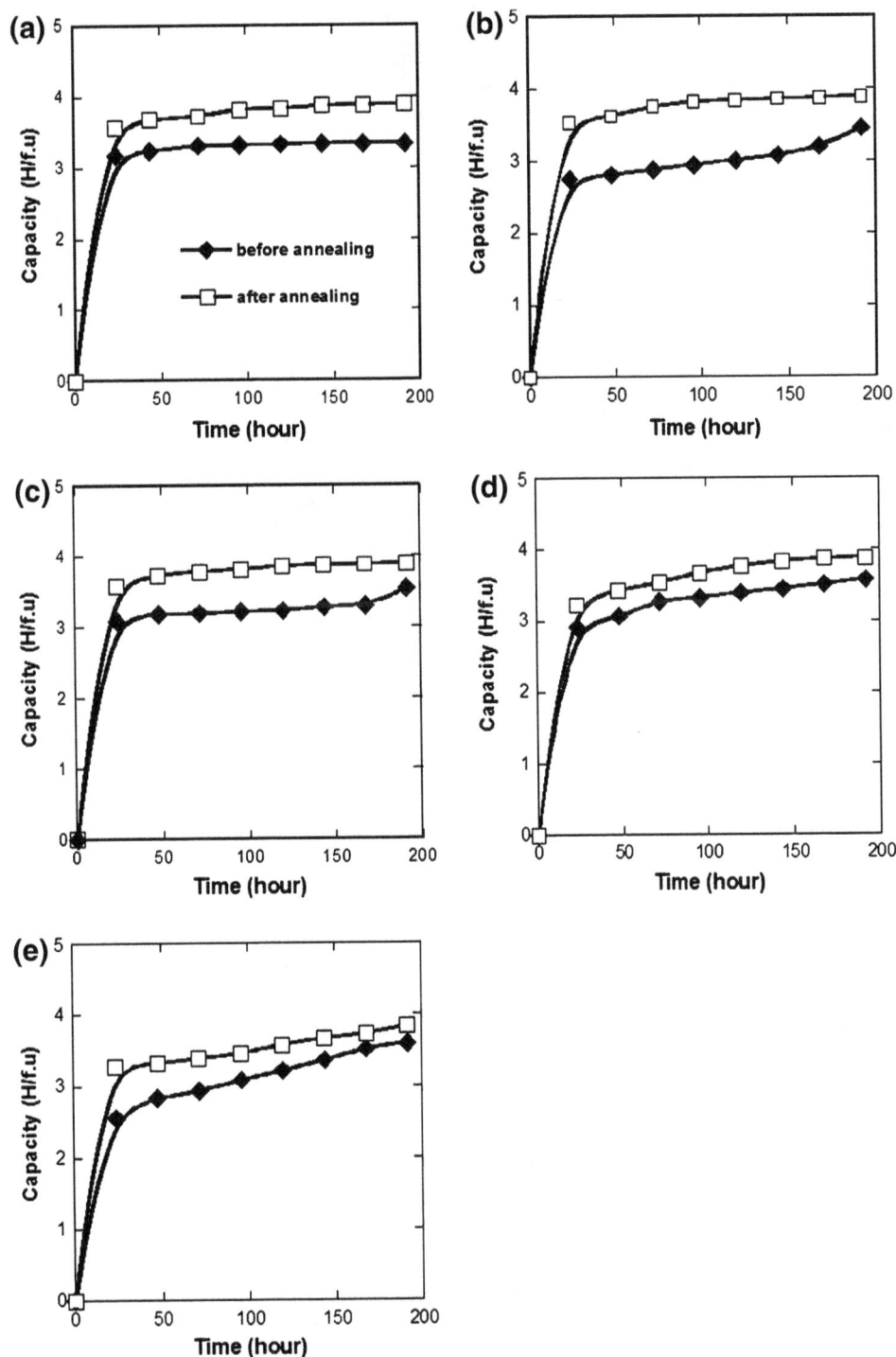

duration, the absorption capacity of Mg_2Ni expands significantly after 25 h of absorption time. For a longer time, the capacity value fairly increases and reaches a plateau. Nearly 90 % of the maximum capacity is reached after an absorption time of 25 h for all Mg_2Ni powders. This behavior reflects a fast kinetic of the hydrogen absorption reaction in the beginning of the absorption process. The maximum capacity values obtained with heat-treated and unheated Mg_2Ni-based powders are presented in Table 2. For all samples, experimental capacities are lower than the theoretical value expected for crystalline Mg_2Ni, i.e. 4 H/f.u. This is mainly due to the heterogeneous chemical composition of prepared composites including crystallized and amorphous Mg_2Ni, together with elemental Ni (Fig. 1). It was already demonstrated that amorphous Mg_2Ni exhibits lower hydrogen absorption capacity than crystallized Mg_2Ni [18]. Moreover, hydrogen atoms are poorly absorbed by the Ni phase in the present conditions. For these reasons, the hydrogen absorption capacity displayed by as-prepared powders is mainly related to Mg_2Ni crystalline phase. Table 2 indicates that the absorption capacity increases by increasing the milling time for unheated compounds. This is explained by the decrease in the amount of residual Ni and the simultaneous formation of Mg_2Ni phase during the mechanical alloying (Fig. 1b).

On the other hand, annealing Mg_2Ni-based compounds enhances the capacity values for all prepared powders (Fig. 4). The absorption capacities of annealed samples vary with time following the same shape than unheated compounds. Again, the capacity value rises during the first 25 h and then stabilizes. The influence of the heat-treatment on the capacity values is indicated in Table 2. Annealed powders exhibit nearly the same absorption capacity with an average value of 3.47 ± 0.01 H/f.u. This similarity derives from the formation of comparable Mg_2Ni microstructures (Fig. 1c) and morphologies upon heat-treating powders. The increase in the capacity value observed upon annealing powders is caused by the transformation of the amorphous Mg_2Ni phase into a highly crystallized one, as evidenced by XRD characterizations (Fig. 1). In the same way, the annealing step promotes further conversion of residual Ni and amorphous Mg into Mg_2Ni. The latter phase is characterized by an absorption capacity higher than the ones of Ni and Mg. However, experimental capacities of annealed products are still lower than theoretical values (Table 2). Westlake et al. [29] reported that the radius of octahedral sites must be larger than 0.4 Å in order to accommodate hydrogen atoms. In the same way, Switendick et al. [30] demonstrated that tetrahedral sites separated by a distance lower than 2.1 Å are unable to absorb hydrogen. Thus, the catalytic activity of as-prepared Mg_2Ni samples may be limited by the narrow size of some unoccupied tetrahedral and octahedral sites.

Table 2 Maximum absorption capacity obtained for ball-milled Mg_2Ni-based powders before and after annealing

Milling time (h)	Maximum capacity (H/f.u)	
	Before annealing	After annealing
10	3.02	3.49
12	3.11	3.48
13	3.19	3.46
15	3.21	3.48
17	3.28	3.46

The hydrogen absorption was performed at a pressure of 11 bars H_2 and a temperature of 280 °C

Annealing temperature: 350 °C, annealing duration: 24 h

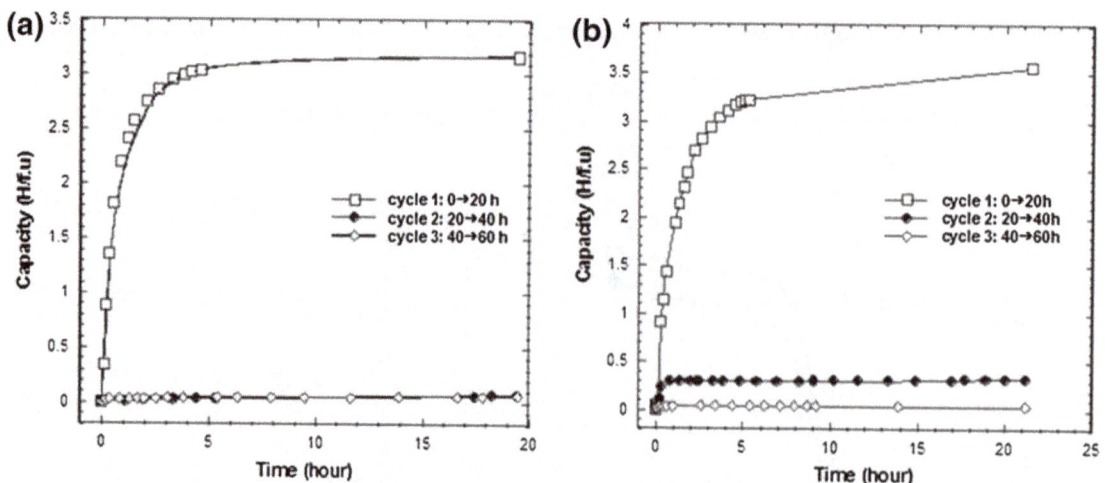

Fig. 5 Evolution of the hydrogen absorption capacity with time for the first, second and third cycles of Mg_2Ni powder ball-milled during 10 h before (**a**) and after subsequent annealing at 350 °C for 24 h (**b**).

The hydrogen absorption was performed at a pressure of 11 bars H_2 and a temperature of 280 °C. The absorption cycles were performed successively without any desorption step

Figure 5 focuses on the evolution of the hydrogen absorption capacity during the first 20 h of absorption time. This experiment is designated as the first absorption cycle and was consecutively repeated two times, without any hydrogen desorption step. Curves are presented for Mg_2Ni-based powders milled during 10 h before and after their heat-treatment. During the first cycle, Mg_2Ni-based composites quickly absorbs 3.05 H/f.u at 5.5 h, which constitutes 96 % of the maximum capacity obtained at the end of this cycle, i.e. 3.17 H/f.u (Fig. 5a). After 5.5 h, the amount of absorbed hydrogen increases weakly, reflecting a slower kinetic of the absorption mechanism. For the second and third cycles, the absorption kinetic is significantly slow as the amount of absorbed hydrogen is too small: 0.070 and 0.067 H/f.u in the second and third cycle, respectively (Fig. 5a). During the first cycle, unoccupied sites of Mg_2Ni microstructure are able to accommodate and absorb hydrogen atoms easily. For further cycles, the hydrogen absorption only occurs inside highly energetic sites and only a small amount of hydrogen may be absorbed. In the case of annealed Mg_2Ni powders (Fig. 5b), most hydrogen is also absorbed during the first cycle. Figure 5b shows that 3.22 H/f.u is absorbed after 5.5 h, which constitutes 85 % of the maximum capacity (3.5 H/f.u). The amounts of absorbed hydrogen in heat-treated powders are 0.29 H/f.u (second cycle) and 0.043 H/f.u (third cycle) and constitute 7.4 and 1.1 % of the maximum absorption capacity, respectively.

Conclusions

Mechanical Alloying of Mg_2Ni-based alloys were performed using a high injected shock power mill. The formation of Mg_2Ni/Ni compound was obtained after only 10 h milling, which corresponds to a cumulated energy of 57 Wh/g. The as-prepared composite also contains Mg–Ni amorphous phase and residual Ni. When the cumulated energy increases from 57 to 98 Wh/g, less residual Ni and higher Mg_2Ni contents are obtained. Consequently, an enhancement of the hydrogen absorption capacity was observed from 3.02 to 3.28 H/f.u upon the increase in the cumulated energy of milling. The combination of mechanical alloying and subsequent annealing at 350 °C for 24 h was able to improve the hydrogen absorption capacity and the hydrogenation kinetics. Independently on the preparation conditions of Mg_2Ni/Ni powders, nearly 95 % of the hydrogen absorption capacity was reached after 5 h of absorption time. For longer absorption durations, the hydrogenation kinetics becomes very slow due to a drastic decrease of unoccupied sites, available to absorb hydrogen atoms, inside Mg_2Ni lattice.

References

1. Abe, T., Tachikawa, T., Hatano, Y., Watanabe, K.: Electrochemical behavior of amorphous MgNi as negative electrodes in rechargeable Ni–MH batteries. J. Alloys Compds. 332, 792–795 (2002)
2. Abdellaoui, M., Mokbli, S., Cuevas, F., Latroche, M., Percherongegan, A., Zarrouk, H.: Structural, solid–gas and electrochemical characterization of Mg_2NiMg_2Ni-rich and $Mg_xNi_{100-x}Mg_xNi_{100-x}$ amorphous-rich nanomaterials obtained by mechanical alloying. Int. J. Hydrog. Energy 31, 247–250 (2006)
3. Nohara, S., Hamasaki, K., Zhang, S.G., Inoue, H., Iwakura, C.: Electrochemical characteristics of an amorphous $Mg_{0.9}V_{0.1}Ni$ alloy prepared by mechanical alloying. J. Alloys Compds. 280, 104–106 (1998)
4. Lenain, C., Aymard, L., Dupont, L.: A new $Mg_{0.9}Y_{0.1}Ni$ hydride forming composition obtained by mechanical grinding. J. Alloys Compds. 292, 84–89 (1999)
5. Dehouche, Z., Djaozandry, R., Goyette, J., Bose, T.K.: Evaluation techniques of cycling effect on thermodynamic and crystal structure properties of Mg_2Ni alloy. J. Alloys Compds. 288, 269–276 (1999)
6. Nohara, S., Inoue, H., Fukumoto, Y.: Compositional and structural characteristics of MgNi alloy prepared by mechanical alloying for use as negative electrodes in nickel-metal hydride batteries. J. Alloys Compds. 259, 183–185 (1997)
7. Kim, J.-H., Kim, J.-H., Hwang, K.-T., Kang, Y.-M.: Hydrogen storage in magnesium based-composite hydride through hydriding combustion synthesis. Int. J. Hydrog. Energy 35, 9641–9645 (2010)
8. Abdellaoui, M., Cracco, D., Percheron-Guegan, A.: Structural characterization and reversible hydrogen absorption properties of Mg_2Ni rich nanocomposite materials synthesized by mechanical alloying. J. Alloys Compds. 268, 233–240 (1998)
9. Abdellaoui, M., Cracco, D.: Structural investigation and solid-H_2 reaction of Mg_2Ni rich nanocomposite materials elaborated by mechanical alloying. J. Alloys Compds. 295, 501–507 (1999)
10. Tojo, T., Yamamoto, I., Zhang, Q., Saito, F.: Discharge properties of Mg_2Ni–Ni alloy synthesized by mechanical alloying. Adv. Powder Technol. 16, 649–658 (2005)
11. Zaluski, L., Zaluska, A., Ström-Olsen, J.O.: Hydrogen absorption in nanocrystalline Mg_2Ni formed by mechanical alloying. J. Alloys Compds. 217, 245–249 (1995)
12. Wang, S., Li, C., Yong, W., Hou, X., Geng, H., Xu, F.: Formation of La-modified $L1_2$-Al_3Ti by mechanical alloying and annealing. Mater. Charact. 59, 440–446 (2008)
13. Mohammadnezhad, M., Shamanian, M, Enayati, M.H., Salehi, M.: Influence of annealing temperature on the structure and properties of the nanograined NiAl intermetallic coatings produced by using mechanical alloying. Surf Coat Technol.

14. Obregón, S.A., Andrade-Gamboa, J.J., Esquivel, M.R.: Synthesis of Al-containing $MmNi_5$ by mechanical alloying: milling stages, structure parameters and thermal annealing. Int. J. Hydrog Energy 37, 14972–14977 (2012)
15. Spassov, T., Solsona, P., Suriñach, S., Baró, M.D.: Optimisation of the ball-milling and heat treatment parameters for synthesis of amorphous and nanocrystalline Mg_2Ni-based alloys. J. Alloys Compds. 349, 242–254 (2003)

16. Spassov, T., Solsona, P., Bliznakov, S., Suriñach, S., Baró, M.D.: Synthesis and hydrogen sorption properties of nanocrystalline $Mg_{1.9}M_{0.1}Ni$ (M = Ti, Zr, V) obtained by mechanical alloying. J. Alloys Compds. **356–357**, 639–643 (2003)

17. Rojas, P., Ordoñez, S., Serafini, D., Zúñiga, A., Lavernia, E.: Microstructural evolution during mechanical alloying of Mg and Ni. J. Alloys Compds. **391**, 267–276 (2005)

18. Gennari, F.C., Esquivel, M.R.: Structural characterization and hydrogen sorption properties of nanocrystalline Mg_2Ni. J. Alloys Compds. **459**, 425–432 (2008)

19. Ebrahimi-Purkani, A., Kashani-Bozorg, S.F.: Nanocrystalline Mg_2Ni-based powders produced by high-energy ball milling and subsequent annealing. J. Alloys Compds. **456**, 211–215 (2008)

20. Ordoñez, S., Rojas, P., Bustos, O., Martínez, V., Serafini, D., San Martín, Y.A.: Crystalline Mg_2Ni obtained from Mg–Ni amorphous precursor produced by mechanical alloying. J. Mater. Sci. Lett. **22**, 717–720 (2003)

21. Abdellaoui, M., Gaffet, E.: A mathematical and experimental dynamical phase diagram for ball-milled Ni10Zr7. J. Alloys Compds. **209**, 351–361 (1994)

22. Abdellaoui, M., Gaffet, E.: The physics of mechanical alloying in a planetary ball mill: mathematical treatment. Acta Metal. Mater. **43**, 1087–1098 (1995)

23. Krill III, C.E., Haberkorn, R., Birringer, R.: Fabrication and spectroscopic characterization of organic nanocrystals. In: Nalwa, H.S. (ed.) Handbook of Nanostructured Materials and Nanotechnology, chap. 8. Academic Press, San Diego (2000)

24. Modak, S., Karan, S., Roy, S.K., Mukherjee, S., Das, D., Chakrabarti, P.K.: Preparation and characterizations of SiO_2-coated nanoparticles of $Mn_{0.4}Zn_{0.6}Fe_2O_4$. J. Magn. Magn. Mater. **321**, 169–174 (2009)

25. Niu, H., Northwood, D.O.: Enhanced electrochemical properties of ball-milled Mg_2Ni electrodes. Int. J. Hydrog. Energy **27**, 69–77 (2002)

26. Poter, D.A., Easterling, K.E.: Phase Transformations in Metals and Alloys. Chapman and Hall, London (1992)

27. Ruggeri, S., Lenain, C., Roué, L., Liang, G., Huot, J., Schulz, R.: Mechanically driven crystallization of amorphous MgNi alloy during prolonged milling: applications in NiMH batteries. J. Alloys Compd. **339**, 195–201 (2002)

28. Yamamoto, K., Orimo, S., Fujii, H., Kitano, Y.: Hydriding properties of the heat-treated MgNi alloys with nanostructural designed multiphase. J. Alloys Compd. **293–295**, 546–551 (1999)

29. Westlake, D.G.: Site occupancies and stoichiometries in hydrides of intermetallic compounds: geometric considerations. J. Less Common Met. **90**, 251–273 (1983)

30. Switendick, A.C.: Band structure calculations for metal hydrogen systems. Z. Phys. Chem. Neue Folge **117**, 89–112 (1979)

18

Synthesis of nanosized MnO_2 prepared by the polyol method and its application in high power supercapacitors

E. Goikolea · B. Daffos · P. L. Taberna ·
P. Simon

Abstract Over the last years, the different polymorphs of MnO_2 have been intensely studied as alternative compounds to amorphous hydrous RuO_2 as supercapacitor electrode materials. In the present work, nanosized birnessite-type MnO_2 platelets were synthesized via the polyol method followed by a subsequent ligand removal with NaOH. The resulting compound is easily dispersible in polar solvents, thus allowing the preparation of stable dispersions (2–3 days) that could be used in thin electrode preparation. The capacitance of the synthesized product was 130 F g^{-1} in a potential window of 0.8 V at a scan rate of 2 mV s^{-1}. The synthesized material was also studied using a cavity microelectrode to evaluate the electrochemical performance of the nanostructured oxide at high scan rates. Cyclic voltammetry measurements were successfully carried out both in the previous potential window between 0.1 and 0.9 V (vs. SHE) and in a larger potential window between −0.6 and 1.1 V (vs. SHE) from 0.1 and up to 1,000 mV s^{-1}. Therefore, herein prepared material could be potentially used for high power applications, although further work should be carried out to upscale such compound without compromising the performance.

Keywords Electrochemical capacitors ·
Pseudocapacitance · Manganese dioxide ·
Nanostructured materials · Cavity microelectrode

E. Goikolea
CIC Energigune, Arabako Parke Teknologikoa,
Albert Einstein 48, 01510 Miñano, Spain

E. Goikolea · B. Daffos · P. L. Taberna · P. Simon (✉)
Université Paul Sabatier, CIRIMAT UMR CNRS 5085,
31062 Toulouse Cedex 4, France
e-mail: simon@chimie.ups-tlse.fr

Introduction

Now that our energy consumption and production culture is changing toward a more sustainable model, the development of improved energy storage systems (EESs) is essential for the consolidation of the different renewable energy sources. For all those applications where high power is a requirement, small electrical devices such as screwdrivers or camera flashes and acceleration/braking units in electric vehicles for instance, supercapacitors or electrochemical capacitors have emerged as a promising type of EES [1, 2]. The advantages of these systems over batteries are indeed their higher power density, much longer cycle-life and faster charging-discharging times. However, since the charge is stored electrostatically by adsorbing the ions from the electrolytic solution on the surface of the active material by charging the so-called double layer, the specific energy of supercapacitors is nearly 1 order of magnitude lower than in batteries (about 10 vs. 100 Wh kg^{-1}, respectively) where the charge storing mechanism involves bulk faradic or redox reactions [3].

Material wise, the most widely used supercapacitors are those based on highly conductive, high surface area carbon materials [4–6]. The specific energy of these materials, ranging from activated carbons to the more in demand graphene [7, 8], can be increased when using organic electrolytes or ionic liquids instead of aqueous electrolytes, i.e., when enlarging the cell potential. A second type of materials comprises conducting polymers and metal oxides, in which the main charge storage mechanism is associated with fast and reversible surface faradic reactions, although double layer capacitance contribution can also play a minor role in the total charge storage capability. Consequently, these materials, also referred as pseudocapacitive materials, can exhibit higher capacitance and, ultimately, higher specific energy than the carbon counterparts in a given electrolyte [9].

Among the major requirements that a metal oxide should fulfill in supercapacitor applications, high electrical conductivity and accessible oxidation states over a range of potentials with no irreversible transformations are of great importance. Regarding these properties and the resulting electrochemical performance, RuO_2 still seems to be the best choice [10], with capacitance values that in the case of amorphous $RuO_2 \cdot nH_2O$ can reach about 900 F g^{-1} in a potential window of 0.8 V [11]. However, the high cost of ruthenium limits the use of its oxides to military, spatial or medical applications where the budget is not an issue. In the search of a synergy among cost, safety and electrochemical performance, MnO_2 polymorphs have turned out to be good candidates [12–14], among which the layered birnessite has emerged as a good compromise between electrochemical performance and sample preparation. In any case, hindered by its poor conductivity and with the only exception of very thin layers of material [15], the specific capacitance of MnO_2 is far from the theoretical 1,250 F g^{-1}, calculated for a redox process involving one electron per manganese atom in a potential window of 0.9 V, and instead it rarely exceeds 400 F g^{-1} [16–20]. The large fluctuations in specific capacitance values that can be found in the literature come principally from the very diverse synthetic routes followed by each research group, which affect the chemical composition, crystallinity, morphology and particle size, and eventually the pseudocapacitive behavior.

In this sense, we have chosen the decomposition of an organometallic precursor in a polyalcohol medium at mild temperature as the method to synthesize birnessite-type nanostructured MnO_2. This so-called polyol method is considered one of the most facile and efficient methods to synthesize nanosized materials. It has already been successfully used to prepare various metal oxide nanoparticles [21–24] and nanosized $LiFePO_4$ for Li-ion battery applications [25, 26]. Regarding manganese oxides, Mn_2O_3 and Mn_3O_4 nanoplates of different shapes [27] and 5–13 nm size nanocrystals of Mn_3O_4 for catalytic [28] or magnetic [29] applications have been synthesized. However, as far as we are concerned, no MnO_2 has been prepared following this synthetic route.

Herein, we present a novel method to obtain nanostructured MnO_2 with birnessite-type structure via the polyol method. The material was studied as a possible candidate for aqueous supercapacitor electrodes both in a conventional three-electrode cell and also by means of cavity microelectrode (CME) technique. Electrochemical tests were performed both in the conventional potential window of 0.8 V and also in a larger potential window of 1.9 V to evaluate the potential usage of the material for high current applications.

Experimental

Materials

Manganese(II) acetylacetonate (Mn(acac)2) and triethylene glycol (TREG) (≥99 %) were purchased from Sigma-Aldrich and used as received without further purification. Sodium hydroxide (98.9 %) was from Fisher Scientific, ethanol and acetone were purchased from Panreac S.A., whereas super 65 conductive black additive was from TimCal.

Synthesis of nanostructured MnO_2

Nanostructured MnO_2 was prepared by reducing Mn(acac)2 in a polyol medium according to a modification of previously reported protocols [23, 30]. Mn(acac)$_2$ (2 mmol) was dispersed in triethylene glycol (30 mL) and stirred at room temperature (18 °C) under a flow of N_2 for 30 min. Under continuously stirring (150 rpm), the mixture was first heated at 180 °C for 30 min and then heated to reflux (~260 °C) for another 30 min. Subsequently, the heating of the mixture was stopped and the black-brown mixture was cooled down to room temperature (~2 h). Under ambient conditions, a 50:50 mixture of ethanol and deionized water was added to the organic mixture (~50 mL) along with a lower volume of an aqueous NaOH solution (1 M, 1 mL). The whole mixture was first stirred for ~1 min and later centrifuged (3,500 rpm, 10 min) to precipitate and separate a dark solid. The collected sample was redispersed in the ethanol–deionized water mixture and precipitated two more times by adding aqueous NaOH to the previous mixture. Finally, the basic mixture was repeatedly washed with acetone until pH ~7, and after removing the solvent by centrifugation (3,500 rpm, 10 min), the final solid product was isolated. The so-obtained nanostructured MnO_2 dark solid could be redispersed in water or in ethanol.

Characterization

Thermogravimetric/differential thermal analysis (TGA) was performed in a Netzsch STA 449 F3 analyzer with a heating rate of 10 °C min^{-1} in Ar atmosphere. Fourier transform infrared (FTIR) spectrum was obtained with a Perkin Elmer Spectrum 400 FT-IR/FT-FIR spectrometer. Analytical measurements were carried out in an atomic emission spectrophotometer with inductively coupled plasma (ICP-AES) (Horiba Jobin Yvon, Activa) using a quartz Meinhard concentric nebulizer, a Scott-type spray chamber and a standard quartz sheath connection between the spray chamber and the torch. Transmission electron microscopy (TEM) images were obtained using a FEI Tecnai G2 microscope at an acceleration voltage of 200 kV. Samples were prepared by drop-casting previously ethanol dispersed powders onto a copper grid. The crystallinity was analyzed by X-ray diffraction (XRD) using a Bruker D8 ADVANCE.

Electrochemical characterization

The electrochemical tests of the nanostructured MnO_2 were carried out with a BioLogic VMP3 or a SP-200 (for ultra-low current) Potentiostat/Galvanostats in a three-electrode configuration using two different working electrodes: a CME and a circular 1.5 cm^2 electrode, both in 0.1 M K_2SO_4 (aq) at 25 °C.

CME technique was used to study the electrochemical performance of the nanostructured MnO_2 powder at different scan rates, starting from 0.1 mV s^{-1} and going up to 1 V s^{-1}. One of the strong points of using the CME technique lays indeed in the ability to go up to several V s^{-1} due to the very low amount of active material studied in the cavity (ohmic drop arising from the bulk of the electrolyte can be neglected). The CME used in this study has already been described in a previous study [31]. Briefly, a hole of 30 μm in diameter and 40 μm in depth was produced by a laser beam at the end of a Pt wire of 60 μm in diameter sealed in a 50 mm thick glass wire. Inside the glass support, the Pt wire was connected to a Cu wire collector via the addition of carbon graphite powder.

For the sake of an optimal electrical contact, nanostructured MnO_2 powder was mixed with acetylene black in the 70/30 weight ratio. This dry composite was further milled in a mortar and manually pressed into the cavity. The so-prepared electrode was integrated to the electrochemical cell along with a Pt electrode (0.25 mm^2 plate) used as counter electrode and a Ag/AgCl (sat KCl) or, in the case of low scan rates, a Hg/Hg_2SO_4 (sat K_2SO_4) as reference electrode.

The electrochemical cell used in the three-electrode system comprised an activated carbon counter electrode, an Ag/AgCl (sat KCl) as a reference electrode and a MnO_2 composite electrode as the working electrode. 1.5 cm^2 circular working electrodes were prepared by mixing nanostructured MnO_2 powder, acetylene black and PTFE in the 60/30/10 weight ratio in the presence of ethanol. Like in the previous method, carbon black additive is required to improve the poor electronic conductivity of the manganese oxide. The mixture was continuously stirred until the evaporation of the solvent and the resulting black paste was subsequently rolled into

films. After drying at 120 °C under vacuum, the ~190 μm thick films were cut into 1.5 cm^2 circular electrodes. Activated carbon electrodes were prepared using the same technique but mixing the active carbon (Norit DLC Super 30) with PTFE in the 95/5 weight ratio. The thickness of the so-prepared 1.5 cm^2 circular electrodes was ~200 μm.

The specific capacitance, C (F g^{-1}) of the MnO_2 composite electrode was determined by integrating the cyclic voltammogram to obtain the corresponding voltammetric charge, Q, and by dividing the latter by the weight of the active material in the composite electrode ($0.6m_{electrode}$) and by the width of the potential window (ΔE).

$$C = \frac{Q}{\Delta E \times 0.6m_{electrode}} \quad (1)$$

For the calculation, the voltammetric charge due to the presence of acetylene black (12 F g^{-1}) was subtracted from the Q value.

Results and discussion

Structural characterization

The XRD pattern of the sample synthesized using the polyol process and presented in Fig. 1 shows very broad diffraction maxima as expected for particle sizes within the nano domain. The peaks at 2θ values of 12.9°, 28.5°, 36.9° and 65.5° can be indexed as the layered 2D allotropic form of MnO_2 (birnessite) [13] (JCPDS 43-1456) with an average basal spacing of 6.8 Å. This interlayer distance depends strongly on the cationic (Na^+ or K^+) and H_2O content between the MnO_6 octahedra sheets. Thus, even if most of the synthetic process took place under nitrogen atmosphere and using organic solvents, the last step with NaOH (aq), which comprises solvent exchange and removal of TREG coating, seems to be crucial for the formation of the MnO_2 layered phase.

The presence of Na in the MnO_2 structure was confirmed by ICP-AES. The calculated ratio between Na and Mn (Na:Mn) was found to be 0.32:1, which also agrees with the layered birnessite structure [32]. TGA result shows a total weight loss of 18 % until about 500 °C that corresponds, first, to the evaporation of adsorbed water molecules (14 %, until 300 °C) and then to the loss of structural water (another 4 %) (Fig. 1b). Taking into account the Na content, the calculated ratio between structural H_2O and Mn (H_2O:Mn) was 0.26:1, that is, the final ratio among Na, H_2O and Mn was found to be 0.32:0.26:1. Above 500 °C, a third weight loss of 4.4 % occurs, most probably caused by the oxygen release

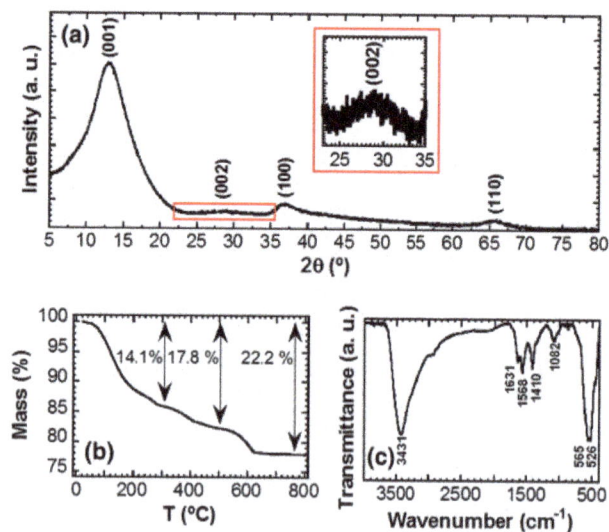

Fig. 1 a XRD pattern of the nanostructured MnO_2 synthesized by the thermal decomposition of $Mn(acac)_2$ along with the **b** FTIR spectra and **c** TGA measurement of the same sample

during the partial reduction of Mn(IV) as also reported for similar type of compounds [33]. Overall, the results suggest that the formula of our compound is $Na_{0.32}MnO_2 \cdot 0.26\ H_2O$ and the average oxidation state of Mn is 3.68.

The FTIR spectrum of the synthesized phase is shown in Fig. 1c. The broad and intense bands at $3,341\ cm^{-1}$ and the sharper one at $1,631\ cm^{-1}$ are characteristic of the stretching and bending modes of water and hydroxyl groups, respectively. These bands along with those appearing between 900 and $1,600\ cm^{-1}$ and associated with different vibration modes of Mn with –OH, O and H^+ groups corroborate the presence of hydronium cations in the structure. The intense doublet at 565 and $526\ cm^{-1}$ is characteristic of tetravalent manganese oxides with birnessite structure and corresponds to the Mn–O stretching vibrations [34, 35].

Figure 2 shows TEM images of the synthesized nanostructured MnO_2 along with the electron diffraction pattern. Contrary to the well-defined and dispersed nanoparticles usually obtained by the polyol method, herein the sample is made of agglomerated small particles of 3–5 nm packed in ~ 50 nm size clusters. Due to the high degree of agglomeration of the particles, it is difficult to give a more accurate size distribution. The nanoparticles display a platelet-type morphology which is in good accordance with the proposed 2D structure of the sample. Moreover, the structure rapidly decomposes and undergoes a phase transition under the electron beam due to the rapid evaporation of the H_2O molecules located in the interlayer of the MnO_6 octahedra sheets that hold the structure (Fig. 2a, b, d).

Electrochemical characterization

Conventional three-electrode cell

Figure 3a presents the cyclic voltammetry (CV) curve of the MnO_2 composite (MnO_2:acetylene black:PTFE) in 0.1 M K_2SO_4 (aq) using a three-electrode cell configuration with a $1.5\ cm^2$ working electrode. The CV displays almost a rectangular shape and a fast current response on potential reversal indicative of the pseudocapacitive nature of the sample. Note that the depicted curve is already the 20th one recorded and that the shape and the current were kept constant after the 2nd cycle. Very weak and broad oxidative and reductive peaks can also be observed at 0.6 and 0.4 V (vs. SHE) respectively, which could be due to a partial K^+ intercalation and deintercalation upon charge and discharge [36]. This cation diffusion seems to be favored by the layered structure of MnO_2. The calculated specific capacitance at $2\ mV\ s^{-1}$ is $130\ F\ g^{-1}$, which is within the usual capacitance value range found in the literature for similar type of samples [19].

When the whole available potential window was explored, that is, from the overpotential for O_2 evolution (1.1 V vs. SHE) to the overpotential for H_2 evolution (−0.8 V vs. SHE), an intense reduction peak was observed at low potential values (−0.10 V vs. SHE) which corresponds to the dissolution of MnO_2 due to the Mn(IV) → Mn(II) reaction (Fig. 3b). At low scan rates, very few of the Mn(II) is retrieved again into the electrode, as indicated by the low intense and broad oxidative peak at 0.3 V (vs. SHE). Both Mn(IV) ↔ Mn(II) reactions start at ~ 0.1 V (vs. SHE) which is in good agreement with the potential value obtained in the Mn Pourbaix diagram for a solution of pH 7.06 like that of a 0.1 M K_2SO_4 (aq). As a result of this loss of material, the specific capacitance of MnO_2 is significantly reduced to almost half of the initial value of $177\ mA\ h\ g^{-1}$ ($335\ F\ g^{-1}$) only after one cycle and to $34\ mA\ h\ g^{-1}$ ($65\ F\ g^{-1}$) after 25 cycles (inset of Fig. 3b). Similar faradic processes, being almost 0.5 electron per unit formula reversible, had also been previously observed when using birnessite in alkaline electrolytes [37–39].

Cavity microelectrode (CME)

The solubility of Mn(II) in the electrolytic solution is well known and has been reported elsewhere [40]. One way to avoid this non-desirable faradic reaction is to reduce the operational potential window, just as shown in Fig. 3. To gain insight into the intrinsic properties of the MnO_2, the oxide was also studied using a CME in the same three-electrode configuration. Thus, parallel to the work done using a conventional cell, the CVs of the nanostructured MnO_2 were first recorded between 0.1 and $1,000\ mV\ s^{-1}$,

Fig. 2 a TEM image of the nanostructured MnO$_2$ synthesized using the polyol method. **b** The electron diffraction pattern of the sample. **c, d** TEM images of the sample after interacting with the electron beam

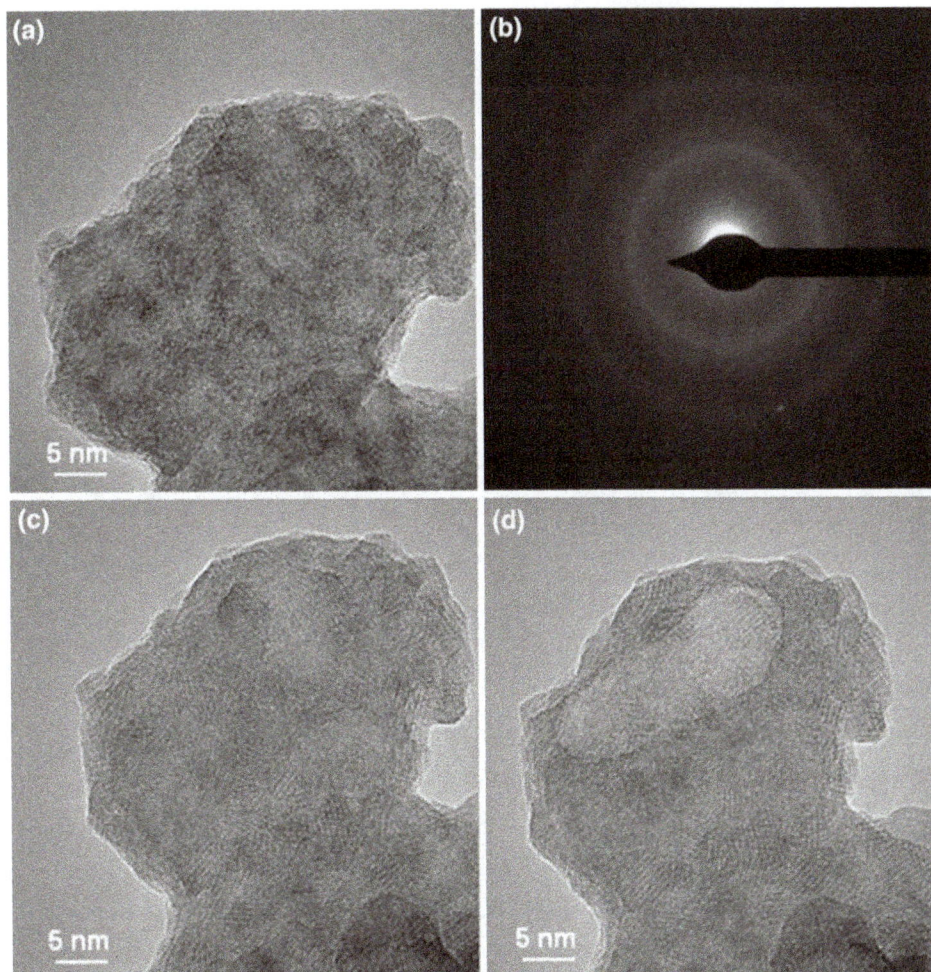

while the cell potential was controlled between 0.1 and 0.9 V (vs. SHE) using a CME cell. Figure 4a shows the CVs measured from 10 and up to 1,000 mV s^{-1} within the mentioned potential window. Note that the current intensity increases proportionally to the scan rate in the whole studied potential range following the $I = kv$ power law, where I is the current intensity in A at 0.5 V and v is the scan rate in V s^{-1} (Fig. 4b). Although slight distortions at high and low scan rates can be identified, in the full studied scan rate range, the curves show the typical rectangular shape as previously observed using conventional electrochemical cells (Fig. 3a). However, note that by CME and using very similar acetylene black to active material ratio, i.e., 30/70 in the conventional cell and 30/60 in the CME, the weak oxidative and reductive peaks observed at 0.6 and 0.4 V are no longer distinguished in all the studied scan rate range. At low scan rates (<10 mV s^{-1}), the plots slightly deviate from the ideal capacitive profile at about 0.1 V because the slower kinetics allow a faradic process to occur along with the redox pseudocapacitive charge storage. Besides, when going up to high scan rates

(>200 mV s^{-1}), the CVs are also slightly distorted due to the increasing ohmic drop in the bulk electrolyte.

Owing to the good mechanical and electrochemical stabilities of the material within the 0.1–0.9 V (vs. SHE) potential window, all the previous CVs were obtained using the same batch of active material. However, for the subsequent electrochemical characterizations in a wider potential window (from −0.6 to 1.1 V vs. SHE) where the Mn dissolution in the electrolyte occurs in the form of Mn^{2+}, another procedure was followed. A different batch of active material was used for each of the scan rates and, after each measurement, the CME was washed and refilled again with the MnO$_2$–carbon black composite. The CME was also cycled at 100 mV s^{-1} in the 0.1–0.9 V (vs. SHE) potential range just before and after measuring any of the subsequent CVs. Although the precise mass inside the CME remained unknown, these reference cycles at 100 mV s^{-1} allowed us (1) to perform the normalization in mass of the CVs showing the faradic reactions (large potential window), (2) to estimate the standard deviation in the filling of the CME ($\sigma < 9.3$ %) and also (3) to evaluate the loss of material due the redox processes.

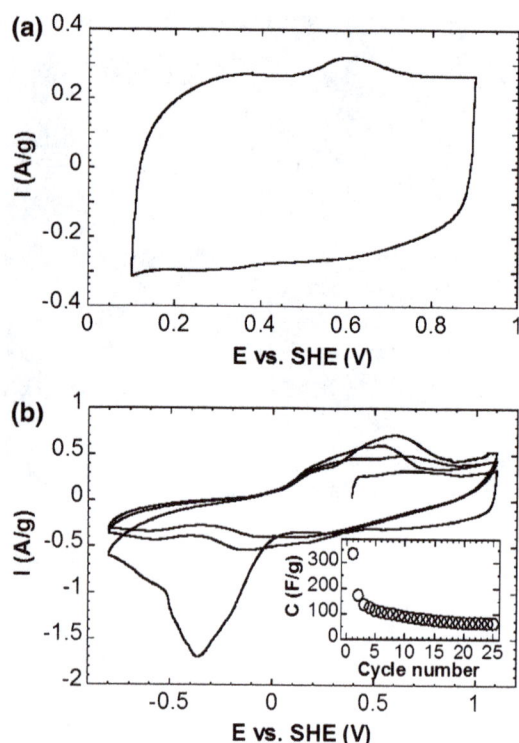

Fig. 3 CV curves of the nanostructured MnO₂ birnessite-type sample at 2 mV s⁻¹ **a** in the 0.1–0.9 V vs. SHE potential window and **b** in the −0.8 to 1.1 V vs. SHE potential window. The *inset* shows the capacitance change in the large potential window after 25 cycles

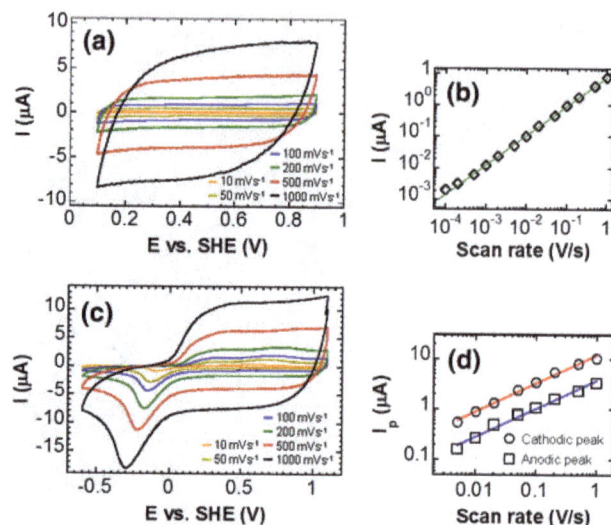

Fig. 4 **a** CV curves of the nanostructured MnO₂ birnessite-type sample at different scan rates in the 0.1–0.9 V vs. SHE potential window and **b** plot of the current intensity at 0.5 V as a function of the scan rate in a logarithm scale. **c** CV curves of the nanostructured MnO₂ birnessite-type sample at different scan rates in the −0.6 to 1.1 V vs. SHE potential window and **d** plot of the anodic (*white squares*) and cathodic (*white circles*) peak intensities as a function of the scan rate in a logarithm scale

Figure 4c shows the CVs obtained at different scan rates between −0.6 and 1.1 V (vs. SHE) after normalizing in mass following the procedure explained above. At intermediate scan rates such as 50 mV s⁻¹, the profile of the CV is similar to that depicted in Fig. 3b for the first cycle, that is, two current peaks can be well distinguished, one corresponding to the reduction of Mn(IV) to Mn(II) at −0.10 V (vs. SHE) and a second oxidative peak at 0.6 V (vs. SHE) corresponding to the oxidation reaction. However, as the scan rate increases, the CV becomes progressively more capacitive and already at 500 mV s⁻¹ only a reduction peak can be observed at −0.20 V (vs. SHE). In this case, the peak intensities follow the linear law $I = kv^{1/2}$, which confirms their faradic nature (Fig. 4d).

This shape variation is highlighted while normalizing the CVs in current, which is performed by simply dividing the current by the scan rate just as shown in Fig. 5. This normalized CV profiles also emphasize the relation between peak separation (ΔE) and scan rate (v). Overall, from 0.5 to 200 mV s⁻¹, the peak separation increases almost logarithmically with increasing scan rate. At 0.5 mV s⁻¹, the peak separation is 480 mV, too large for a reversible system, and it increases about 150 mV every tenfold increase in v. Below 10 mV s⁻¹, the separation shift remains almost equal for both peaks, while just above 5 mV s⁻¹ the shift in potential of the anodic peak is larger

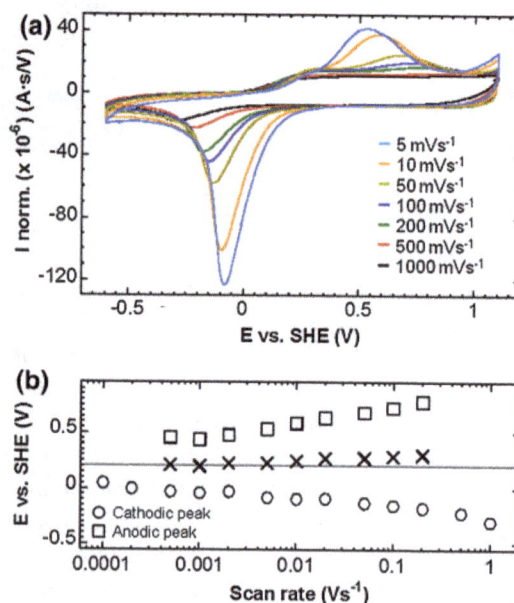

Fig. 5 **a** Normalized CV curves of the nanostructured MnO₂ birnessite-type sample at different scan rates in the −0.6 to 1.1 V vs. SHE potential window and **b** anodic (*white squares*) and cathodic (*white circles*) peak potential evolution at different scan rates. The *black crosses* show the mean peak potential

than that of the cathodic peak (40 mV larger every tenfold increase in v). At scan rates larger than 200 mV s⁻¹ and lower than 0.5 mV s⁻¹, the oxidative peak diminishes and the peak separation cannot longer be studied.

For fast scan rates, the anodic peak is obscured by the capacitive envelope of the CV and the relative intensity of the reduction peak decreases about 10 times with respect to low scan rate CVs. In this scan rate regime, faradic reactions are hampered while capacitive processes are dominant which results in a low Mn dissolution. On the contrary, at low kinetics (<0.5 mV s^{-1}), Mn^{2+} cations generated in the cathodic process cannot be retrieved into the CME. Thus, the anodic peak corresponding to the Mn(II) $\rightarrow$ Mn(IV) reaction is no longer observed, but in this case, due to the undesired loss of active material. Although very recently specific capacitance values obtained by CME have been reported for MnO$_2$ [41], in general, CME technique does not allow performing precise quantitative measurements. In the present work, the active material loss was estimated from the discharge current fading. In this sense, while at 1,000 mV s^{-1} the discharge current remains almost constant from one cycle to the subsequent and, thus, the mass loss can be considered negligible ($\sim$2 %), at 1 mV s^{-1} the current drops 25 % and already at 0.1 mV s^{-1} 84 %. Interestingly, this suggests that at high current rate, our nanoparticulate MnO$_2$ material can sustain an overvoltage with very small damage. This point could be really interesting for applications requiring high level of safety, which together with the high scan rate ability makes this compound a promising candidate for power applications. Note also that due to the good stability of the dispersions prepared in polar solvents such as water or ethanol (2–3 days), this nanostructured MnO$_2$ is well suited for thin electrode preparation. Overall, a particular effort should be made on the formulation of the nanostructured MnO$_2$ obtained by the polyol route for upscaling such compound for supercapacitor applications.

Conclusions

We report a simple and potentially scalable procedure to prepare nanostructured MnO$_2$ with birnessite-type layered structure starting from a polyol medium of Mn(acac)2 and followed by the reflux of the mixture in a two-step process first at 180 °C and finally at 260 °C under a flow of N$_2$. The layered structure of the oxide is kept stable thanks to the intercalated Na$^+$ and H$_2$O during the last stage of the synthetic process, which gives an average basal spacing of 6.8 Å. The sample consists of small platelet-type particles of 3–5 nm packed in $\sim$50 nm size clusters. When cycling the material at 2 mV s^{-1} in the usual potential window of 0.8–0.9 V, a capacitance value of 130 F g^{-1} was obtained, whereas in a larger potential window of 1.9 V, the initial capacitance was 335 F g^{-1} but faded to 65 F g^{-1} after 25 cycles due to the loss of active material through the reduction of Mn(IV) to the cationic state of Mn(II) at

-0.10 V (vs. SHE). However, the reduced size of the particles offers shorter transport and diffusion path lengths for ions and electrons that can ultimately improve charge/discharge capacities at high current densities. CME technique allowed performing cycling tests from 0.1 up to 1,000 mV s^{-1} and showed that at high scan rates MnO$_2$ can be cycled without a significant Mn dissolution even within the largest potential window. These results along with the chemical and structural properties of the material open up the possibility of using the herein prepared nanostructured MnO$_2$ for high power requirement applications.

Acknowledgments The present work was supported by the Basque Government through Etortek project energiGUNE'10. The authors thank V. Roddatis (CIC Energigune) for TEM imaging, M. Casas-Cabanas (CIC Energigune) for her assistance with XRD measurements and SGIker Advanced Research Facilities from the University of the Basque Country for ICP-AES measurements.

References

1. Miller, J.R., Burke, A.F.: Electrochemical capacitors: challenges and opportunities for real-world applications. Electrochem. Soc. Interface **17**, 53–57 (2008)
2. Miller, J.R., Simon, P.: Electrochemical capacitors for energy management. Science **321**, 651–652 (2008)
3. Simon, P., Gogotsi, Y.: Materials for electrochemical capacitors. Nat. Mater. **7**, 845–854 (2008)
4. Pandolfo, A.G., Hollenkamp, A.F.: Carbon properties and their role in supercapacitors. J. Power Sources **157**, 11–27 (2006)
5. Zhai, Y., Dou, Y., Zhao, D., Fulvio, P.F., Mayers, R.T., Dai, S.: Carbon materials for chemical capacitive energy storage. Adv. Mater. **23**, 4828–4850 (2011)
6. Presser, V., Heon, M., Gogotsi, Y.: Carbide-derived carbons—from porous networks to nanotubes and graphene. Adv. Funct. Mater. **21**, 810–833 (2011)
7. Miller, J.R., Outlaw, R.A., Holloway, B.C.: Graphene double-layer capacitor with ac line-filtering performance. Science **329**, 1637–1639 (2010)
8. Zhu, Y., Murali, S., Stoller, M.D., Ganesh, K.J., Cai, W., Ferreira, P.J., Pirkle, A., Wallace, R.M., Cychosz, K.A., Thommes, M., Su, D., Stach, E.A., Ruoff, R.S.: Carbon-based supercapacitors produced by activation of graphene. Science **332**, 1537–1541 (2011)
9. Wang, G., Zhang, L., Zhang, J.: A review of electrode materials for electrochemical supercapacitors. Chem. Soc. Rev. **41**, 797–828 (2012)
10. Hadzi-Jordanov, S., Angerstein-Kozlowska, H., Vukoviff, M., Conway, B.E.: Reversibility and growth behavior of surface oxide films at ruthenium electrodes. J. Electrochem. Soc. **125**, 1471–1480 (1978)
11. Long, J.W., Swider, K.E., Merzbacher, C.I., Rolison, D.R.: Voltammetric characterization of ruthenium oxide-based aerogels and other RuO$_2$ solids: the nature of capacitance in nanostructured materials. Langmuir **15**, 780–785 (1999)

12. Feng, Q., Kanohb, H., Ooi, K.: Manganese oxide porous crystals. J. Mater. Chem. **9**, 319–333 (1999)

13. Ghodbane, O., Pascal, J.-L., Favier, F.: Microstructural effects on charge-storage properties in MnO$_2$-based electrochemical supercapacitors. ACS Appl. Mater. Interfaces **1**, 1130–1139 (2009)

14. Ghodbane, O., Pascal, J.-L., Fraisse, B., Favier, F.: Structural in situ study of the thermal behavior of manganese dioxide materials: toward selected electrode materials for supercapacitors. ACS Appl. Mater. Interfaces **2**, 3493–3505 (2010)

15. Toupin, M., Brousse, T., Belanger, D.: Charge storage mechanism of MnO$_2$ electrode used in aqueous electrochemical capacitor. Chem. Mater. **16**, 3184–3190 (2004)

16. Lee, H.Y., Goodenough, J.B.: Supercapacitor behavior with KCl electrolyte. J. Solid State Chem. **144**, 220–223 (1999)

17. Toupin, M., Brousse, T., Bélanger, D.: Influence of microstucture on the charge storage properties of chemically synthesized manganese dioxide. Chem. Mater. **14**, 3946–3952 (2002)

18. Wang, Y., Yuan, A., Wang, X.: Pseudocapacitive behaviors of nanostructured manganese dioxide/carbon nanotubes composite electrodes in mild aqueous electrolytes: effects of electrolytes and current collectors. J. Solid State Electrochem. **12**, 1101–1107 (2008)

19. Bélanger, D., Brousse, T., Long, J.W.: Manganese oxides: battery materials make the leap to electrochemical capacitors. Electrochem. Soc. Interface **17**, 49–52 (2008)

20. Li, W., Liu, Q., Sun, Y., Sun, J., Zou, R., Li, G., Hu, X., Song, G., Ma, G., Yang, J., Chen, Z., Hu, J.: MnO$_2$ ultralong nanowires with better electrical conductivity and enhanced supercapacitor performances. J. Mater. Chem. **22**, 14864–14867 (2012)

21. Sun, S., Zeng, H.: Size-controlled synthesis of magnetite nanoparticles. J. Am. Chem. Soc. **124**, 8204–8205 (2002)

22. Feldmann, C.: Polyol-mediated synthesis of nanoscale functional materials. Adv. Funct. Mater. **13**, 101–107 (2003)

23. Sun, S., Zeng, H., Robinson, D.B., Raoux, S., Rice, P.M., Wang, S.X., Li, G.: Monodisperse MFe$_2$O$_4$ (M = Fe, Co, Mn) nanoparticles. J. Am. Chem. Soc. **126**, 273–279 (2004)

24. Salado, J., Insausti, M., Lezama, L., Gil de Muro, I., Goikolea, E., Rojo, T.: Preparation and characterization of monodisperse Fe$_3$O$_4$ nanoparticles: an electron magnetic resonance study. Chem. Mater. **23**, 2879–2885 (2011)

25. Kim, D.-H., Kim, J.: Synthesis of LiFePO$_4$ nanoparticles in polyol medium and their electrochemical properties. Electrochem Solid State Lett. **9**, A439–A442 (2006)

26. Oh, S.W., Huang, Z.-D., Zhang, B., Yu, Y., He, Y.-B., Kim, J.-K.: Low temperature synthesis of graphene-wrapped LiFePO$_4$ nanorod cathodes by polyol method. J. Mater. Chem. **22**, 17215–17221 (2012)

27. Liu, L., Yang, Z., Liang, H., Yang, H., Yang, Y.: Shape-controlled synthesis of manganese oxide nanoplates by a polyol-based precursor route. Mater. Lett. **64**, 891–893 (2010)

28. Rhadfi, T., Piquemal, J.-Y., Sicard, L., Herbst, F., Briot, E., Benedetti, M., Atlamsani, A.: Polyol-made Mn$_3$O$_4$ nanocrystals as efficient Fenton-like catalysts. Appl. Cat. A **386**, 132–139 (2010)

29. Sicard, L., Le Meins, J.-M., Méthivier, C., Herbst, F., Ammar, S.: Polyol synthesis and magnetic study of Mn$_3$O$_4$ nanocrystals of tunable size. J. Magn. Magn. Mater. **322**, 2634–2640 (2010)

30. Wan, J., Cai, W., Feng, J., Meng, X., Liu, E.: In situ decoration of carbon nanotubes with nearly monodisperse magnetite nanoparticles in liquid polyols. J. Mater. Chem. **17**, 1188–1192 (2007)

31. Come, J., Taberna, P.-L., Hamelet, S., Masquelier, C., Simon, P.: Electrochemical kinetic study of LiFePO$_4$ using cavity microelectrode. J. Electrochem. Soc. **158**, A1090–A1093 (2011)

32. Hu, Y., Zhu, H., Wang, J., Chen, Z.: Synthesis of layered birnessite type manganese oxide thin films on plastic substrates by chemical bath deposition for flexible transparent supercapacitors. J. Alloys Comp. **509**, 10234–10240 (2011)

33. Yang, X., Makita, Y., Liu, Z.-H., Sakane, K., Ooi, K.: Structural characterization of self-assembled MnO$_2$ nanosheets from birnessite manganese oxide single crystals. Chem. Mater. **16**, 5581–5588 (2004)

34. Potter, R.M., Rossman, G.R.: The tetravalent manganese oxides: identification, hydration, and structural relationships by infrared spectroscopy. Am. Miner. **64**, 1199–1218 (1979)

35. Anant, M.V., Pethkar, S., Dakshinamurthi, K.: Distortion of MnO$_6$ octahedra and electrochemical activity of Nstutite-based MnO$_2$ polymorphs for alkaline electrolytes-an FTIR study. J. Power Sources **75**, 278–282 (1998)

36. Brousse, T., Toupin, M., Dugas, R., Athouël, L., Crosnier, O., Bélanger, D.: Crystalline MnO$_2$ as possible alternatives to amorphous compounds in electrochemical supercapacitors. J. Electrochem. Soc. **153**, A2171–A2180 (2006)

37. Donne, S.W., Lawrance, G.A., Swinkels, D.A.J.: Redox processes at the manganese dioxide electrode.1. Constant-current intermittent discharge. J. Electrochem. Soc. **144**, 2949–2953 (1997)

38. Donne, S.W., Lawrance, G.A., Swinkels, D.A.J.: Redox processes at the manganese dioxide electrode 2. Slow-scan cyclic voltammetry. J. Electrochem. Soc. **144**, 2954–2961 (1997)

39. Donne, S.W., Lawrance, G.A., Swinkels, D.A.J.: Redox processes at the manganese dioxide electrode 3. Detection of soluble and solid intermediates during reduction. J. Electrochem. Soc. **144**, 2961–2967 (1997)

40. Raymundo-Piñero, E., Khomenko, V., Frackowiak, E., Béguin, F.: Performance of manganese oxide/CNTs composites as electrode materials for electrochemical capacitors. J. Electrochem. Soc. **152**, A229–A235 (2005)

41. Athouël, L., Arcidiacono, P., Ramirez-Castro, C., Crosnier, O., Hamel, C., Dandeville, Y., Guillemet, P., Scudeller, Y., Guay, D., Bélanger, D., Brousse, T.: Investigation of cavity microelectrode technique for electrochemical study with manganese dioxides. Electrochim. Acta **86**, 268–276 (2012)

19

Synthesis parameter dependence of the electrochemical performance of solvothermally synthesized Li$_4$Ti$_5$O$_{12}$

Qian Yang · Hailei Zhao · Jie Wang ·
Jing Wang · Chunmei Wang · Xinmei Hou

Abstract Pure Li$_4$Ti$_5$O$_{12}$ with high crystallinity was successfully synthesized by a solvothermal process. The effects of initial Li/Ti ratio and post-heating temperature on the phase evolution, particle morphology and electrochemical properties were systematically investigated. Excess lithium, compared to the theoretical value in Li$_4$Ti$_5$O$_{12}$, was required to get pure Li$_4$Ti$_5$O$_{12}$ due to the condensation reaction. Low Li/Ti ratio led to the appearance of secondary phase rutile TiO$_2$, while high heat-treatment temperature easily resulted in particle agglomeration of Li$_4$Ti$_5$O$_{12}$ powder. The existence of rutile TiO$_2$ decreased the specific capacity, and the particle agglomerate had a strong negative effect on the rate capability of electrode. The sample synthesized at the optimized condition exhibited a stable specific capacity of 150 mAh/g and a good rate performance.

Keywords Li$_4$Ti$_5$O$_{12}$ · Solvothermal synthesis · Heat treatment · Electrochemical performance

Q. Yang · H. Zhao (✉) · J. Wang · J. Wang · C. Wang
School of Materials Science and Engineering, University
of Science and Technology Beijing, Beijing 100083, China
e-mail: hlzhao@ustb.edu.cn

H. Zhao
Beijing Key Lab of Advanced Energy Materials,
Beijing 100083, China

X. Hou
School of Metallurgical and Ecological Engineering, University
of Science and Technology Beijing, Beijing 100083, China

Introduction

Lithium-ion batteries have attracted much attention as important energy supply in portable electronic devices, hybrid electrical vehicles and electrical vehicles because of their high power and energy density [1–3]. At present, new electrode materials exhibiting excellent rate capability and high safety performance are urgently demanded to meet the requirement of electrical vehicles. The spinel lithium titanate Li$_4$Ti$_5$O$_{12}$ is being considered as an ideal anode material in lithium-ion batteries due to its unique characteristics, including very flat charge/discharge voltage plateaus and a small structural change during charge/discharge processes. The zero-strain insertion characteristic provides material with an excellent cycling performance [4, 5], while that of the flat operating voltage at 1.55 V (versus Li$^+$/Li) can avoid the deposition of dendritic metallic lithium, therefore a high operational safety can be expected [6, 7]. Despite the high Li deintercalation/intercalation potential, it can, in principle, be coupled with high-voltage cathodes such as LiNi$_{0.4}$Mn$_{1.6}$O$_4$ to provide a cell with an operating voltage of approximately 3 V [8].

However, Li$_4$Ti$_5$O$_{12}$ is an insulator, its rate capability is greatly limited by its inherently low lithium-ion diffusivity and electronic conductivity. Typical approaches to resolve this problem include employing nanoparticles to reduce the diffusion length of lithium ions, and increase the contact area between the electrode and the electrolyte [9–11], doping Li$_4$Ti$_5$O$_{12}$ with aliovalent cation (Al^{3+}, Ga^{3+}, Co^{3+}, Mg^{2+}, Ta^{5+}) [12–14] in Li and Ti sites to produce mixed valence of Ti^{3+}/Ti^{4+}, and thus increase the electronic conductivity, and incorporating directly the conductive second phase (carbon, Ag and so on) [7, 15, 16].

Actually, the particle size and the crystalline ordering degree have strong impacts on the electrochemical properties of electrode. Small-sized active material can not only reduce the lithium-ion diffusion distance, but also increases the contact area with conductive reagent and electrolyte solution, thus can decrease the local current density and mitigate the electrode polarization. The high crystallinity is believed to be beneficial to the good cycling stability of electrode [9]. Compared to the doped materials, the pure material is easier to be synthesized and handled in practical operations. Many methods, including conventional solid-state reaction [12–14], sol–gel method [6, 17, 18], solvothermal technique [19–22], combustion synthesis [23], rheological phase reaction [11] and other synthesis routes, have been exploited to prepare $Li_4Ti_5O_{12}$ materials. Among them, solvothermal technique with simple and flexible controls has spurred considerable interests. Although $Li_4Ti_5O_{12}$ powders prepared by solvothermal method have been investigated extensively [19–22], the work concerning the effect of the synthesis parameters on the electrochemical properties is very limited. Considering that the practical composition of the synthesized material via solvothermal route is usually different from the nominal composition, in this work, the effect of initial Li/Ti ratio in starting solution on the phase purity and the electrochemical properties was investigated. The influence of the post-heat-treatment temperature on the electrochemical performance of $Li_4Ti_5O_{12}$ electrode was also addressed. The synthesized $Li_4Ti_5O_{12}$ exhibited excellent rate capability and cycling performance, showing the solvothermal synthesis is a promising method to obtain high-performance $Li_4Ti_5O_{12}$ anode material.

Experimental

Materials synthesis

The spherical precursors of $Li_4Ti_5O_{12}$ powders were synthesized by solvothermal method using lithium acetate (LiAc, AR $\geq$99.0 %, Beijing Yili Fine Chemicals Co., Ltd.) and tetrabutyl titanate [$Ti(O(CH_2)_3CH_3)_4$, denoted as $Ti(OR)_4$, AR $\geq$99.0 %, Beijing Jinlong Chemical Reagent Co., Ltd.) as Li and Ti cation sources, respectively. The molar ratios of the mixtures were fixed at different proportions (Li/Ti ratio = 0.8–1.4). $Ti(OR)_4$ was dissolved in ethanol under magnetic stirring, and then LiAc was added into the mixtures with further stirring to obtain a homogeneous dispersion system. The concentration of $Ti(OR)_4$ in ethanol was 1.4×10^{-4} mol/ml. The transparent solution was then transferred into a 100 ml teflon-lined stainless steel autoclave and kept at 180 °C for 24 h. After cooling down to room temperature, a milky white precursor was prepared. The produced powder was washed and filtered with ethanol to eliminate the unreacted reagents and

the partial organic compounds. The precipitate was dried at 80 °C in air for 3 h. To obtain well-crystallized $Li_4Ti_5O_{12}$, the precursor was calcined at 800 °C for 2 h in air with a heating rate of 5 °C/min. At last, the effect of heat-treatment temperature on the particle morphology and electrochemical properties was investigated. The precursor with the optimal Li/Ti ratio based on the above results was subjected to calcination at temperatures of 400, 600, and 800 °C, respectively.

Characterization

Phase purity and crystallinity of the synthesized samples were identified by means of powder X-ray diffraction (XRD) performed on a Rigaku D/MAX-A diffractometer with Cu Kα radiation source ($\lambda = 1.54056$ Å) in the range of $10° \leq 2\theta \leq 90°$, while the morphology and size distribution of precursors and post-treated powders were observed on a LEO-1450 scanning electron microscope (SEM). The actual molar ratio of Li/Ti in the precursor was determined by inductively coupled plasma atomic emission spectrometer (ICP-AES) (IRIS Intrepid II XSP). The thermal behavior of the precursor powders was examined by a thermogravimetry–differential thermal analysis (TG–DTA) instrument (Netzsch STA 449 C) with a heating rate of 10 °C/min from room temperature to 900 °C under air.

Electrochemical measurement

Half-cells were used to evaluate the electrochemical performance. Celgard 2400 microporous membrane was used as separator, a lithium foil as negative electrode, and 1 M $LiPF_6$ dissolved in a mixture of ethyl methyl carbonate (EMC), dimethyl carbonate (DMC) and ethylene carbonate (EC) with a volume ratio of 1:1:1 as the electrolyte. The working electrode was made from the mixture of active material $Li_4Ti_5O_{12}$, acetylene black (AB) and polyvinylidene fluoride (PVdF) in a weight ratio of 85:10:5. The slurry of the mixture was uniformly pasted on aluminum foil, dried and cut into disks. Then the working electrode was dried under vacuum at 120 °C for 24 h before electrochemical evaluation. Cell assembly was carried out in a glove box filled with high-purity argon where the oxygen and water vapor contents were each <1 ppm.

The galvanostatic charging–discharging test was employed to evaluate the cycle stability and electrochemical capacity of the samples by a computer-controlled Land CT 2001A battery test system. The cell was cycled at different current densities, and the cut-off voltage for charging and discharging processes was 1.0–2.5 V, respectively. The specific capacity was calculated based on the whole weight of the synthesized samples, including $Li_4Ti_5O_{12}$ and possible impurity TiO_2. AC electrochemical impedance spectroscopy was measured by a Solartron 1260/1287 (UK) impedance analyzer in the frequency range from 1 MHz to 0.01 Hz at the state of fully lithiated. Experiments were carried out at room temperature.

Results and discussion

Effect of Li/Ti molar ratio

Due to the complicated coordination process of tetrabutyl titanate [$Ti(OR)_4$] with LiAc in the solvothermal condition, the chemical composition of the resultant is usually different from the nominal composition. To prepare high-purity $Li_4Ti_5O_{12}$, it is essential to control the starting molar ratio of Li/Ti in the reaction mixtures. Different solutions with Li/Ti ratio = 0.8, 1.0, 1.2 and 1.4 were solvothermally treated in an autoclave, and the precipitates after heat treatment were subjected to the phase identification by XRD. The results are shown in Fig. 1. The samples with Li/Ti molar ratios of 0.8, 1.0, 1.2, 1.4 are named as LTO0.8, LTO1.0, LTO1.2 and LTO1.4, respectively. For all samples, diffraction peaks indexed on the cubic spinel phase $Li_4Ti_5O_{12}$ with $Fd\bar{3}m$ space group (JCPDS No. 49-0207) are observed. However, some additional diffraction peaks corresponding to rutile TiO_2 with P42/mnm space group (JCPDS No. 21-1276) are also detected at the same time except for sample LTO1.4. Although the theoretical Li/Ti ratio of $Li_4Ti_5O_{12}$ is 0.8, sample LTO0.8 shows strong TiO_2 peaks, demonstrating that a significant amount of lithium remained in the solution. With increasing Li/Ti ratio, the relative peak intensity of $Li_4Ti_5O_{12}$ increases gradually, while that of TiO_2 decreases

remarkably and finally tends to disappear. When the molar ratio of Li/Ti reaches 1.4, the single-phase spinel type $Li_4Ti_5O_{12}$ without any impurity is obtained.

Compared to the theoretical ratio, the excess lithium required to obtain pure single-phase $Li_4Ti_5O_{12}$ in the solvothermal synthesis route is related to the reaction mechanism of $Ti(OR)_4$ and LiAc. Under solvothermal condition, part of tetrabutyl titanate $Ti(OR)_4$ may first resolve in ethanol and take alcoholysis reaction to form $Ti(OR)_{4-x}(OH)_x$, as expressed by reaction (1). The $Ti(OR)_{4-x}(OH)_x$ monomers then condense with LiAc to produce $Ti(OR)_{4-x}(OH)_{x-y}(OLi)_y$ via reaction (2). Different condensation reactions occur among the clusters of $Ti(OR)_{4-x}(OH)_{x-y}(OLi)_y$. The condensation may occur between Ti–(OH) groups, producing H_2O as by-product, while the reaction may also occur between Ti–OH and Ti–OLi producing LiOH, corresponding to reaction (3) and (4), respectively. The product of the solvothermal reaction was composed of mainly Li–Ti–O amorphous material with some remained organic radicals that already show the basic lattice structure of $Li_4Ti_5O_{12}$, as shown in Fig. 6 (sample LTO), and small amount of LiOH as by-product. Therefore, part of the lithium remains in the solution. If Li source LiAc is not excess in the starting materials than the theoretically required amount, then the attached lithium in the solid particles after condensation reactions (3) and (4) will be inadequate to form $Li_4Ti_5O_{12}$, leading to the generation of trace of TiO_2 after heat treatment, as evidenced in Fig. 1.

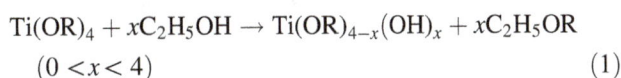

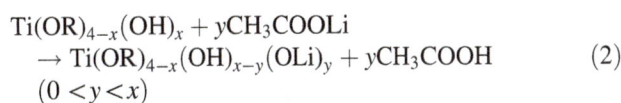

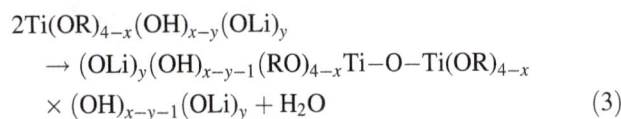

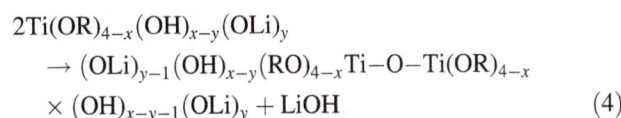

$$Ti(OR)_4 + xC_2H_5OH \rightarrow Ti(OR)_{4-x}(OH)_x + xC_2H_5OR$$
$$(0 < x < 4) \tag{1}$$

$$Ti(OR)_{4-x}(OH)_x + yCH_3COOLi$$
$$\rightarrow Ti(OR)_{4-x}(OH)_{x-y}(OLi)_y + yCH_3COOH \tag{2}$$
$$(0 < y < x)$$

$$2Ti(OR)_{4-x}(OH)_{x-y}(OLi)_y$$
$$\rightarrow (OLi)_y(OH)_{x-y-1}(RO)_{4-x}Ti-O-Ti(OR)_{4-x}$$
$$\times (OH)_{x-y-1}(OLi)_y + H_2O \tag{3}$$

$$2Ti(OR)_{4-x}(OH)_{x-y}(OLi)_y$$
$$\rightarrow (OLi)_{y-1}(OH)_{x-y}(RO)_{4-x}Ti-O-Ti(OR)_{4-x}$$
$$\times (OH)_{x-y-1}(OLi)_y + LiOH \tag{4}$$

Analyzed by means of ICP-AES, the actual Li/Ti molar ratio in the precursor of sample LTO1.4 is 1.0, indicating that some lithium remained in solution, which is in good agreement with reaction (4). Considering that the Li/Ti is 0.8 in $Li_4Ti_5O_{12}$, the high Li/Ti (1.0) in the precursor of LTO1.4 implies that some lithium is lost during calcination [17, 24, 25].

To investigate the thermal decomposition behavior of the precursor, TG–DTA examination was performed on the

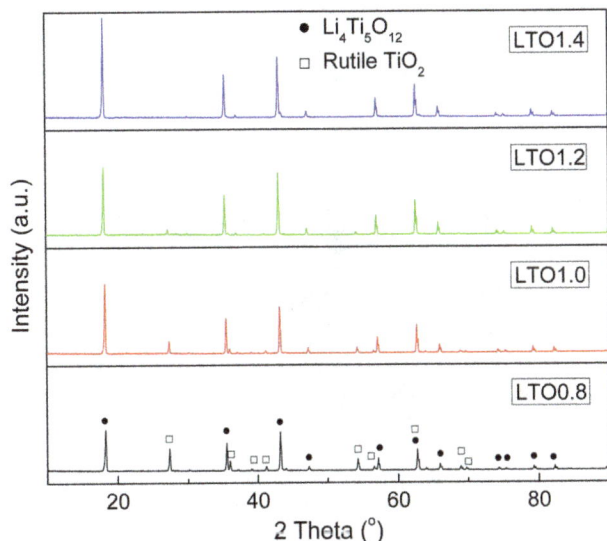

Fig. 1 XRD patterns of $Li_4Ti_5O_{12}$ samples after heat treated at 800 °C in air

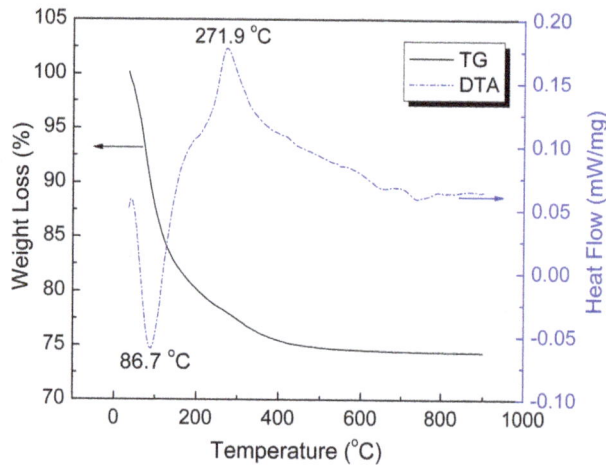

Fig. 2 TG–DTA curves of precursor with initial Li/Ti = 1.4

precursor of sample LTO1.4. The result is shown in Fig. 2. In the TG curve, the total mass loss obtained in the temperature range from room temperature to 900 °C is approximately 25.64 %. The first step of mass loss about 19.06 % occurred between room temperature and 180 °C, corresponding to the endothermic peak at 86.7 °C in the DTA curve, is due to the removal of adsorbed ethanol and water molecules. The second step of mass loss occurred in 180–600 °C, associating with the exothermic peak at 271.9 °C, is attributed to the loss of the organics and the formation of $Li_4Ti_5O_{12}$ phase. When the temperature is above 600 °C, no major weight loss was examined, indicating that the decomposition of organic groups was completed.

The particle morphologies of $Li_4Ti_5O_{12}$ precursors with different molar ratios of Li/Ti are shown in Fig. 3. The samples are all composed of well-dispersed spherical particles with some small particles adhering to their surface. With increasing Li/Ti ratio, the number of small particles decreases remarkably, the average particle size reduces from 3 to 2 μm and the particle distribution tends to be more uniform.

After heat treatment at 800 °C in air, the powders exhibit even smaller particle size of about 1–1.5 μm and smoother particle surface, as presented in Fig. 4. This is considered to be mainly resulted from the decomposition of organic groups on the particle surface of $Li_4Ti_5O_{12}$ precursor.

With $Li_4Ti_5O_{12}$/Li half-cell, the cycling performances of samples LTO0.8, LTO1.0, LTO1.2 and LTO1.4 at 0.2 C were examined, and the results are shown in Fig. 5a. All the synthesized active materials of LTO0.8, LTO1.0, LTO1.2 and LTO1.4 display a stable cycling performance. The specific capacity of these samples increases with increasing Li/Ti ratio, and sample LTO1.4 exhibits the highest specific capacity among these samples. The existence of rutile TiO_2 is apparently detrimental to lithium storage capacity of the samples. This is in good agreement with the literature results that only small amounts of lithium ions can be intercalated in rutile TiO_2 at room temperature [26, 27].

The initial discharge–charge potential curves of samples LTO0.8, LTO1.0, LTO1.2 and LTO1.4 are shown in Fig. 5b. All the samples exhibit only one typical discharge/charge potential plateau of $Li_4Ti_5O_{12}$ representing a two-phase reaction between $Li_4Ti_5O_{12}$ and $Li_7Ti_5O_{12}$ [28], no plateau corresponds to the lithiation/delithiation process of rutile TiO_2 [29].

Fig. 3 SEM images of $Li_4Ti_5O_{12}$ precursors obtained by solvothermal reaction with **a** Li/Ti = 0.8, **b** Li/Ti = 1.0, **c** Li/Ti = 1.2, **d** Li/Ti = 1.4

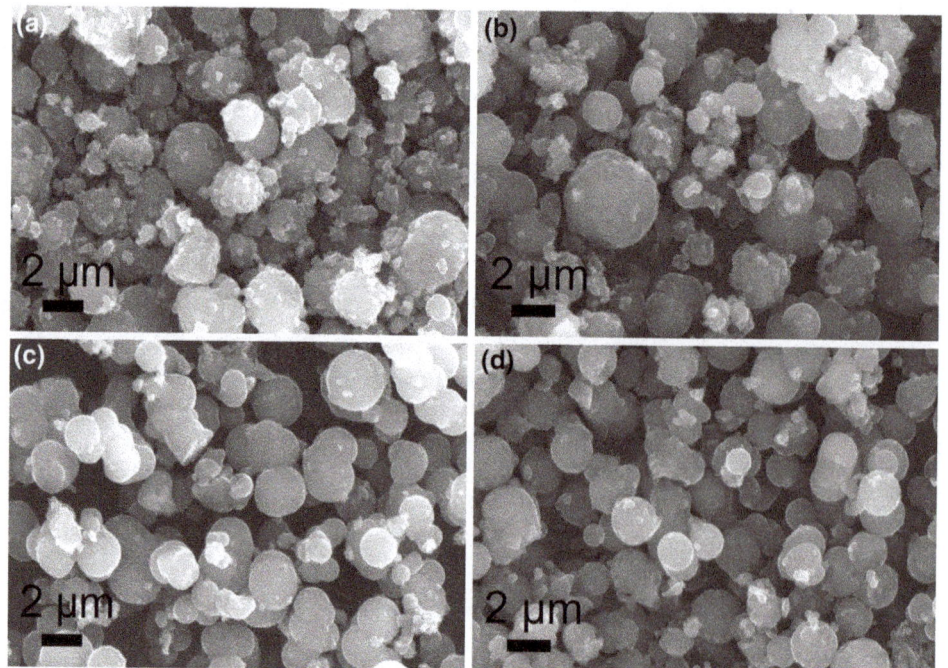

Fig. 4 SEM images of $Li_4Ti_5O_{12}$ samples after heat treated at 800 °C in air **a** LTO0.8, **b** LTO1.0, **c** LTO1.2, **d** LTO1.4

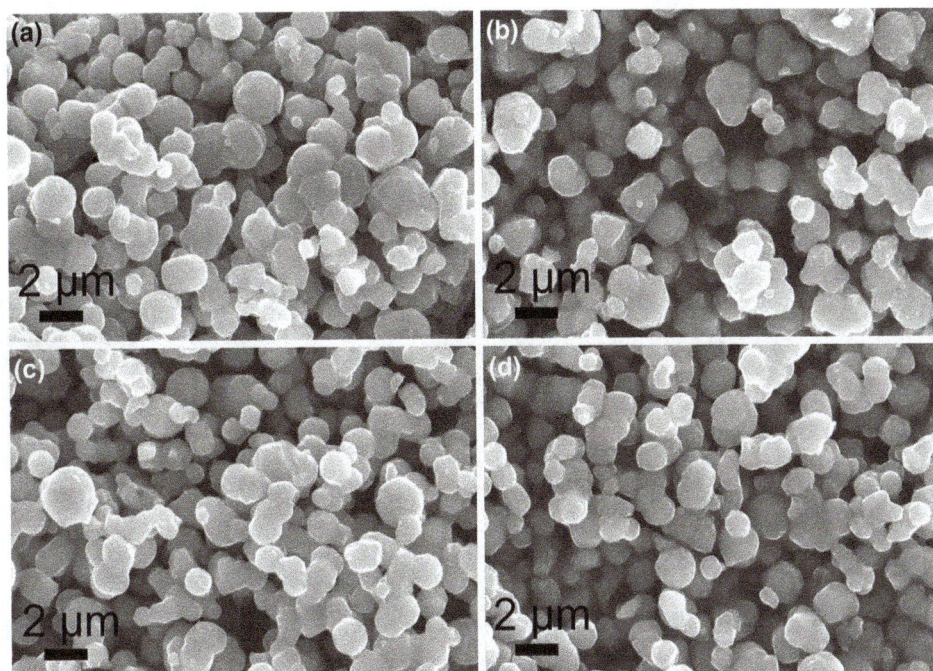

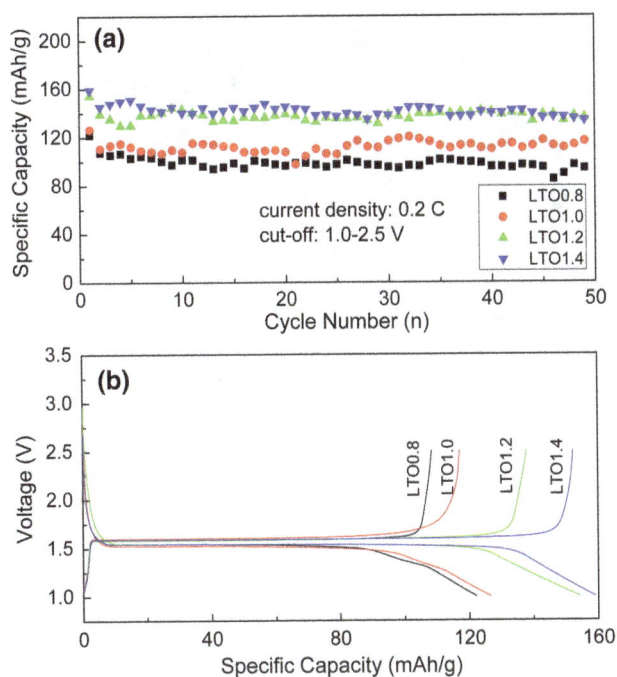

Fig. 5 a Discharge specific capacity of $Li_4Ti_5O_{12}$ prepared from starting materials with different Li/Ti ratios; **b** initial discharge–charge curves of the samples LTO0.8, LTO1.0, LTO1.2 and LTO1.4

The results indicate that the impurity rutile TiO_2 in the samples does not have electrochemical activity in this condition.

Effect of heat-treatment temperature

Considering that the electrochemical performances of $Li_4Ti_5O_{12}$ are closely related with its crystallinity, a further heat-treatment investigation was conducted. Precursor of sample LTO1.4, showing pure phase and high specific capacity after 800 °C-treatment, was selected to subject the heat-treatment test to clarify the effect of treating temperature on the structure and electrochemical performance of synthesized $Li_4Ti_5O_{12}$. The precursor and the samples heat treated at 400, 600 and 800 °C are denoted as LTO, LTO4, LTO6 and LTO8, accordingly. Figure 6 shows the XRD patterns of these samples. They are all identified with a pure cubic phase $Li_4Ti_5O_{12}$. The precursor features amorphous structure. The appearance of messy background and the broad peaks with weak intensities indicate the poor crystallinity of the formed precursor. The peak intensities of the spinel phase significantly enhance when the heat-treatment temperature increases, exhibiting the improvement of crystallinity. When the heat-treating temperature is up to 600 °C, good crystallinity and pure spinel phase $Li_4Ti_5O_{12}$ have been identified from the XRD data. The lattice parameters calculated according to the XRD data are 8.3395 (3), 8.3661 (9), 8.3695 (9) Å for LTO4, LTO6 and LTO8, respectively, which are in good agreement with previous reported values [6, 30, 31]. The average crystallite sizes calculated from the full width at half maximum (FWHM) of peak (111) are 11.47, 44.70 and 50.04 nm for LTO4, LTO6 and LTO8, respectively. The crystallite size increases with the increase of heat-treatment temperature. The results demonstrate that pure $Li_4Ti_5O_{12}$ powders with high crystallinity, small crystallite size can be successfully synthesized by solvothermal method. Additionally, the synthesized $Li_4Ti_5O_{12}$ powders display high dispersity and good sphericity without particle agglomeration, which are

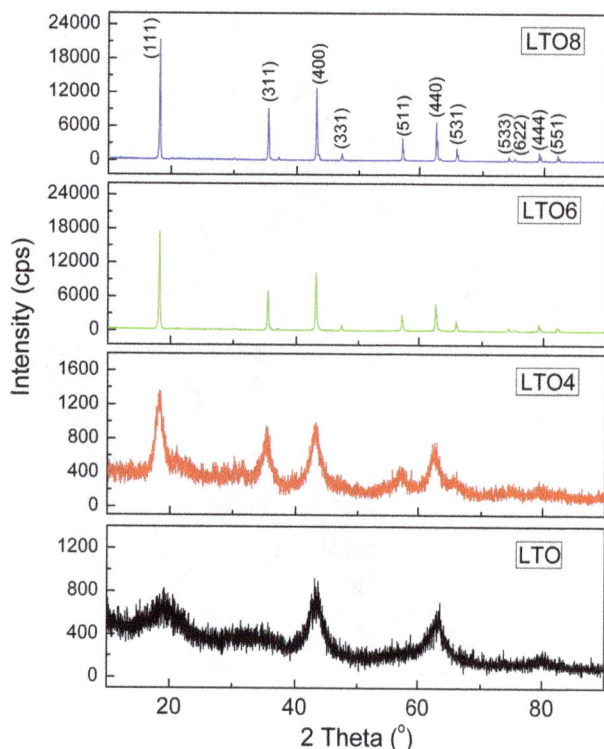

Fig. 6 XRD patterns of the precursor and the final powders calcined at different temperatures

beneficial for both of electrochemical performance and practical electrode preparation. In most previous studies, spinel $Li_4Ti_5O_{12}$ was synthesized by solid-state reaction at 800–1000 °C for 5–24 h [12–14]. Compared with the conventional solid-state technique, the calcination temperature of the solvothermally synthesized products was greatly decreased and the dwelling time was significantly shortened.

The morphologies of the $Li_4Ti_5O_{12}$ powders after heat treatment are shown in Fig. 7. With increasing temperature, the particle size decreases slightly and the particle surface becomes much smoother. However, the 800 °C-treated sample LTO8 shows an obvious particle agglomeration, several small particles stick together to form a large one, showing a poor dispersity.

To investigate the influence of heat-treatment temperature on the electrochemical performance, the cycling performances of the samples LTO4, LTO6 and LTO8 were examined. As shown in Fig. 8, sample LTO4 displays extremely high irreversible capacity compared to other samples, mainly resulting from the remained organic groups on the particle surface due to its lower heating temperature. All samples show good cycling stability since the second cycle, while sample LTO6 exhibits the highest electrochemical capacity of 150 mAh/g among all the samples. To understand the dependence of electrochemical capacity of LTO on the heating temperature, several

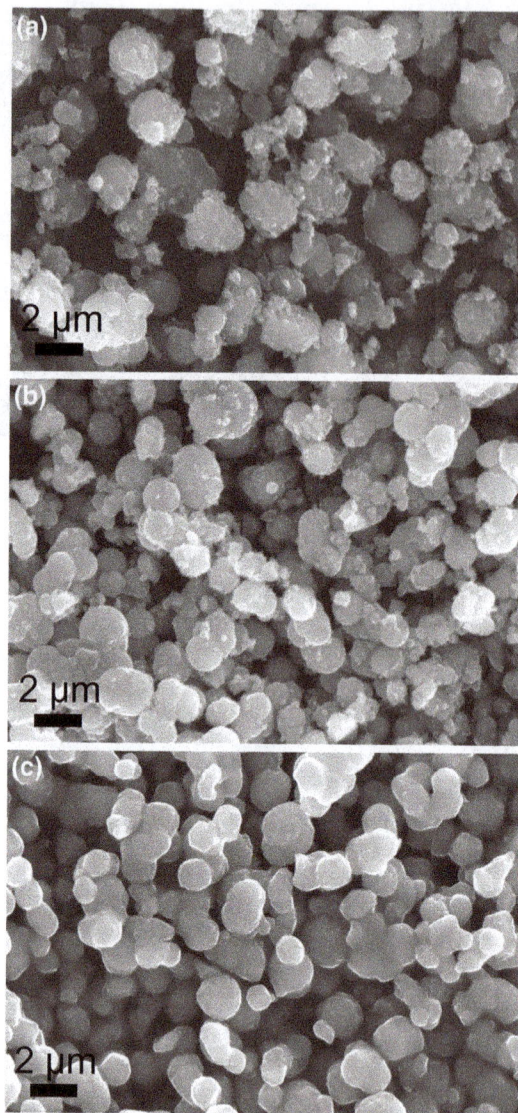

Fig. 7 SEM images of samples **a** LTO4, **b** LTO6, **c** LTO8

aspects should be taken into account. The first is the crystallinity of powders. High crystallinity is usually beneficial to the good electrochemical performance of $Li_4Ti_5O_{12}$ negative electrode [9]. On the other hand, particle agglomeration is unfavorable for the diffusion of lithium ions due to the elongated diffusion distance. The characteristics of good crystallinity and small particle size of sample LTO6 guarantee its high specific capacity. Due to the obvious particle agglomeration, sample LTO8 exhibits the lowest specific capacity, even worse than sample LTO4. Besides the long diffusion path of lithium ions of the agglomerated particles that impedes the kinetics of lithium intercalation into the LTO8 host structure, the lowered specific surface area should be taken into consideration because it can reduce the contact area between the electrode and the AB, and therefore deteriorate the electronic conduction. Furthermore, the lowered specific

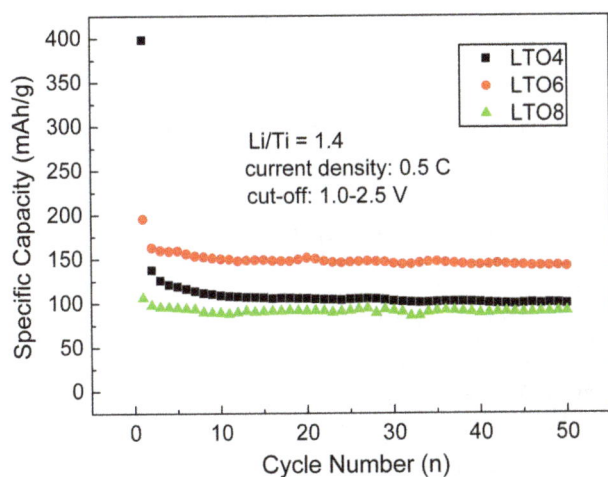

Fig. 8 Discharge capacities of samples LTO4, LTO6 and LTO8

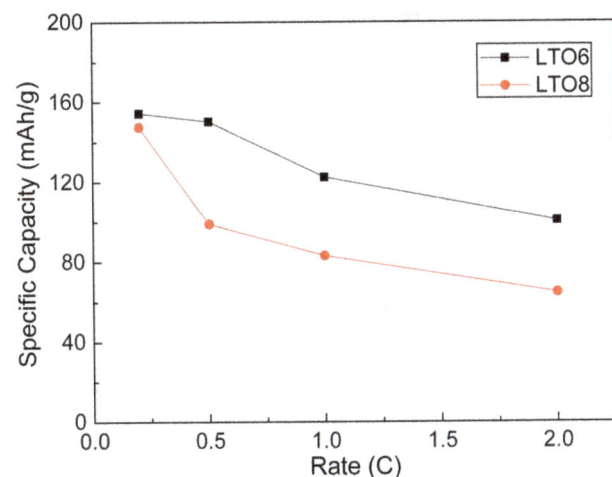

Fig. 9 Specific capacities of samples LTO6 and LTO8 at different rates between 1.0 and 2.5 V

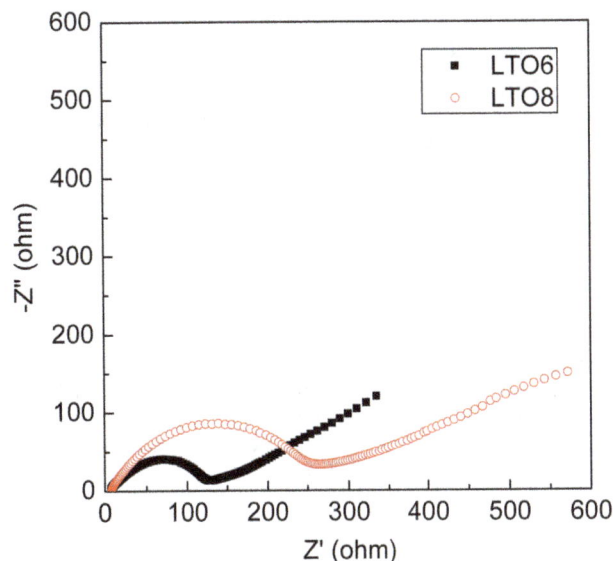

Fig. 10 AC impedance spectra of Li/Li$_4$Ti$_5$O$_{12}$ cells with LTO6 and LTO8 as active material

AC impedance measurement was performed on LTO6 and LTO8 electrodes, the results are shown in Fig. 10. Each curve is comprised of a depressed semi-circle in the high-medium frequency region and an oblique straight line in the low frequency region. The semi-circle in the high-medium frequency region is mainly related to the charge-transfer process, while the inclined line in the low frequency region is attributed to the Warburg impedance that reflects lithium-ion diffusion behavior in the Li$_4$Ti$_5$O$_{12}$ electrode [34]. Sample LTO6 displays a significantly lower resistance of charge transfer than that of sample LTO8, which is certainly associated with the larger specific surface area of active material LTO6. These results can interpret well the experimental results in Figs. 8 and 9.

Conclusion

Pure and well-crystallized Li$_4$Ti$_5$O$_{12}$ powders were synthesized from Ti(O(CH$_2$)$_3$CH$_3$)$_4$ and LiAc by solvothermal route involving a further heat treatment at relatively low temperature with short dwelling time. The Li/Ti molar ratio in starting materials and post-heat-treatment temperature have strong impacts on the electrochemical performance of Li$_4$Ti$_5$O$_{12}$ anode material. Excess lithium, compared to the theoretical value in Li$_4$Ti$_5$O$_{12}$, is required for the synthesis of pure-phase Li$_4$Ti$_5$O$_{12}$ in this preparation condition. Low Li/Ti ratio (<1.4 in atom) easily results in the coexistence of the secondary phase TiO$_2$, while high treatment temperature leads to particle agglomeration of Li$_4$Ti$_5$O$_{12}$ powders. The former causes the decrease in specific capacity, but the latter deteriorates the rate capability of

surface area can result in the increase in actual current density on particle surface, and thus cause a large polarization of the electrode, which is another reason for the low specific capacity of sample LTO8 [6, 31–33].

Figure 9 illustrates the rate capabilities of LTO6 and LTO8 electrodes. The discharge capacity of both samples decreases gradually with increasing current density. Nevertheless, the sample LTO6 maintains a higher reversible capacity at the same current density as compared to the sample LTO8, indicating the good rate capability of sample LTO6. At lower current density of 0.2 C, both the samples reveal relatively good capacity characteristics due to the sufficient insertion/extraction of lithium ions. High current density makes their specific capacity different. The capacity ratio of $C_{2\,C}/C_{0.2\,C}$ is 0.65 and 0.44 for samples LTO6 and LTO8, respectively. The good rate capability as well as the high specific capacity of sample LTO6 makes it a promising material for a better commercial application.

electrode. The 600 °C-treated sample LTO6 with starting $LiAc/Ti(OR)_4 = 1.4$ exhibits the highest specific capacity and best rate performance due to its high purity, good crystallinity and excellent dispersity. At 0.5 C, LTO6 displays a stable reversible capacity of ca. 150 mAh/g. EIS measurement reveals that LTO6 has lower charge-transfer resistance compared to the 800 °C-treated LTO8, which is mainly attributable to the high specific surface area and small particle size of sample LTO6.

Acknowledgments This work was financially supported by National Basic Research Program of China (2013CB934003), "863" program (2013AA050902), National Nature Science Foundation of China (21273019) and the Foundamental Research Funds for the Central Universities (FRF-MP-13-002B).

References

1. Tarascon, J.-M., Armand, M.: Issues and challenges facing rechargeable lithium batteries. Nature **414**, 359–367 (2001)
2. Sato, K., Noguchi, M., Demachi, A., Oki, N., Endo, M.: A mechanism of lithium storage in disordered carbons. Science **264**, 556–558 (1994)
3. Scrosati, B., Garche, J.: Lithium batteries: status, prospects and future. J. Power Sources **195**, 2419–2430 (2010)
4. Ohzuku, T., Ueda, A., Yamamoto, N.: Zero-strain insertion material of $Li[Li_{1/3}Ti_{5/3}]O_4$ for rechargeable lithium cells. J. Electrochem. Soc. **142**, 1431–1435 (1995)
5. Ariyoshi, K., Yamato, R., Ohzuku, T.: Zero-strain insertion mechanism of $Li[Li_{1/3}Ti_{5/3}]O_4$ for advanced lithium-ion (shuttlecock) batteries. Electrochim. Acta **51**, 1125–1129 (2005)
6. Jiang, C., Ichihara, M., Honma, I., Zhou, H.: Effect of particle dispersion on high rate performance of nano-sized $Li_4Ti_5O_{12}$ anode. Electrochim. Acta **52**, 6470–6475 (2007)
7. Yuan, T., Cai, R., Shao, Z.: Different effect of the atmospheres on the phase formation and performance of $Li_4Ti_5O_{12}$ prepared from ball-milling-assisted solid-phase reaction with pristine and carbon-precoated TiO_2 as starting materials. J. Phys. Chem. C **115**, 4943–4952 (2011)
8. Patoux, S., Daniel, L., Bourbon, C., Lignier, H., Pagano, C., Cras, F.L., Jouanneau, S., Martinet, S.: High voltage spinel oxides for Li-ion batteries: from the material research to the application. J. Power Sources **189**, 344–352 (2009)
9. Guerfi, A., Sévigny, S., Lagacé, M., Hovington, P., Kinoshita, K., Zaghib, K.: Nano-particle $Li_4Ti_5O_{12}$ spinel as electrode for electrochemical generators. J. Power Sources **119–121**, 88–94 (2003)
10. Prakash, A.S., Manikandan, P., Ramesha, K., Sathiya, M., Tarascon, J.-M., Shukla, A.K.: Solution-combustion synthesized nanocrystalline $Li_4Ti_5O_{12}$ as high-rate performance Li-ion battery anode. Chem. Mater. **22**, 2857–2863 (2010)
11. Yin, S.Y., Song, L., Wang, X.Y., Zhang, M.F., Zhang, K.L., Zhang, Y.X.: Synthesis of spinel $Li_4Ti_5O_{12}$ anode material by a modified rheological phase reaction. Electrochim. Acta **54**, 5629–5633 (2009)
12. Huang, S., Wen, Z., Zhu, X., Lin, Z.: Effects of dopant on the electrochemical performance of $Li_4Ti_5O_{12}$ as electrode material for lithium ion batteries. J. Power Sources **165**, 408–412 (2007)
13. Chen, C.H., Vaughey, J.T., Jansen, A.N., Dees, D.W., Kahaian, A.J., Goacher, T., Thackeray, M.M.: Studies of Mg-substituted $Li_{4-x}Mg_xTi_5O_{12}$ spinel electrodes ($0 \leq x \leq 1$) for lithium batteries. J. Electrochem. Soc. **148**, A102–A104 (2001)
14. Wolfenstine, J., Allen, J.L.: Electrical conductivity and charge compensation in Ta doped $Li_4Ti_5O_{12}$. J. Power Sources **180**, 582–585 (2008)
15. Cheng, L., Yan, J., Zhu, G.N., Luo, J.Y., Wang, C.X., Xia, Y.Y.: General synthesis of carbon-coated nanostructure $Li_4Ti_5O_{12}$ as a high rate electrode material for Li-ion intercalation. J. Mater. Chem. **20**, 595–602 (2010)
16. Huang, S., Wen, Z., Zhang, J., Yang, X.: Improving the electrochemical performance of $Li_4Ti_5O_{12}$/Ag composite by an electroless deposition method. Electrochim. Acta **52**, 3704–3708 (2007)
17. Sorensen, E.M., Barry, S.J., Jung, H.-K., Rondinelli, J.R., Vaughey, J.T., Poeppelmeier, K.R.: Three-dimensionally ordered macroporous $Li_4Ti_5O_{12}$: effect of wall structure on electrochemical properties. Chem. Mater. **18**, 482–489 (2006)
18. Bach, S., Pereira-Ramos, J.P., Baffier, N.: Electrochemical properties of sol–gel $Li_{4/3}Ti_{5/3}O_4$. J. Power Sources **81–82**, 273–276 (1999)
19. Fattakhova, D., Petrykin, V., Brus, J., Kostlánová, T., Dĕdeček, J., Krtil, P.: Solvothermal synthesis and electrochemical behavior of nanocrystalline cubic Li–Ti–O oxides with cationic disorder. Solid State Ion. **176**, 1877–1885 (2005)
20. Kostlánová, T., Dĕdeček, J., Krtil, P.: The effect of the inner particle structure on the electronic structure of the nano-crystalline Li–Ti–O spinels. Electrochim. Acta **52**, 1847–1856 (2007)
21. Li, J., Tang, Z., Zhang, Z.: Controllable formation and electrochemical properties of one-dimensional nanostructured spinel $Li_4Ti_5O_{12}$. Electrochem. Commun. **7**, 894–899 (2005)
22. Lee, S.C., Lee, S.M., Lee, J.W., Lee, J.B., Lee, S.M., Han, S.S., Lee, H.C., Kim, H.J.: Spinel $Li_4Ti_5O_{12}$ nanotubes for energy storage materials. J. Phys. Chem. C **113**, 18420–18423 (2009)
23. Yuan, T., Wang, K., Cai, R., Ran, R., Shao, Z.: Cellulose-assisted combustion synthesis of $Li_4Ti_5O_{12}$ adopting anatase TiO_2 solid as raw material with high electrochemical performance. J. Alloys Compd. **477**, 665–672 (2009)
24. Zhang, B., Liu, Y., Huang, Z., Oh, S., Yu, Y., Mai, Y.-W., Kim, J.-K.: Urchin-like $Li_4Ti_5O_{12}$-carbon nanofiber composites for high rate performance anodes in Li-ion batteries. J. Mater. Chem. **22**, 12133–12140 (2012)
25. Huang, S., Wen, Z., Gu, Z., Zhu, X.: Preparation and cycling performance of Al^{3+} and F^- co-substituted compounds Li_4Al_x-$Ti_{5-x}F_yO_{12-y}$. Electrochim. Acta **50**, 4057–4062 (2005)
26. Koudriachova, M.V., Harrison, N.M., de Leeuw, S.W.: First principles predictions for intercalation behavior. Solid State Ion. **175**, 829–834 (2004)
27. Zachau-Christiansen, B., West, K., Jacobsen, T., Atlung, S.: Lithium insertion in different TiO_2 modifications. Solid State Ion. **28–30**, 1176–1182 (1988)
28. Liu, J., Li, X., Yang, J., Geng, D., Li, Y., Wang, D., Li, R., Sun, X., Cai, M., Verbruggeb, M.W.: Microwave-assisted hydrothermal synthesis of nanostructured spinel $Li_4Ti_5O_{12}$ as anode materials for lithium ion batteries. Electrochim. Acta **63**, 100–104 (2012)
29. Chen, J.S., Lou, X.W.: The superior lithium storage capabilities of ultra-fine rutile TiO_2 nanoparticles. J. Power Sources **195**, 2905–2908 (2010)
30. Colin, J.F., Godbole, V., Novák, P.: In situ neutron diffraction study of Li insertion in $Li_4Ti_5O_{12}$. Electrochem. Commun. **12**, 804–807 (2010)

31. Jung, H.G., Kim, J., Scrosati, B., Sun, Y.K.: Micron-sized, carbon-coated $Li_4Ti_5O_{12}$ as high power anode material for advanced lithium batteries. J. Power Sources **196**, 7763–7766 (2011)

32. Zhang, N., Liu, Z., Yang, T., Liao, C., Wang, Z., Sun, K.: Facile preparation of nanocrystalline $Li_4Ti_5O_{12}$ and its high electrochemical performance as anode material for lithium-ion batteries. Electrochem. Commun. **13**, 654–656 (2011)

33. Aricò, A.S., Bruce, P., Scrosati, B., Tarascon, J.-M., van Schalkwijk, W.: Nanostructured materials for advanced energy conversion and storage devices. Nat. Mater. **4**, 366–377 (2005)

34. Yoshikawa, D., Kadoma, Y., Kim, J.-M., Ui, K., Kumagai, N., Kitamura, N., Idemoto, Y.: Spray-drying synthesized lithium-excess $Li_{4+x}Ti_{5-x}O_{12-\delta}$ and its electrochemical property as negative electrode material for Li-ion batteries. Electrochim. Acta **55**, 1872–1879 (2010)

Transesterification of peanut and rapeseed oils using waste of animal bone as cost effective catalyst

Ali A. Jazie · H. Pramanik · A. S. K. Sinha

Abstract Heterogeneous catalysts were developed from goat animal bones for biodiesel production via transesterification process. Desirable feedstock like peanut and rapeseed oils were chosen as raw material for the transesterification process. The bone catalysts calcined at 900 °C shows low crystallite size (41.47434 nm) and higher surface area (90.6523 m^2/g) compared to catalysts calcined at other temperatures. The maximum biodiesel yield of 94 % for peanut oil and 96 % for rapeseed oil were obtained at 20:1 molar ratio of methanol to oil, addition of 18 wt% of bone catalyst (calcined at 900 °C, 2 h), 60 °C reaction temperature and reaction time of 4 h. The fuel properties of biodiesel produced were compared with ASTM standards for biodiesel. Reusability of the catalyst was also tested.

Keywords Biodiesel · Peanut oil · Rapeseed oil · Animal bone waste · Heterogeneous catalysis

Introduction

Biodiesel seems to be a realistic fuel for future; it has become more attractive recently because of its environmental benefits. Biodiesel is an environmentally friendly fuel that can be used in any diesel engine without modification [1]. Recently, heterogeneous catalysts derived from renewable materials, such as oyster shell [2], egg shell [3], mud crab shell [4], and mollusk shells [5] have been employed for conversion of oils to biodiesel. Previously, those catalysts source were generally considered as waste. Normally, disposal of these waste materials from seafood processing are an economic or environmental problem for entrepreneurs and local governments. However, biodiesel production catalysts prepared from these "wastes" are a promising "green" technology. Hydroxyapatite the main component of bones and teeth, attracts considerable interests in many areas because of acid–base properties, ion-exchange ability, and adsorption capacity [6]. Thus, bone waste of animal which contain hydroxyapatite (HA) can also be used as raw material to develop heterogeneous green catalysts for biodiesel production. HA is an efficient solid base catalyst used for many reactions such as the Michael addition [7] and the reaction of ring opening [8]. Bone does have 65–70 % hydroxyapatite and 30–35 % organic compounds (on a dry weight basis) [9]. Recently, Obadiah et al. worked on the biodiesel production from palm oil using calcined waste animal bone as catalyst. The biodiesel yield was 96.78 % under optimal reaction conditions of 20 wt% of catalysts, 1:18 oil to methanol ratio, 200 rpm of stirring of reactants and at a temperature of 65 °C [10]. Desirable feedstock characteristics include adaptability to local growing conditions, regional availability, high oil content, favorable FA composition, compatibility with existing farm practices, low agricultural inputs, definable growing season, uniform seed maturation rates, markets for byproducts, compatibility with fallow lands, and rotational adaptability with commodity crops [11]. Biodiesel prepared from feedstocks that meet all or most of these criteria hold the greatest promise as alternatives to mineral diesel. Feed stocks of interest in the current study included peanut and rapeseed oils. The peanut or groundnut belongs to the Fabaceae family and China and India represent 56 % of the world's cultivated area. The percentage of oleic acid in traditional peanut oil ranges

A. A. Jazie (✉) · H. Pramanik · A. S. K. Sinha
Department of Chemical Engineering and Technology,
Indian Institute of Technology, (Banaras Hindu University),
Varanasi 221 005, UP, India
e-mail: jazieengineer@yahoo.com

from 41 to 67 %, whereas high-oleic cultivars contain close to 80 % of this constituent [12]. Rapeseed is now second largest oilseed crop after soybean and the third largest vegetable oil and characterized by high level of erucic acid which may cause serious damage to heart and lever [13].

The main objective of the present study is to develop heterogeneous catalyst from waste animal bone which is available in plenty amount. We report our studies on the effect of process variables, i.e., catalyst concentration, catalyst type, temperature; methanol to oil molar ratio and reaction time on the yield of biodiesel produced by the transesterification of peanut and rapeseed oils. Reusability of the catalyst was tested. The properties of biodiesel were also measured and reported in this paper.

Experimental

Materials and catalyst preparation

Commercial peanut and rapeseed oils were used in the present study. Methanol was procured from Fisher Scientific, India. Methyl ester and triolien were obtained from Sigma-Aldrich, USA. All chemicals used were analytical reagents. Bones of goat animal were obtained from butcher shops in Varanasi, UP, India. Firstly, the bones were crushed into small chips. To remove impurity and undesirable material, the bone chips were rinsed several times with hot water to remove tissue and fat. The clean bone chips were subsequently dried at 378 K for 24 h in a hot air oven. The bone chips were grounded to fine powdered and another drying for the same conditions were performed before calcinations. The fine bone powder was calcined in the muffle furnace at different temperatures for 2 h under static air.

Catalyst characterizations

The crystalline phases of calcined samples were analyzed by X-ray diffraction (XRD). The samples were characterized by N_2 adsorption–desorption (Micromeritics, ASAP 2020) for their BET surface area, pore volume and pore size. FTIR spectra were obtained with FTIR (Thermo-Nicolet 5700 model). The spectra were obtained in the 500–4,000 cm^{-1} region, with a resolution of 4 cm^{-1}. Averages of 32 scans were recorded. The elemental compositions were determined by X-ray Photoelectron spectroscopy (XPS) (Kratos Amicus, Shimadzu, Japan) under vacuum mode.

Experimental setup

The transesterification reaction was carried out in a batch reactor. A 500 mL three necked round bottom glass flask

was used. It had provisions for a water-cooled condenser, thermometer, and mechanical stirrer. The flask was kept inside a water bath with thermostat which maintained the temperature from 30 to 70 °C. The reaction mixture was stirred at 600 rpm for proper mixing of catalyst and reaction mixtures for all test runs.

Transesterification reaction

The oils were heated at 378 K for 1 h in N_2-purge to evaporate water and other volatile impurities. Heated oils were allowed to cool to room temperature. Subsequently, a mixture of methanol and catalyst at a designated amount was added to the oil. Each experiment was allowed to continue for a set period of time. The reaction mixture was allowed to cool down and equilibrate which resulted in separation of two layers. The upper layer consisted of methyl esters (biodiesel) and unconverted triglycerides. The lower layer contained glycerol, excess methanol, catalyst and any soap formed during the reaction and possibly some entrained methyl esters. After separation of the two layers by sedimentation the upper methyl esters layer was dried at 378 K for 4 h to remove water content from biodiesel layer. The catalyst was separated from lower layer by centrifugation and filtration. The recovered catalysts were regenerated to check the reusability.

Testing of vegetable oil and biodiesel (methyl esters) properties

In the present work, vegetable oil and methyl esters (biodiesel) were analyzed by FTIR (Thermo-Nicolet 5700 model). The spectra were obtained in the 500–4,000 cm^{-1} region, with a resolution of 4 cm^{-1}. Averages of 32 scans were recorded using a multi bounce ATR. The method developed by Giuliano et al. [14] was used for quantitative analysis. The height of absorbance band at wave number 1,741 cm^{-1} was used to calculate the concentration of ester in the biodiesel layer. A calibration curve was obtained by measuring the height of the 1,741 cm^{-1} bands for samples of ester and oil of known compositions (methyl ester and triolien). A calibration curve (Fig. 1) between concentration of FAME and peak height is developed to determine the yield of biodiesel using Eq. (1). The yield of biodiesel was calculated using the following formula:

$$\text{Yield} = \frac{E_1}{W_0} \times E_c \qquad (1)$$

where: E_1, E_c, W_0 are the biodiesel layer volume in (mL), ester concentration in (g mL^{-1}), and weight of vegetable oil used in (g) respectively. Figure 2 shows the FTIR of vegetable oil and product biodiesel. The samples of vegetable oil and biodiesel were tested for their fuel properties

Fig. 1 The calibration curve of ester

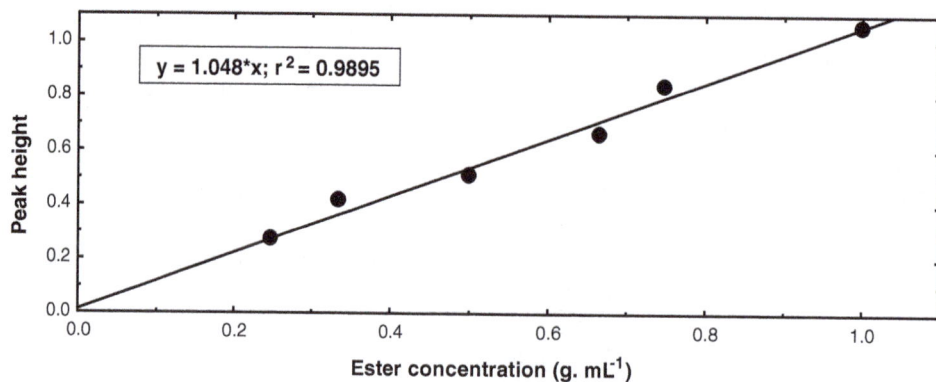

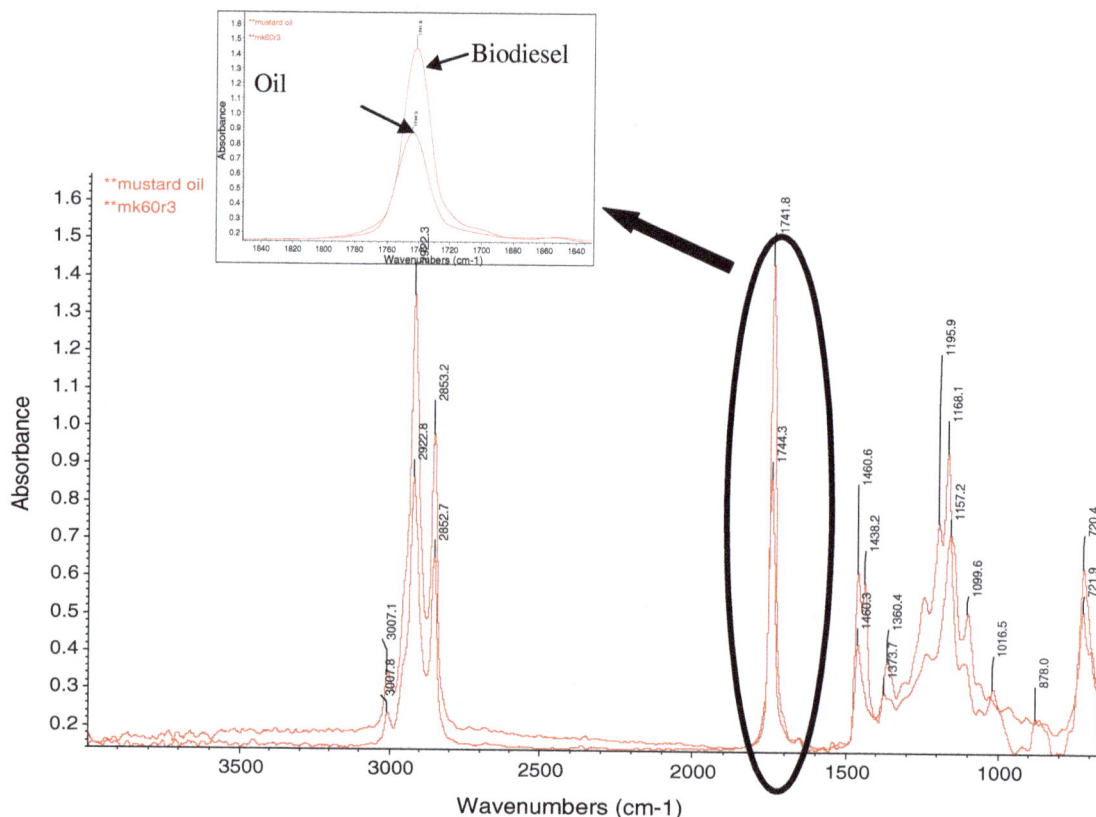

Fig. 2 The FTIR spectra of vegetable oil and vegetable oil methyl ester samples at wave number about 1,740 cm^{-1}

also. The flash point was determined by Cleveland open cup method using ASTM D92-53. The cloud point and the pour point determinations were made using cloud and pour point apparatus as specified in IP15/60. The kinematic viscosity was determined at 313 K, using a Redwood viscometer as specified in ASTM D445. Calorific value was measured using a bomb calorimeter (IP12/63T). The acid value was determined by a standard titration method as specified in ASTM D664 [15]. The density at room temperature of the biodiesel was measured as specified in ASTM D4052. The Cetane index was estimated according to ASTM D976.

Results and discussion

XRD analysis

XRD spectra of calcined animal bone samples were obtained with Cu radiation ($\lambda = 0.154178$ nm) at 40 kV, 30 mA, a scan speed of 0.1 °/s, and a scan range of 10–80°. Indexing of the diffraction peaks was done using a Joint Committee on Powder Diffraction Standards (JCPDS) file. The XRD patterns of calcined animal bone at 800, 900 and 1,000 °C, respectively were presented in Fig. 3. It can be seen that the XRD pattern of calcined animal bone at

900 °C shows sharper peaks, indicating better crystallinity. The peak positions for hydroxyapatite are in good agreement with the JCPDS (09-0432) having lattice parameters $a = b = 0.942$ nm, $c = 0.688$ nm, and no pattern indicating the presence of impurities was observed. Therefore, standard HA with hexagonal structure is formed. As shown in Fig. 3, all the peaks corresponding to the standard hydroxyapatite are obvious in the spectra of the calcined animal bone. Therefore, it can be concluded that the calcination process have eliminated the collagen and organic compounds from the animal bone and did not affect the molecular skeleton of the hydroxyapatite. Table 1 shows the planar spacings (estimated by Bragg's law) and the intensities at the strongest peaks in the XRD spectra. These results have been compared with the standard HA data (JCPDS). The HA obtained by the calcinations at 900 °C does have planar spacings and intensities very close to the standard HA.

Table 2 BET surface area (S_{BET}), total pore volume and crystallite size of the bone

Catalyst type	Calcinations temperature (°C)	Surface area (m²/g)	Total pore volume (cm³/g)	Crystallites size (nm)
Bone	800	4.0173	0.016895	57.24773
Bone	900	90.6523	0.050995	41.47434
Bone	1,000	1.2008	0.002773	95.0028

The crystalline size of animal bone catalysts were also calculated from the XRD data using Scherrer's formula given by Qin et al. [16] after correction for instrumental broadening (Eq. 2):

$$D \approx 0.9\lambda/\beta \cos 2\theta \qquad (2)$$

The crystallite size of 800, 900 and 1,000 °C calcined catalysts were calculated and the results were shown in Table 2. Whereas the crystallite size of the catalyst upon calcinations at 900 °C reduced to 41 nm. This shows that crystallinity of the animal bone decreased on calcination. Yoosuk et al. [17] also observed that calcination of CaO decreased its crystallinity with the increase in temperature.

FTIR analysis

The FTIR patterns of bone with respect to calcinations at 800, 900 and 1,000 °C are presented in Fig. 4. The presence of OH and PO_4 functional groups were confirmed by FTIR spectra. The peaks at 471.8, 568.3, 603.6, 962.4, 1,035.1 and 1,094.9 cm^{-1} were attributed to the PO_4 group, while the peaks at 1,630.2 and 3,571 cm^{-1} denote the OH group [18, 19]. Presence of carbonate group in the carbonated hydroxyapatite is either at the phosphate tetrahedron (B-type) or at the hydroxyl site (A-type). Biological apatites do have both types however; B-type is more abundant [20]. The sharp band at 3,573 cm^{-1} in all

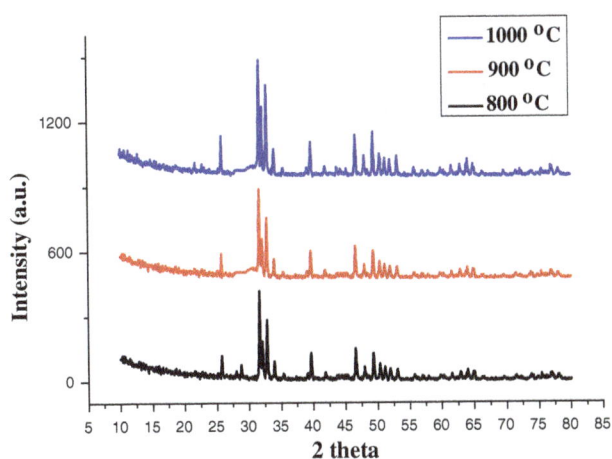

Fig. 3 XRD of the calcined bone catalyst at 800, 900 and 1,000 °C respectively

Table 1 Planar spacings and intensities obtained from X-ray diffraction for the obtained ones from calcined bone at 800, 900 and 1,000 °C the results are compared with the standard HA (JCPDS)

(hkl)	d (nm)				Relative intensity (%)			
	JCPDS	800 °C	900 °C	1,000 °C	JCPDS	800 °C	900 °C	1,000 °C
002	0.3440	0.3455	0.3445	0.3444	40	31	42	48
211	0.2814	0.2824	0.2818	0.2819	100	100	100	100
112	0.2778	0.2789	0.2783	0.2785	60	63	61	66
202	0.2631	0.2639	0.2633	0.2636	25	22	24	32
310	0.2261	0.2270	0.2265	0.2267	20	25	23	27
222	0.1943	0.1949	0.1945	0.1946	30	27	30	37
213	0.1841	0.1845	0.1843	0.1845	40	30	36	33
321	0.1806	0.1810	0.1807	0.1809	20	18	19	30
004	0.1722	0.1724	0.1722	0.1728	20	14	20	25

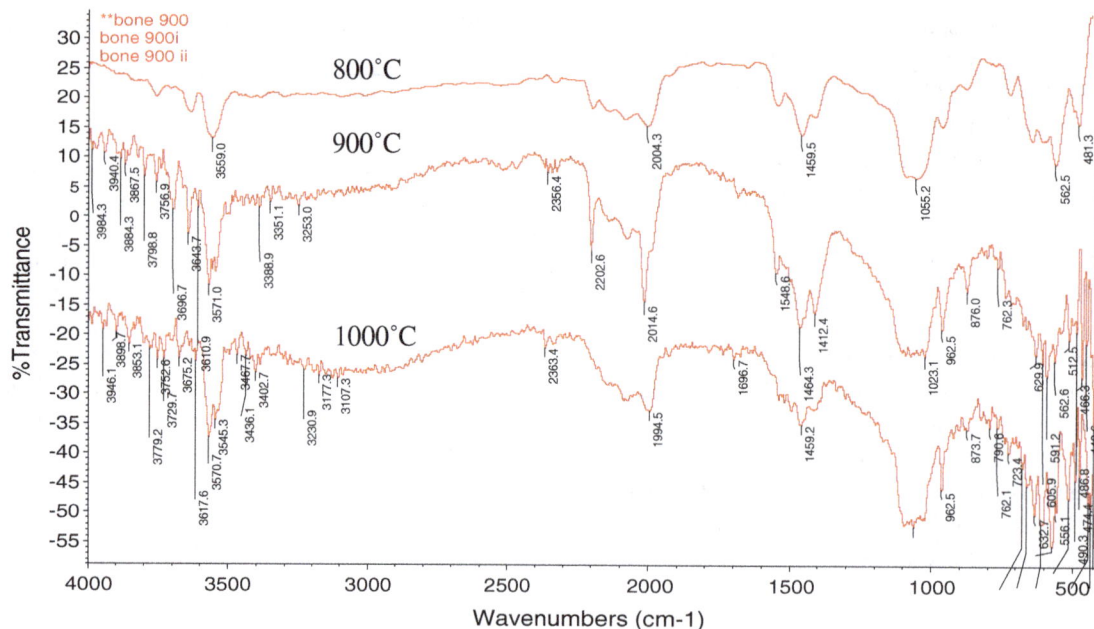

Fig. 4 FTIR spectra of bone calcined at 800, 900 and 1,000 °C

the spectra of Fig. 4 is assigned to the OH stretching mode [21]. As shown in Fig. 4, the intensity of the OH stretching band is moderate in the spectra of calcined animal bone at 800 °C and 1,000 °C and high in the spectra of calcined animal bone at 900 °C. However, the OH stretching peak is located at the same wave number for all the obtained apatites. Fleet and Liu [22] have reported that, decrease in the intensity of this band means increase in carbonate content and absence of structurally bound OH. Figure 4 shows the bands of the carbonate v_2 CO_3 (peak at 850 cm^{-1}) and v_3 CO_3 (peaks at 1,412 cm^{-1}) for both types. As shown in this figure, both types of v_2 CO_3 and v_3 CO_3 are present in the obtained apatites, however, A-type is meager. From Fig. 4, one can say all the obtained apatites have B-type carbonate.

BET surface area

As shown in Table 2, the BET surface area of catalyst synthesized from bone at 900 °C was 90.6523 m^2 g^{-1} which is higher than that of other two catalysts calcined at 800 and 1,000 °C (Table 2). The BET studies confirmed that the particle size decreased as the calcinations temperature increased from 800 to 900 °C, leading to an increase in surface area and with further increase in temperature the particle size increases due to sintering. These should be due to a severe reduction of unit cell after the complete decarbonation [23]. The bone catalysts calcined at 800 and 1,000 °C were found to be less-porous materials due to their trace pore volume. While catalyst calcined at 900 °C exhibit slightly better porous structure. The surface fraction was reflected by a significant decrease in the average pore volume and pore size (Table 2). The elemental compositions of bone catalyst calcined at 900 °C as following: C (0.3 wt%), O (37 wt%), Na (0.6 wt%), Mg (0.5 wt%), P (16 wt%) and Ca (35 wt%).

Effect of calcination temperature on biodiesel yield

Figure 5 shows effect of calcination temperature on the yield of biodiesel. When the calcination temperature was 800 °C, a yield of 80 % was achieved, whereas, a low level of activity was observed when the calcination temperature was <800 °C. The maximum biodiesel yield of 95 % was achieved for the bone calcined at 900 °C with a reaction time of 4 h at 60 °C and stirring speed of 600 rpm. It may be due to decarbonation of hydroxyapatite at different calcination temperatures, providing the different amounts of active basic sites for the transesterification. However, the biodiesel yield was decreased to 82 % for the bone calcined at 1,000 °C. The reason may be, due to sintering of the catalysts at high temperature (1,000 °C). Thus, the best catalytic performance was obtained for calcination temperatures at 900 °C, when the XRD patterns showed a prominent crystalline peak for hydroxyapatite.

Effect of catalyst concentration on biodiesel yield

Figure 6 shows the effects of catalyst concentration on the transesterification of peanut and rapeseed oils. The biodiesel yield was obtained for different catalysts concentrations from 3 to 24 wt%. Temperature was maintained at

60 °C, stirring speed of 600 rpm and methanol to oil ratio 20:1. The biodiesel yield did not reach the maximum when the catalyst concentration was maintained below 18 wt% for the two types of oils. The maximum biodiesel yield of 96 % for rapeseed oil and 94 % for peanut oil was obtained using catalyst calcined at 900 °C. However, the yield did not increase when the catalyst loading was above 18 wt%. The reason may be, at higher loading of catalysts the reaction mixture becomes high viscous due to formation of slurries which resulted in mass transport limitation. Therefore, the optimum catalyst loadings was found to be 18 wt% in this system.

Effect of methanol to oil molar ratio

Figure 7 shows the effect of methanol to oil molar ratio on biodiesel yield. The maximum biodiesel yield was obtained for 4 h of reaction under excess methanol at temperature of 60 °C. The maximum biodiesel yield of 96 % for rapeseed oil and 94 % for peanut oil was obtained using catalyst calcined at 900 °C. Biodiesel yield increases with the increase in methanol to oil molar ratio from 3:1 to 20:1. Further increase in methanol to oil molar ratio beyond 20:1, the biodiesel yield decreased (Fig. 7). The high amount of methanol (methanol to oil ratio of 20:1)

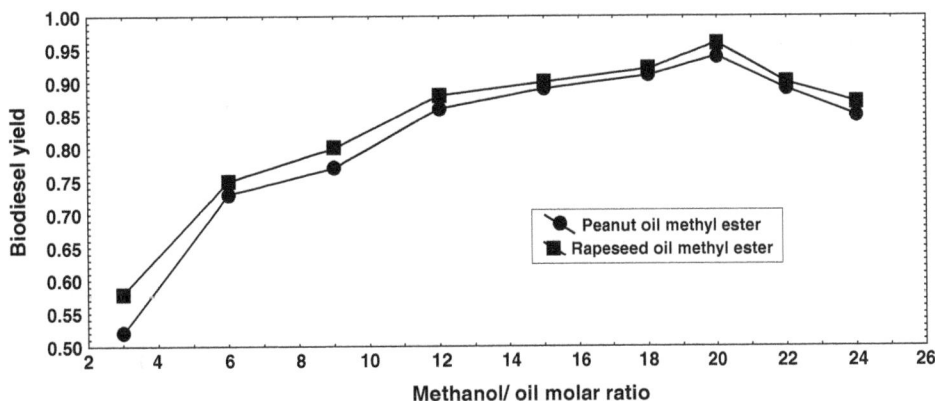

Fig. 5 Effect of calcination temperature

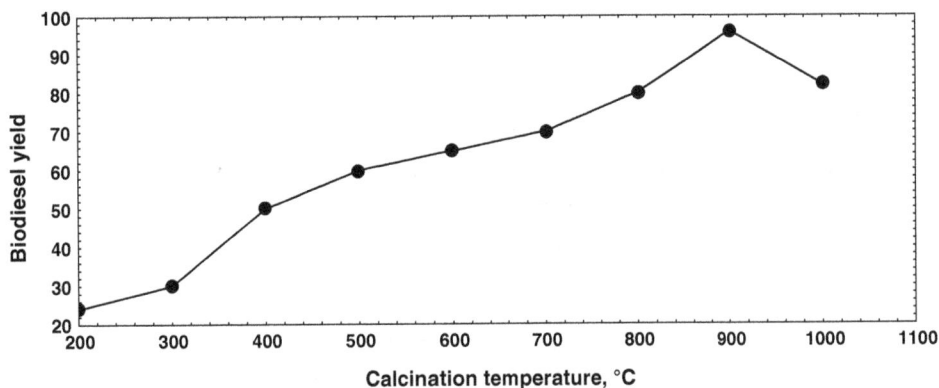

Fig. 6 Effect of catalyst concentration

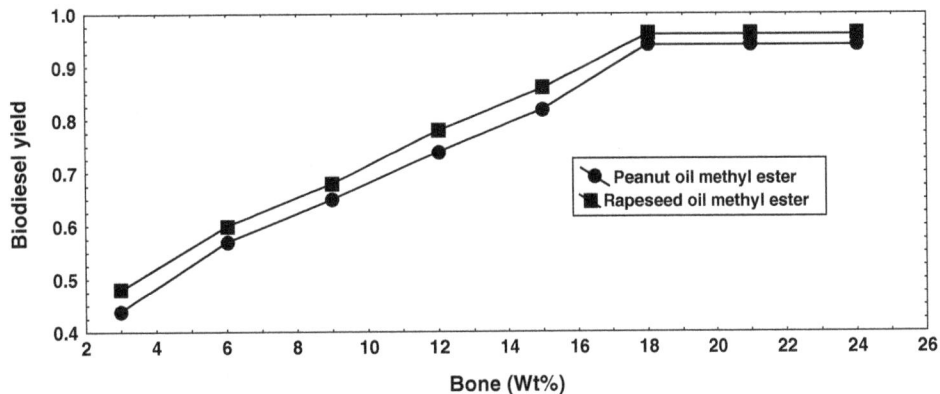

Fig. 7 Effect of methanol to oil molar ratio

promoted the formation of methoxy species on the Catalyst surface, leading to a shift in the equilibrium in the forward direction, thus increasing the biodiesel yield. However, further increase in the methanol to oil ratio (>20:1), did not promote the reaction because the catalyst content decreased to result in a hindrance for the access of glyceride molecules to active sites. Also it may be understood that the glycerol would largely dissolve in excessive methanol and subsequently inhibit the reaction of methanol to the reactants and catalyst, thus interfering with the separation of glycerin, which in turn lowers the biodiesel yield by shifting the equilibrium in the reverse direction [24]. However, considering the biodiesel yield and catalyst separation, the selected optimum molar ratio of methanol to oil is 20:1.

Effect of reaction temperature

A maximum conversion of 96 % for rapeseed oil and 94 % for peanut oil was obtained in 4 h at 60 °C, just below the boiling point of methanol (Fig. 8). This result parallels that of Obadiah et al. [10] who found a comparatively higher temperature (65 °C) to be optimum to obtain a high yield (96.78 %). When the reaction was carried out at 65 °C, which is above the boiling point of methanol, the solvent

vaporized and remained in the vapor phase in the reactor causing a reduction in the methanol in the reaction media. Whereas, biodiesel yield of 80 and 82 % are obtained for the catalysts calcined at 800 and 1,000 °C respectively.

Effect of reaction time and stirring speed

The change of product distribution on time in the transesterification of peanut and rapeseed oils over bone calcined at 900 °C are shown in Fig. 9. The biodiesel yield increased significantly by increasing the reaction time from 0.5 to 5 h and a maximum yield was obtained at 4 h. The results showed that the yield increased with time, reaching maximum value of 96 % for rapeseed oil and 94 % for peanut oil after 4 h using catalyst calcined at 900 °C. The yield of approximately 45 % was attained within 30 min for the two type of oils used. The biodiesel yield increased as the stirring rate was increased and reached a maximum at a rate of 600 rpm as shown in Fig. 10.

Reusability of waste catalysts

The catalyst was recovered by centrifuge and thoroughly washed with methanol. Figure 11 shows yields after reuse of the catalyst. The results indicated that the catalyst can

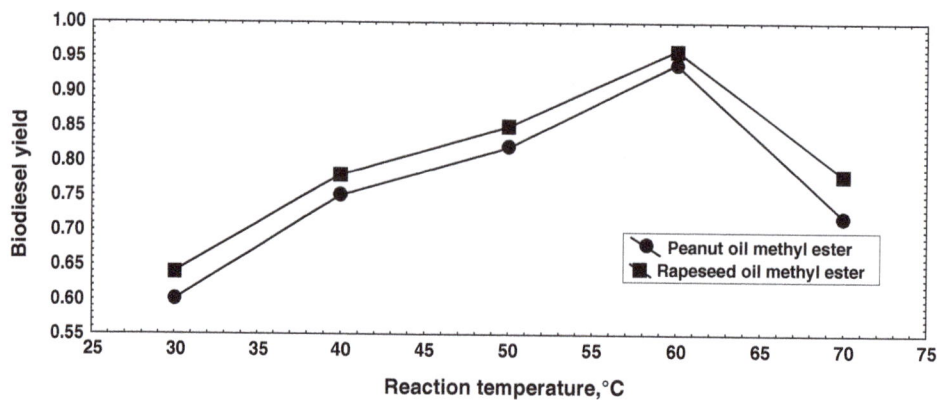

Fig. 8 Effect of reaction temperature

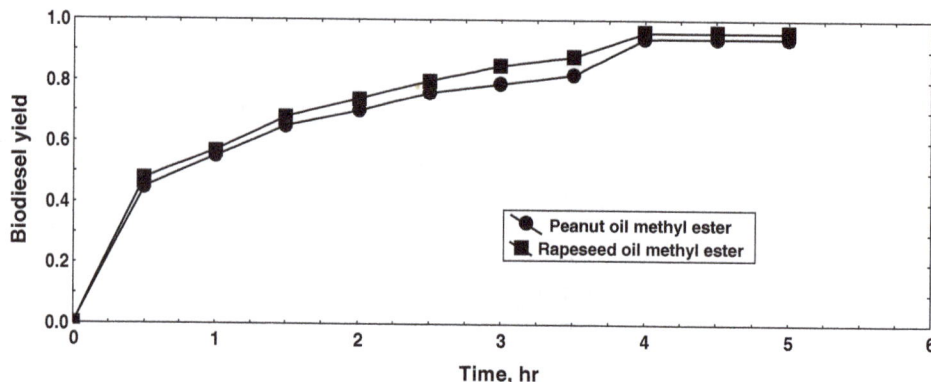

Fig. 9 Effect of reaction time

be repeated use for six times with no apparent loss of activity. After the 6th cycle of transesterification, the yield was still 93 %. After being used for more than 6 times, the catalyst lost activity gradually and was completely deactivated after being used more than 15 times. The deactivation of catalyst may be ascribed to its structure change. When compared to the previous work [25], the bone catalyst can be reused in fewer repetition times than CaO. This reflects the difference in the number of the active site and the reaction conditions. Also, the regeneration procedure largely influences the performance of the catalyst in the next use [26].

Properties of vegetable oil and methyl ester

The fuel properties of vegetable oils and corresponding biodiesels are given in Table 3. The peanut and rapeseed methyl esters have fuel property values relatively closer to that of mineral diesel. As compared to the mineral diesel specific gravity of 0.85, biodiesel specific gravity in the present analysis were 0.868, 0.88 respectively. The ASTM standard D6751 prescribed an acceptable viscosity at 40 °C range for biodiesel to be 1.9–6.0 mm^2/s, which was satisfied by biodiesel produced in the present work. The calorific value of methyl ester was same as

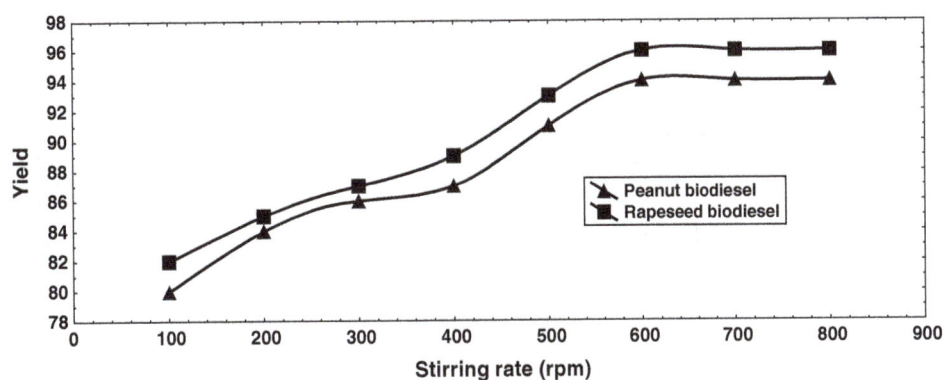

Fig. 10 Effect of stirring rate

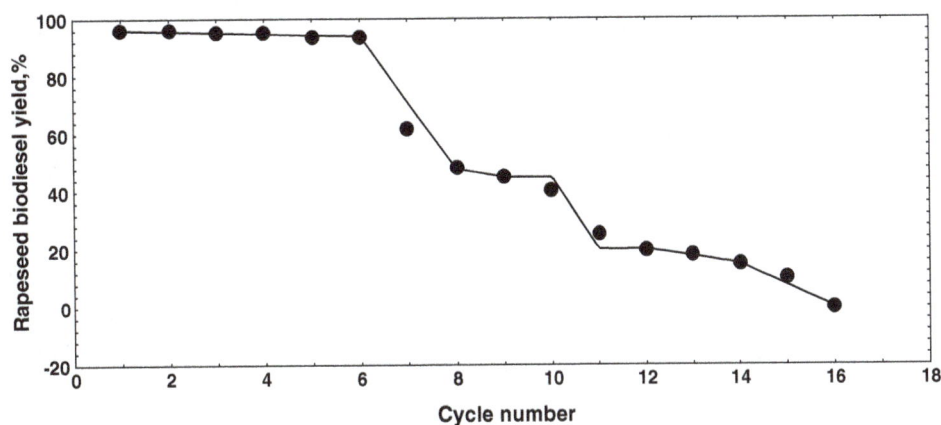

Fig. 11 Effect of reusability of the catalyst

Table 3 The fuel properties of vegetables oils and corresponding biodiesel with testing methods

Properties	Peanut oil	Rapeseed oil	Peanut methyl ester	Rapeseed methyl ester	Biodiesel standard ASTM D 6751-02	Mineral diesel	Test method
Specific gravity at 15/15 °C	0.89	0.91	0.868	0.88	0.87–0.90	0.85	ASTM D4052
Viscosity at 38 °C (mm^2/s)	39.6	51	4.9	4.15	1.9–6.0	1.9–4.1	ASTM D445
Calorific value (MJ/kg)	39.782	40.17	44.8	45	–	45	IP12/63T
Pour point (°C)	−6.7	−30	−1	−9	–	–	IP15/60
Cloud point (°C)	12.8	−4	5	−3	–	–	IP15/60
Flash point (°C)	271	246	172	170	130 °C	52	ASTM D92-53
Acid value (mg KOH/g)	0.71	0.62	0.45	0.37	0.8 max.	–	ASTM D 664
Cetane index	41.8	37.6	54	61.2	47 min.	45	ASTM D976

that of diesel. Cold flow properties of biodiesel are important indicators of the commercial applicability of the fuel. The key flow properties for biodiesel fuel specification are cloud and pour point. The values of cloud and Pour points of peanut methyl ester produced in the present work were found −1 and 5 °C and for rapeseed methyl ester were found −9 and −3 °C. Biodiesel produced from rapeseed oil is more satisfactory than that from peanut oil because peanut oil biodiesel may have a problem in winter. A possible solution for this problem would be the use of pour and cloud point depressors [27]. According to ASTM standard D6751, no value is given for cloud and pour point. The flash and fire points were higher than those for no. 2 diesel (85–95 °C). A higher value of both points decreases the risk of fire. The acid value was determined using the ASTM D664. The acid value of the biodiesels produced from peanut oil and rapeseed oil were 0.45, 0.37 mg KOH/g respectively. It is well within the specified limit of 0.8 mg KOH/g (biodiesel standard ASTM D6751). In the present analysis of peanut and rapeseed methyl esters, cetane index was 54.2 and 61.2 respectively. A typical value for mineral diesel is about 46. The cetane index is higher in biodiesel obtained from peanut oil and rapeseed oil. This parameter guarantees good control of the combustion, increasing performance and improving cold starts [28]. Thus, most of the fuel properties of peanut and rapeseed methyl esters were quite comparable to those of ASTM biodiesel standards, and therefore, the biodiesel produced from peanut oil and rapeseed oil can be used as substitute for mineral diesel.

Conclusion

Our present study revealed that bone waste is a good heterogeneous catalyst for the transesterification of rapeseed oil and peanut oil. The calcination of bone at 900 °C resulted in an increase in surface area, leading to better catalytic activity for the formation of methyl esters. Under the optimum conditions of 20:1 molar ratio of methanol to oil, addition of 18 wt% of bone catalyst (calcined at 900 °C, 2 h), 60 °C reaction temperature, the biodiesel yield was 96 % for rapeseed oil and 94 % for peanut oil at 4 h. This high efficient and low-cost waste catalyst could make the process of biodiesel production from vegetable oils used in the present process economic and fully eco-friendly making it competitive with petroleum diesel. The experimental results showed that the biodiesel produced in the present work has fuel property values relatively closer to that of mineral diesel and comparable to those of ASTM biodiesel standards.

References

1. Atabani, A.E., Silitonga, A.S., Badruddin, I.A., Mahlia, T.M.I., Masjuki, H.H., Mekhilef, S.: A comprehensive review on biodiesel as an alternative energy resource and its characteristics. Renew. Sustain Energy Rev. **16**(4), 2070–2093 (2012)
2. Nakatani, N., Takamori, H., Takeda, K., Sakugawa, H.: Transesterification of soybean oil using combusted oyster shell waste as a catalyst. Bioresour. Technol. **100**, 1510–1513 (2009)
3. Wei, Z., Xu, C., Li, B.: Application of waste eggshell as low-cost solid catalyst for biodiesel production. Bioresour. Technol. **100**, 2883–2885 (2009)
4. Boey, P.L., Maniam, G.P., Hamid, S.A.: Biodiesel production via transesterification of palm olein using waste mud crab (Scylla serrata) shell as a heterogeneous catalyst. Bioresour. Technol. **100**, 6362–6368 (2009)
5. Viriya-empikul, N., Krasae, P., Puttasawat, B., Yoosuk, B., Chollacoop, N., Faungnawakij, K.: Waste shells of mollusk and egg as biodiesel production catalysts. Bioresour. Technol. **101**(10), 3765–3767 (2010)
6. Mori, K., Yamaguchi, K., Hara, T., Mizugaki, Ebitani, K.T., Kaneda, K.: Controlled synthesis of hydroxyapatite-supported palladium complexes as highly efficient heterogeneous catalysts. J. Am. Chem. Soc. **124**(39), 11572 (2002)
7. Zahouily, M., Abrouki, Y., Bahlaouan, B., Rayadh, A., Sebti, S.: Hydroxyapatite: new efficient catalyst for the Michael addition. Catal. Commun. **4**(10), 521 (2003)
8. Onaka, M., Ohta, A., Sugita, K., Izumi, Y.: New application of solid base to regioselective ring openings of functionalized epoxides and oxetanes with cyanotrimethylsilane. Appl. Catal. A **125**(2), 203 (1995)
9. Turek S.L., Buckwalter J.A.: Orthopaedics: Principles and applications (1994)
10. Obadiah, A., Swaroopa, G.A., Kumar, S.V., Jeganathan, K.R., Ramasubbu, A.: Biodiesel production from Palm oil using calcined waste animal bone as catalyst. Bioresour. Technol. **116**, 512–516 (2012)
11. Moser, B.R., Vaughn, S.F.: Evaluation of alkyl esters from Camelina sativa oil as biodiesel and as blend components in ultra low-sulfur diesel fuel. Bioresour. Technol. **101**, 646–653 (2010)
12. Davis, J.P., Dean, L.O., Faircloth, W.H., Sanders, T.H.: Physical and chemical characterizations of normal and high-oleic oils from nine commercial cultivars of peanut. J. Am. Oil Chem. Soc. **85**, 235–243 (2008)
13. Lang, X., Dalai, A.K., Bakhshi, N.N., Reaney, M.J., Hertz, P.B.: Biodiesel esters as lubricity additives: effects of process variables and evaluation of low-temperature properties. Fuels Int. 207–227 (2001)
14. Zagonel, G.F., Peralta-Zamora, P., Ramos, L.P.: Multivariate monitoring of soybean oil ethanolysis by FTIR. Talanta **63**, 1021–1025 (2004)
15. ASTM. American standards for testing of materials (2003). (D 189–01, D 240–02, D4052–96, D 445–03, D 482–74, D 5555–95, D 6751-02, D 93–02a, D 95–990, D 97–02)
16. Qin, C., Li, C., Hu, Y., Shen, J., Ye, M.: Facile synthesis of magnetite iron oxide nanoparticles using 1-methyl-2-pyrrolidone as a functional solvent. Colloids Surf. A **336**, 130–134 (2009)

17. Yoosuk, B., Udomsap, P., Puttasawat, B., Krasae, P.: Improving transesterification activity of CaO with hydration technique. Bioresour. Technol. **101**, 3784–3786 (2010)

18. Michael, S.L., Shanthi, P., Ashok, M., Balasubramanian, T., Riyasdeen, A., Akbarsha, M.A.: Synthesis and characterization of nano-hydroxyapatite at ambient temperature using cationic surfactant. Mater. Lett. **63**(24), 2123 (2009)

19. Wang, Y.J., Chen, J.D., Wei, K., Zhang, S.H., Wang, X.D.: Surfactant-assisted synthesis of hydroxyapatite particles. Mater. Lett. **60**(27), 3227 (2006)

20. Antonakos, A., Liarokapis, E., Leventouri, T.: Micro-Raman and FTIR studies of synthetic and natural apatites. Biomaterials **28**(19), 3043 (2007)

21. Elliot, J.C.: Structure and Chemistry of the Apatites and other Calcium Orthophosphates. Elsevier, Amsterdam (1994)

22. Fleet, M.E., Liu, X.: Coupled substitution of type A and B carbonate in sodium-bearing apatite. Biomaterials **28**(6), 916 (2007)

23. Boynton, R.S.: Chemistry and technology of lime and limestone, 2nd edn. Wiley, New York (1980)

24. Yang, L., Zhang, A., Zheng, Z.: Shrimp shell catalyst for biodiesel production. Energy Fuels **23**, 3859–3865 (2008)

25. Liu, X., He, H., Wang, Y., Zhu, S., Piao, X.: Transesterification of soybean oil to biodiesel using CaO as a solid base catalyst. Fuel **87**, 216–221 (2008)

26. Ngamcharussrivichai, C., Totarat, P., Tanachai, S., Bunyakiat, K.: Biodiesel production through transesterification over natural calciums. Fuel Process. Technol. **91**, 1409–1415 (2010)

27. Reed, T.B., Graboski, M.S., Gaur, S.: Development and commercialization of oxygenated diesel fuels from waste vegetable oils. Biomass Bioenerg. **3**, 11 (1992)

28. Shuit, S.H., Lee, K.T., Kamaruddin, A.H., Yusup, S.: Reactive extraction of *Jatropha curcas* L. seed for production of biodiesel: process optimization study. Environ. Sci. Technol. **44**, 4361–4367 (2010)

21

Potential vegetable sources for biodiesel production: cashew, coconut and cotton

Jennifer Judith Lafont · Amelia Andrea Espitia ·
José Ricardo Sodré

Abstract This work presents a study on crude oil and biodiesel obtained from the seeds of the tropical plants *Anacardium occidentale L* (cashew), *Cocos nucifera* (coconut palm) and *Gossypium hirsutum* (upland cotton). The following crude oil and biodiesel physical–chemical properties were determined: acid number, iodine value, copper corrosivity, density and viscosity at different temperatures. Also, the chemical composition of the fatty acid methyl esters was measured using gas chromatography and a comparison was made with biodiesel from other sources reported in the literature. The analysis pointed out that cashew, coconut palm and upland cotton are potential sources for biodiesel production. Among the biodiesel types tested, cashew showed the highest oxidation stability.

Keywords Biodiesel · Cashew · Coconut · Cotton · Fatty acid

Introduction

The world energy demand and environmental contamination have both increased throughout the years from the increase of global warming gases and particles suspension in the atmosphere. The main reason is the use of fossil fuels as energy source for the development of industrial, technological, economical, transport and agricultural activities, among others [1]. This situation produced the search for alternative and renewable energy sources, such as biodiesel. Biodiesel is a biofuel, the chemical composition of which is a blend of alkyl esters produced together with glycerin from esterification or transesterification reactions of free fatty acids and triglycerides present in vegetable oils or animal fat [2].

Vegetable oils are the most studied ones for biodiesel production, among them palm, soybean, rapeseed and sunflower [3]. Other oleaginous plants have recently been considered for biodiesel production, such as *Anacardium occidentale L*. This is popularly known as cashew, belonging to the Anacardiaceous group, typical from tropical zone, and it presents excellent nutritional and medical properties. Its fruit contains A and C vitamins and it is used to elaborate juices. The fruit is united to a pseudo fruit of gray color that has two types of oils, one of them, of black color, is viscous and corrosive due to the presence of anacardic acids, used as insect repellent. The other is light amber, with high concentration of fatty acids, and it has started to be studied as a potential biodiesel source [4]. The fatty acid composition of *Anacardium occidentale L* has previously been reported by [5, 6].

Cocos nucifera is a palm tree commonly known as coconut, belonging to the Arecaceae group. It grows in tropical climate and humid atmosphere. Its thin trunk can reach 20 m tall, and its large fruit can have 20–30 cm of diameter, with a fibrous skin, inside which is the fruit of nut of oval shape, light brown color and hollow, that is filled with a juice during its growth. This fruit is used to obtain oil and its juice is used in refreshing drinks [7]. Previous work on coconut biodiesel has been presented by [5, 8].

J. J. Lafont · A. A. Espitia
Department of Chemistry, University of Cordoba,
Carrera 6, N° 76-103, Montería, Córdoba, Colombia
e-mail: jenniferlafontmendoza@gmail.com

A. A. Espitia
e-mail: andreaespitia36@gmail.com

J. R. Sodré (✉)
Department of Mechanical Engineering, Pontifical Catholic
University of Minas Gerais, Av. Dom José Gaspar 500, Belo
Horizonte, MG 30535-901, Brazil
e-mail: ricardo@pucminas.br

The *Gossypium hirsutum*, known as upland cotton, is a bush belonging to the malvaceae group that can achieve 2 m tall. Its flowers are yellow and, when fecundated, change to pink. After withering they turn into fruits, which are capsules of 3–5 pistils with 6–10 seeds in each of them. As the seeds are mature, their epidermis generates the cellulosic fibers of cotton, which are important textile products. The seeds are also rich sources of oil and proteins, with high industrial value [9]. From these biodiesel sources, cotton is the only one that has been extensively studied by many researchers, among them [1, 2, 8, 10–13].

This paper presents an analysis of the physical–chemical properties of the crude oil and biodiesel produced from the tropical plants *Anacardium occidentale L, Cocos nucifera* and *Gossypium hirsutum*. The properties evaluated are acid number, iodine value, copper corrosion, density and viscosity. The methyl esters present in each biodiesel type are also investigated to verify if those are real potential sources for high-quality biofuel production. This work aims at giving further insight on the quality of biodiesel produced from those sources, considering that there is still a lack of information particularly for cashew and coconut.

Methodology

Reactants and samples

All solvents used in this work were of analytical grade, manufactured by Merck, distilled at reduced pressure before use. The tested samples were seeds of cashew (Fig. 1), upland cotton (Fig. 2) and coconut (Fig. 3), the plants of which were grown at the location of Montería, in Córdoba, Colombia.

Crude oil and biodiesel production

The tested cashew and cotton seeds were peeled and crunched by a milling machine. Then, they were dried at 105 °C during 30 min in a micro processing muffle model Terrígeno, of rated temperature 1,200 °C, rated power 2.6 kW and maximum volume 8.8 L. After drying, the crunched seeds were weighted and the drying procedure was repeated until the difference between the measured weights was below 0.05 %.

The dry sample (100 g) was submitted to an oil extraction process using hexane as a solvent. The solvent was recovered by a rotary evaporator. The concentrations of crude oil obtained were 77.88 % for cashew and 75.63 % for cotton. The coconut seed was grated and the

Fig. 2 Upland cotton seeds

Fig. 1 Cashew seeds

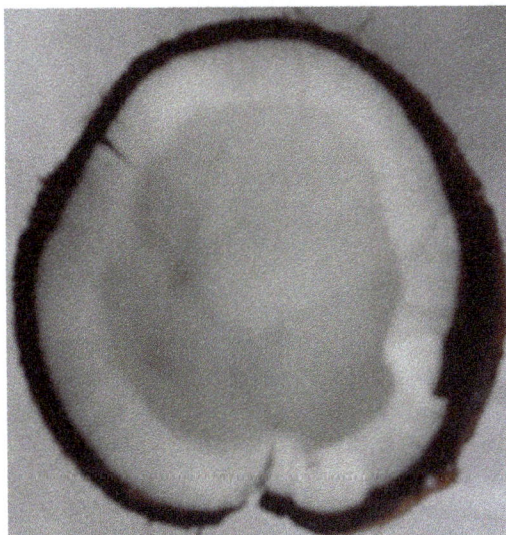

Fig. 3 Coconut seed interior

oil was extracted at a warmed wet medium, from which 76.78 % of crude oil was obtained.

The biodiesel was obtained from an esterification reaction in acid medium, followed by oil transesterification using methanol with a molar ratio of 6:1 and potassium hydroxide (KOH) as a catalyst. With this purpose, 50 mL of oil was used together with a solution of methanol and 1 % of KOH, based on the oil amount. The solution was blended to the oil and the reaction was allowed to take place for 1-h at 210 rev/min and 328 K. This procedure was performed three times. Afterwards, the solvent was recovered and the remaining substance was biodiesel–glycerol blend. This blend was taken to a separation funnel and remained still for 24 h to separate the biodiesel. The biodiesel was washed several times with hot deionized water to reach the neutrality. Finally, the biodiesel was warmed at 383 K during 10 min to eliminate humidity [12].

Identification and quantification of fatty acid methyl esters

The identification and quantification of the fatty acid methyl esters of all samples were performed using a gas chromatograph Agilent model 6,890 N with flame ionization detector (FID) analyzer. A capillary column model Carbowax 20 M was used for separation of the sample components. The column had a nominal bore of 0.32 mm and a length of 30 m, with 0.25 μm wall film thickness and stationary phase poly ethylene glycol. The injected volume amount was 1 μL. The methodology applied is described by EN14103 standard. Methyl heptadecanoate was used as an internal standard substance. The fatty acid methyl esters present in the samples were identified by comparison with the retention times of the fatty acids of the standard substance.

Physical–chemical analysis of crude oil and biodiesel

The iodine value is a chemical property that describes the iodine mass absorbed by 100 g of sample. It is taken as a measure of the degree of unsaturation of the biodiesel. Once the iodine is added to the fatty acid double bonds, the iodine value is increased as the double bonds are increased. As a consequence, the biodiesel with high concentration of saturated fatty acids presents low iodine value, which is increased with mono-, di- and triunsaturates. When the biodiesel iodine value is high and the engine temperature is increased, the fuel is degraded faster and forms solid deposits in the engine. The iodine value was determined according to EN 14111 standard.

The acid number describes the amount of KOH required to neutralize the free fatty acids and other acid compounds present in a mass unit of biodiesel. The acid number was

determined according to ASTM D664 standard. The acid number is considered as an indicator of biodiesel degradation. A high acid number affects engine fuel injection system and increases corrosion of engine components.

The peroxide index measures the initial oxidation of the fresh oil or the rancidity grade in equivalents of active oxygen per unit mass of oil, allowing for oxidation detection before organoleptic detection. The peroxide index is a measure of the oil degree of rancidity evolution, and was here determined according to AOAC Cd 8b-90 standard.

The saponification index is related to the molecular weight through the total fatty acids (free and combined) present in the oil sample. The saponification index was measured by the AOAC Cd 3–25 standard.

Humidity is an important oil factor that must be controlled. High humidity can generate a hydrolysis reaction of the triglycerides, forming free fatty acids that will contribute to a high acid value. Oil humidity was measured according to the AOAC 950.46 standard.

Copper corrosion is a parameter that allows for prediction of biodiesel corrosive action and detection of corrosive components or acids that attack some copper leagues, such as bronze, which are present in some engine components. Copper corrosion was determined following ASTM D130 standard.

Density is a property that can indicate the contamination degree of biodiesel. High biodiesel density can increase fuel consumption and generate higher amounts of gaseous pollutant emissions and particulate matter to the atmosphere. Also, the biodiesel density is an important parameter for storage and transportation as it can vary the fuel volume with varying temperature at a given pressure. The biodiesel viscosity must be minimized to avoid reduced engine power due to difficult flow through the filters and injectors of the fuel system. The biodiesel density and viscosity were determined according to ASTM D7042-12 standard.

The concentrations of total and free glycerin in biodiesel are important parameters to measure the biofuel quality. Low values of free glycerin reveal that high concentrations of methyl esters have been produced. On the other hand, the presence of glycerin with higher values than the standard limits indicates low conversion of the crude oil in biodiesel and poor fuel purification. The total and free glycerin content was measured by the AOCS Ca 14–56 standard.

Methanol content is another parameter that refers to the biodiesel purification degree. Low amounts of free methanol can reduce the fuel flammability index, oxidize the aluminum and zinc engine parts and cause damage to the fuel injection system. Methanol content was determined using EN14111 standard.

The cetane number measures the biodiesel ignition quality, and is determined by ASTM D613 standard. To calculate its value, the correlation developed by Bamgboye and Hansen [14] from fatty acid profiles has been employed:

$$NC = 61.1 + 0.088x_1 + 0.133x_2 + 0.152x_3 - 0.101x_4 - 0.039x_5 - 0.243x_6 - 0.395x_7 \tag{1}$$

where x_1 is the concentration of myristic acid (%), x_2 is the concentration of palmitic acid (%), x_3 is the concentration of stearic acid (%), x_4 is the concentration of palmitoleic acid (%), x_5 is the concentration of oleic acid (%), x_6 is the concentration of linoleic acid (%) x_7 is the concentration of linolenic acid (%).

Results and discussion

The fatty acid profile of cashew biodiesel here measured is compared with the ones reported by [5, 6] (Table 1). The biodiesel chemical composition makes cashew a potential source for high-quality biodiesel production. A predominant concentration of oleic acid (C18:1) is observed. The oleic is a monounsaturated fatty acid that is favorable to oxidation stability, reducing degradation and polymerization. The oxidation process is favored by high number of double bonds to form the epoxides responsible for oil aging and it is typical of the oil types that present high concentrations of polyunsaturated fatty acids, such as linoleic acid (C18:2) and linolenic acid (C18:3).

Cashew biodiesel shows high concentration of oleic acid (C18:1) and low concentrations of linoleic acid (C18:2), linolenic acid (C18:3) and palmitic acid (C16:0), which is a saturated fatty acid not recommended as fuel for cold climate (Table 1). The results here found for cashew biodiesel are in close agreement with those found by [5, 6]. In general, oleic acid (C18:1) was the component that showed the highest concentration (61.36–65.4 %), followed by linoleic acid (C18:2) (16–19.48 %), palmitic acid (C16:0) (10.36–10.5 %) and stearic acid (C18:0) (7.2–9.04 %). Small concentrations of linolenic acid (C18:3) (0.29–0.52 %) have also been identified by this work and by [5]. Those authors also reported low concentrations of myristic (14:0), palmitoleic (16:1), arachidic (C21:0) and behenic acids (C22:0), all varying from 0.07–0.67 %. These components have not been reported by [6].

Coconut biodiesel composition shows a predominant concentration of lauric acid (C12:0) (46.91–64.44 %), followed by myristic acid (C14:0) (18.50–20.45 %) and palmitic acid (C16:0) (7.71–8.90 %) (Table 2). The presence of high concentrations of saturated fatty acids limits the use of coconut biodiesel at low temperatures. The

Table 1 Chemical composition of cashew biodiesel

Fatty acid	CC:DB[a]	Cashew biodiesel composition (%)		
		This work	[6]	[5]
Lauric	C 12:0	0.00	0.00	0.00
Myristic	C 14:0	0.00	0.00	0.07
Palmitic	C 16:0	10.43	10.5	10.36
Palmitoleic	C 16:1	0.00	0.00	0.19
Stearic	C 18:0	8.21	7.2	9.04
Oleic	C 18:1	61.36	65.4	63.38
Linoleic	C 18:2	19.48	16.0	16.17
Linolenic	C 18:3	0.52	0.00	0.29
Arachidic	C 20:0	0.00	0.00	0.67
Eicosenoic	C 20:1	0.00	0.00	0.00
Behenic	C 22:0	0.00	0.00	0.07
Erucic	C 22:1	0.00	0.00	0.00
Lignoceric	C 24:0	0.00	0.00	0.00

[a] Carbon content (CC) per double bonds (DB)

Table 2 Chemical composition of coconut biodiesel

Fatty acid	CC:DB	Coconut biodiesel composition (%)		
		This work	[8]	[5]
Caprylic	C 8:0	6.10	6.46	0.00
Capric	C 10:0	5.93	5.62	0.00
Lauric	C 12:0	47.20	46.91	64.44
Myristic	C 14:0	18.50	18.74	20.45
Palmitic	C 16:0	8.90	9.69	7.71
Palmitoleic	C 16:1	0.00	0.00	0.09
Stearic	C 18:0	3.10	2.83	1.73
Oleic	C 18:1	7.30	6.83	4.61
Linoleic	C 18:2	1.50	2.21	0.96
Linolenic	C 18:3	0.00	0.00	0.00
Arachidic	C 20:0	0.00	0.10	0.04
Eicosenoic	C 20:1	0.00	0.00	0.00
Erucic	C 22:1	0.00	0.00	0.00

results here obtained closely resemble those by [8]. Both show the highest quantitative difference from the concentration of lauric acid (C12:0) found by [5].

Cotton biodiesel presents a chemical composition with high concentration of linoleic acid (C18:2) (53.14 % to 58 %), followed by palmitic acid (C16:0) (24.90–28.70 %) and oleic acid (C18:1) (13–18.93 %) (Table 3). Thus, this biodiesel type is more likely to be oxidized and degraded, requiring storage with high concentrations of antioxidant additives. The component concentrations found in this work are in close agreement with those found by [8, 13] and in reasonable agreement with the results found by [10, 11].

The physical–chemical properties obtained for cashew, coconut and cotton crude oil and biodiesel attend ASTM D6751-12 specifications (Tables 4, 5, 6). The acid number of all crude oil studies was higher than that of the corresponding biodiesel. No significant change of biodiesel iodine value was observed in comparison with the corresponding crude oil. The iodine value was coherent with the fatty acid methyl ester chemical composition, showing the lowest value for coconut biodiesel (saturated), followed by cashew (monounsaturated) and cotton (diunsaturated). Copper corrosion tests showed low values for all crude oils and biodiesel tested.

Cashew biodiesel density and viscosity here obtained have close values to the ones presented by [5] (Table 4). The properties of cashew crude oil have not been found in the literature for comparison purpose. Coconut biodiesel

viscosity here determined is also in agreement with the value presented by [5] (Table 5). However, the iodine value here found for coconut biodiesel is higher than that reported by [5], which is probably a consequence from the different compositions found for the saturated fatty acids (see Table 2).

All properties of cotton biodiesel here determined resemble those found in the literature (Table 6). In general, the properties of cotton crude oil here found are in good agreement with the results obtained by [1], with exception of a much higher acid number, which may be a reflection of the storage process of the crude oil. Both this work and the one by [1] show cotton crude oil viscosity much higher than that presented by [2], and much lower than the one described by [12].

One of the most important biodiesel properties is kinematic viscosity, as it affects fuel injection in the engine. The viscosity of the studied biodiesel types was much lower than that of the corresponding crude oils, thanks to the transesterification process. The high viscosity of crude oils is the main obstacle for their direct use in diesel engines [15]. The viscosity of all biodiesel types tested attended ASTM D6751-12 standard (Tables 4, 5, 6). Increasing temperature causes the fuel to flow more easily, being favorable to fuel injection efficiency and atomization. Low temperatures reduce fuel fluidity, as fuel viscosity is increased (Fig. 4).

The density of the biodiesel types tested presented a slight reduction in comparison with the corresponding crude oils, and is within the recommended range by ASTM D6751-12 standard (Tables 4, 5, 6). In general, biodiesel density is decreased with increasing temperature (Fig. 5). High fuel density causes increased fuel consumption and exhaust emissions of particulate matter, unburned hydrocarbons and carbon dioxide.

Table 3 Chemical composition of upland cotton biodiesel

Fatty acid	CC:DB	Cotton biodiesel composition (%)				
		This work	[8]	[13]	[11]	[10]
Caprylic	C 8:0	0.00	0.00	0.00	0.00	0.00
Capric	C 10:0	0.00	0.00	0.00	0.00	0.00
Lauric	C 12:0	0.00	0.00	0.00	0.00	0.00
Myristic	C 14:0	0.00	0.72	0.00	0.00	0.00
Palmitic	C 16:0	25.43	25.93	24.90	28.70	28.00
Palmitoleic	C 16:1	0.00	0.00	0.00	0.00	0.00
Stearic	C 18:0	2.21	1.74	2.63	0.90	1.00
Oleic	C 18:1	15.36	15.98	18.93	13.00	13.00
Linoleic	C 18:2	54.48	55.12	53.14	57.40	58.00
Linolenic	C 18:3	0.52	0.16	0.00	0.00	0.00
Arachidic	C 20:0	0.10	0.22	0.29	0.00	0.00
Eicosenoic	C 20:1	0.00	0.07	0.00	0.00	0.00
Erucic	C 22:1	0.00	0.00	0.00	0.00	0.00

Table 4 Physical–chemical properties of cashew crude oil and biodiesel

Property	Cashew crude oil	Cashew biodiesel		ASTM D6751
	This work	This work	[5]	
Density @ 20 °C (kg/m^3)	920	890	–	850–900
Viscosity @ 40 °C (mm^2/s)	27.23	4.32	3.69	1.9–6.0
Acid number (mg KOH/g)	1.60	0.50	–	≤0.50
Iodine value (g I$_2$/100 g)	83.0	82.50	86.65	–
Copper corrosion strip	1b	1b	–	Max 3
Peroxide index (meqO$_2$/Kg)	1.39	0.92	–	–
Saponification index (mgKOH/g)	175.4	145.2	201	–
Water content (mg/Kg)	536	312.35	–	Max 500
Free glycerin (% wt/wt)	–	0.01	.	0.02
Total glycerin (% wt/wt)	–	0.12	.	0.24
Methanol content (% wt/wt)	–	0.1	–	0.2
Cetane number	–	56.40	54.03	Min 47

Table 5 Physical–chemical properties of coconut crude oil and biodiesel

Property	Coconut crude oil	Coconut biodiesel		ASTM D6751
	This work	This work	[5]	
Density @ 20 °C (kg/m^3)	910	870	–	850–900
Viscosity @ 40 °C (mm^2/s)	28.80	2.80	2.30	1.9–6.0
Acid number (mg KOH/g)	2.22	0.19	–	≤0.50
Iodine value (g I$_2$/100 g)	8.10	10.12	5.98	–
Copper corrosion strip	1b	1b	–	Max 3
Peroxide index (meqO$_2$/Kg)	1.05	0.86	–	–
Saponification index (mgKOH/g)	155.4	151.2	262	–
Water content (mg/Kg)	487	230.25	–	Max 500
Free glycerin (% wt/wt)	–	0.018	–	0.02
Total glycerin (% wt/wt)	–	0.11	–	0.24
Methanol content (% wt/wt)	–	0.1	–	0.2
Cetane number	–	63.73	65.80	Min 47

Table 6 Physical–chemical properties of upland cotton crude oil and biodiesel

Property	Cotton crude oil				Cotton biodiesel		ASTM D6751
	This work	[1]	[2]	[12]	This work	[13]	
Density @ 20 °C (kg/m^3)	930	915	910	912	880	875	850–900
Viscosity @ 40 °C (mm^2/s)	35.20	33.5	18.20	50.00	4.75	4.07	1.9–6.0
Acid number (mg KOH/g)	8.20	0.16	–	0.11	0.35	0.16	≤0.50
Iodine value (g I$_2$/100 g)	108.80	104.7	–	–	112.40	–	–
Copper corrosion strip	1b	–	–	–	1b	1a	Max 3

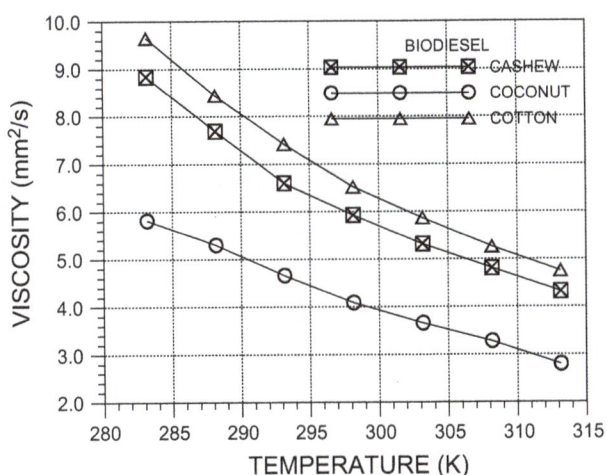

Fig. 4 Decrease of biodiesel viscosity with increasing temperature

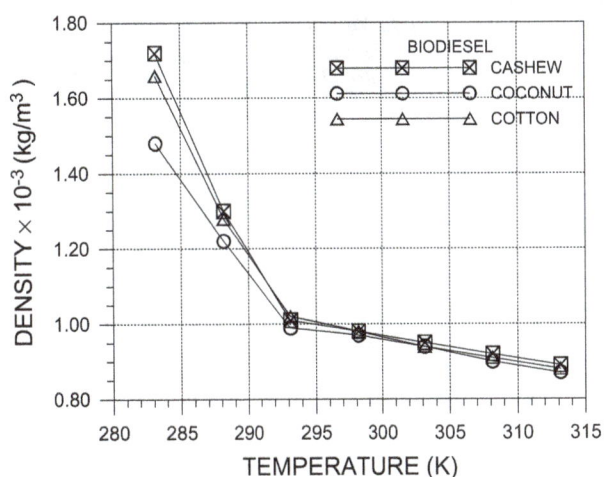

Fig. 5 Decrease of biodiesel density with increasing temperature

The other physical–chemical properties that indicate coconut and cashew biodiesel quality, such as water content, total and free glycerin content, methanol content and cetane number, shown in Tables 4 and 5, attend the ASTM 6751 standard for biodiesel quality. These results indicate that both materials, coconut and cashew, are potential sources to obtain high-quality biofuels.

Conclusions

The potential biodiesel sources studied—cashew (*Anacardium occidentale L*), coconut (*Cocos nucifera*) and upland cotton (*Gossypium hirsutum*)—are all rich in fatty acids and allows for fast biodiesel production processes. The physical–chemical properties of the biodiesel produced from those sources attend the ASTM D6751-12 standard (ASTM, 2012c) for biodiesel specification. From the chemical composition of the fatty acid methyl esters, cashew biodiesel presented the highest oleic acid concentration, being an indication of its high oxidation stability. The presence of high concentrations of saturated fatty acids in coconut biodiesel may limit its use under low temperatures. The composition of cotton biodiesel indicates that, among the sources tested, it is the most likely to be oxidized and degraded.

Acknowledgments The authors thank Universidad de Córdoba for the financial support to this work. Thanks are also due to CNPq and FAPEMIG.

References

1. Borugadda, V.B., Goud, V.V.: Biodiesel production from renewable feedstocks: status and opportunities. Renew Sustain Energy Rev **16**, 4763–4784 (2012)

2. Leung, D.Y., Xuan, W., Leung, M.K.H.: A review on biodiesel production using catalyzed transesterification. Appl Energy **87**, 1083–1095 (2010)

3. Murugesan, A., Umarani, C., Chinnusamy, T.R., Krishnan, M., Subramanian, R., Neduzchezhain, N.: Production and analysis of bio-diesel from non-edible oils—a review. Renew Sustain Energy Rev **13**, 825–834 (2009)

4. Lafont, J.J., Pàez, M.S., Portacio, A.A.: Extracción y caracterización fisicoquímica del aceite de la semilla (almendra) del marañón (A*nacardium occidentale L)*. Inf Tecnol **22**, 51–58 (2011)

5. Winayanuwattikun, P., Kaewpiboon, C., Piriyakananon, K., Tantong, S., Thakernkarnkit, W., Chulalaksananukul, W., Yongvanich, T.: Potential plant oil feedstock for lipase-catalyzed biodiesel production in Thailand. Biomass Bioenergy **32**, 1279–1286 (2008)

6. Adeyeye, E.I.: Fatty acid composition of zonocerus variegatus, macrotermes bellicosus and anacardium occidentale kernel. Int J Pharma Bio Sci **2**, 135–144 (2011)

7. Sun, J., Yu, B., Curran, P., Liu, S.Q.: Quantitative analysis of volatiles in transesterified coconut oil by headspace-solid-phase microextraction-gas chromatography–mass spectrometry. Food Chem. **129**, 1882–1888 (2011)

8. Giakoumis, E.G.: A statistical investigation of biodiesel physical and chemical properties, and their correlation with the degree of unsaturation. Renewable Energy **50**, 858–878 (2013)

9. Anwar, F., Rashid, U., Knothe, G.: Evaluation of biodiesel obtained from cottonseed oil. Fuel Process. Technol. **90**, 1157–1163 (2009)

10. Goering, C.E., Schwab, A.W., Daugherty, M.J., Pryde, E.H., Heakin, A.J.: Fuel properties of eleven oils. Trans ASAE **25**, 1472–1483 (1982)

11. Demirbas, A.: Biodiesel fuels from vegetable oils via catalytic and non-catalytic supercritical alcohol transesterifications and other methods: a survey. Energy Convers. Manage. **44**, 2093–2109 (2003)

12. Meng, X., Chen, G., Wang, Y.: Biodiesel production from waste cooking oil via alkali catalyst and its engine test. Fuel Process Technol. **89**, 851–857 (2008)

13. Rashid, U., Anwar, F., Knothe, G.: Evaluation of biodiesel obtained from cottonseed oil. Fuel Process. Technol. **90**, 1157–1163 (2009)

14. Bamgboye, A., Hansen, A.: Prediction of cetane number of biodiesel fuel from the fatty acid methyl ester (FAME) composition. Int Agrophy **22**, 21–29 (2008)

15. Refaat, A.A.: Correlation between the chemical structure of biodiesel and its physical properties. Int. J. Environ. Sci. Tech. **6**, 677–694 (2009)

22

Characterization, kinetics and statistical screening analysis of gmelina seed oil extraction process

Chigozie F. Uzoh · Okechukwu D. Onukwuli ·
Joseph T. Nwabanne

Abstract The kinetics and optimization of oil extraction from gmelina seed was studied. The effect of various process variables such as temperature, time, volume of solvent, particle size, and their interaction on oil yield was investigated. A predictive model describing the oil yield in terms process variables was derived from multiple regression analysis. Optimum yield of 49.90 % was predicted at extraction temperature of 60 °C, extraction time of 60 min, seed particle size of 150 μm and 150 ml volume of solvent, for the process. It was found that oil yield increased with increase in temperature, time and volume of solvent but decreased with increase in seed particle size. The extract was analyzed to examine its physico-chemical properties (acid value, iodine value, peroxide value, viscosity, saponification value, moisture and ash content, refractive index, smoke, flash and fire points, and specific gravity) and structural elucidation by standard methods and instrumental techniques. Results revealed that the oil is not edible but find potential in biodiesel production. The kinetic study showed that the extraction process follows a second order mechanism with a rate constant of 1.26×10^{11} $(dm^3 \ mol^{-1})^3 \ S^{-1}$.

Keywords Gmelina seed oil · Process optimization · Solvent extraction · Kinetics

Introduction

The rapidly growing global demand for energy and industrial raw materials from crude oil and the consequent depletion of crude oil reserves in addition to adverse environmental concerns and unstable nature of the international market make imperative the need to explore alternative sources of fuel and industrial raw materials. There has been an increase in the world production of oil seeds over the last 30 years. This would appear to be related to the increasing demand for oil seed products and by-products. Most seeds are cultivated primarily for their oil and meal. The oil from most seeds can be used for edible purposes (example is fluted pumpkin seed oil, soya bean seed oil) and about 80 % of the world production of vegetable oil is for human consumption while the remaining 20 % is shared between animals and chemical process industries [1]. As a result of the extensive demand of oil for consumption and industrial uses, analysis of many oils has been carried out. Most vegetable seed oils find wide applications in the production of soaps, paints, varnishes, lacquers, lubricants, hydraulic fluids, printing inks, dyes, pesticides, and insecticides [2–5]. Also, with increase in petroleum prices and uncertainties surrounding petroleum availability, vegetable oil can be employed as an alternative fuel and for biodiesel production which aims to overcome energy crisis problem [6]. Due to the increasing applications of vegetable oils, a number of seed oils have been characterized but the vast majority have not been adequately evaluated and this is particularly valid for gmelina, which falls into this group of under-utilized species of plant. The ability of a particular oil seed to fit into the growing industry depends on its utilization potential, rate of production, and availability of processing technology [7].

Gmelina arborea is a fast growing tree, which grows on different localities and prefers moist fertile valleys, they attain moderate to large height up to 40 m and 140 cm in diameter [8]. It is occurring naturally throughout greater

C. F. Uzoh (✉) · O. D. Onukwuli · J. T. Nwabanne
Chemical Engineering Department, Faculty of Engineering,
Nnamdi Azikiwe University, Awka, Nigeria
e-mail: epio4real@yahoo.com

part of India at altitudes up to 1,500 m. It also occurs naturally in Myanmar, Thailand, Laos, Cambodia, Vietnam, and in southern provinces of China, and has been planted extensively in Sierra Leone, Nigeria and Malaysia [9].

A lot of researches are being carried to find alternative ways of producing oil for process industries and for food industry. It has been found that almost all the seeds contain oil, hence these gives ground for other researchers to consider studies on the possible uses of other oil producing substances found in people's everyday lives. There are various ways of extracting oil from oilseeds but solvent extraction has been reported to be most efficient techniques [10]. Now, gmelina seeds are already proven to produce oil [11], this fact itself is already useful information for researchers who seek to find alternative sources of oil. The ability of the oil to fit depends on its constituents, its composition, the rate of production, and availability of the processing technology. The study of these constituents is important for their effective uses. Recently, Sangay et al. [6], reported that a biodiesel produced from gmelina seed oil met all the necessary biodiesel standard of ASTM D6751 and EN 14214. Gmelina seed oil have been found to be a sustainable material for biodiesel and alkyd resin synthesis in terms of its availability, renewability, and biodegradability. Gmelina based biodiesel has been developed keeping two criteria in mind; the biodiesel met all the technical and industrial standards and, met all the ecologically relevant standards [6, 12].

The solvent extraction process must be carried out within a predetermined space of the controllable factors that are known to have considerable effects on the system response i.e. oil yield. The characterization (or screening) of these major contributing variables to the process response was achieved through design-of-experiment (DOE) and analysis of variance (ANOVA) while the control and optimization tasks were accomplished using the well-known response surface methodology (RSM).

The central focus of this study therefore, is to: explore the most significant factors which influence the overall efficiency of gmelina seed oil extraction process; characterize their effects and predict the possible route(s) to a desired optimal; characterize the extract for structural establishment and physico-chemical properties; evaluate a suitable kinetics mechanism and rate constant for the extraction process. Gmelina seed has been found most suitable in that; it is not edible, widely available, will not interfere in the food chain and can stand out as sustainable material for energy. Overall, this new solvent extraction design will guarantee increased overall process efficiency through;

1 Lower investment costs (more efficient and easy to execute).
2. Lower operating costs (operating at optimal conditions of the process variables).

RSM has been widely adopted in industries such as drug and food industry, chemical and biological processes, for the purpose of either producing high quality products or operating the process in a more economical manner and ensuring that the process operates in a more stable and reliable way [13]. It has also been successfully applied to different processes for achieving its optimization using experimental designs [14–20].

Kinetic modelling

A relevant kinetic data are required to analyze and design an extraction process especially in industrial scale. In these experiments, mass transfer kinetic model was proposed to represent our experimental data because there is no reaction between gmelina seed oil with the organic solvent, n-hexane. For this model, the main mechanism which controls the rate of gmelina seed oil extraction was assumed to be the mass transfer of gmelina seed oil from gmelina seed (solid) to bulk liquid n-hexane. Mass transfer rate of the oil from gmelina seed to organic solvents can be written as;

$$\frac{dW_A}{dt} = K \cdot A(C_{Ai} - C_A) \tag{1}$$

where dW_A/dt is the mass transfer rate of gmelina seed oil (g/s), C_A and C_{Ai} are the concentration of gmelina seed oil in bulk liquid (organic solvents) at time t (g L^{-1}) and at equilibrium (g L^{-1}), respectively. K is mass transfer coefficient and A is surface area for mass transfer process. Since the extraction was taken in batch process and its volume was kept constant during process, therefore

$$\frac{dW_A}{dt} = K \cdot \frac{A}{V} \cdot (W_{Ai} - W_A). \tag{2}$$

$$\frac{dW_A}{dt} = K \cdot a \cdot (W_{Ai} - W_A) \tag{3}$$

where $k \cdot a$ is volumetric mass transfer coefficient. Applying boundary conditions as follows:

1. At the beginning of extraction process ($t = 0$), the mass of gmelina seed oil in bulk liquid is zero, $W_A = 0$.
2. At any time t, mass of gmelina seed oil in bulk liquid is $W_A = W_{Ai}$ [10].

On integration, Eq. (5) gives;

$$W_A = W_{Ai}[1 - \exp(-k \cdot a \cdot t)] \tag{4}$$

Equation (6) can be rearranged in terms of yield per mass of gmelina seed to give:

$$Y_A = Y_{Ai}[1 - \exp(-k \cdot a \cdot t)] \qquad (5)$$

where Y_A and i Y_{Ai} are yield of gmelina seed oil in bulk liquid at time t and at equilibrium per mass of gmelina seed.

If a second-order rate law is considered, the rate of dissolution for the oil contained in the solid to solution can be described as:

$$\frac{dC_t}{dt} = K(C_s - C_s)^2 \qquad (6)$$

where k = the second-order extraction rate constant (L g^{-1} min^{-1}), C_s = the extraction capacity (concentration of oil at saturation in g L^{-1}), C_t = the concentration of oil in the solution at any time (g L^{-1}), t (min).

By considering the boundary condition $t = 0 - t$ and $C_t = 0 - C_t$, the integrated rate law for a second-order extraction was obtained:

$$C_t = \frac{C_s^2 kt}{1 + C_s kt} \qquad (7)$$

Linear form of the Eq. (9) would be:

$$\frac{t}{C_t} = \frac{1}{KC_s^2} + \frac{1}{C_s} \qquad (8)$$

The extraction rate can be written as:

$$\frac{C_s}{t} = \frac{1}{\left(\frac{1}{KC_s^2}\right) + \left(\frac{t}{C_s}\right)} \qquad (9)$$

As $t \to 0$, the initial extraction rate, h, can be written as:

$$h = KC_s^2 \qquad (10)$$

Rearranging Eq. (12), the concentration of oil at any time can be obtained as:

$$C_s = \frac{t}{\frac{1}{h} + \frac{t}{C_s}} \qquad (11)$$

The initial extraction rate, h, the extraction capacity, C_s and the second-order extraction constant, k, can be calculated experimentally by plotting t/C_t vs. t.

The influence of temperature on the rate constant was determined by fitting k to Arrhenius equation,

$$K = A \exp\left(\frac{-E}{RT}\right) \qquad (12)$$

Using plots of ln k as a function of $1/T$

Materials and methods

Materials

The gmelina fruits were collected locally from a forest in ministry of forest reserve, Anambra State in Nigeria. It was soaked in water for 8 days so as to easily separate the fruit pulp from the seed (de-pulp). The seeds were sun-dried and crushed mechanically using corona blender; the crushed samples were then separated into different particle sizes using laboratory test sieves (150, 300, 600, 850 μm and 1 mm). The samples were then dried using the mermmet oven, stored in air tight containers and were labeled adequately.

Ethanol, sodium hydroxide, potassium hydroxide, and hydrochloric acid were obtained from BDH chemical ltd., Poole, England. Petroleum ether, diethyl ether, phenolphthalein indicator, glacial acetic acid, chloroform, distilled water, carbon tetrachloride, wij's iodide solution were purchased from Merck chemicals, Germany. Potassium iodide (KI) and sodium thiosulphate were obtained from M&B, England. The organic solvent used for the oil extraction was n-hexane. All the reagents were commercial grade and were used without further purification.

Instruments and methods

The analysis of the oil was performed with a Thermo Finnigan Trace GC/Trace DSQ/A1300, (E.I Quadropole) equipped with a SGE-BPX5 MS fused silica capillary column (film thickness 0.25 μm) for GC–MS detection, and an electron ionization system with ionization energy of 700 eV was used. Carrier gas was helium at a flow rate of 10 ml min^{-1} injector and MS transfer line temperatures were set at 220 and 290 °C, respectively. The oven temperature was programmed from 50 to 150 °C at 3 °C min^{-1}, then held isothermal for 100 min, and raised to 250 °C at 10 °C min^{-1}. Diluted samples (1/100, v/v, in methylene chloride) of 1.00 μL were injected manually in the slitless mode. The identification of individual components was based on the comparison of their relative retention times with those of authentic samples on SGE-BPX5 capillary column, and by matching their mass spectral of peaks with those obtained from authentic samples and/or the Wiley 7N and TRLIB libraries spectra and published data. The chemical compositions of oil were also confirmed by SHIMADZU FTIR-84008. Viscosity was determined by Brookfield viscometer, RVT Model (#Spindle 3, RPM 20). The physico-chemical properties of the extract were determined by standard methods (ASTM, 1973).

Extraction of oil from gmelina

Ten grams of grinded meal was extracted with n-hexane. The extraction temperature was varied from 20 °C to boiling point of the solvent while the reaction time was varied between 5 and 60 min. The solvent to solid ratio was investigated from 2:1 to 5:1 and particle size was varied from 150 to 1,000 μm. At the end of the extraction, the micelle was filtered using a vacuum filtration (Millipore glass base and funnel) to remove suspended solids. Subsequently, the solvent was separated from the oil using rotary vacuum evaporator (Laborota 4000) and was collected in the receiving flask. The oil which was remained in the sample flask was weighed after the process was completed. The percentage of extracted oil was calculated by dividing the amount of obtained oil by the amount of the seeds multiplied by 100. The percentage oil yield was calculated using the expression below:

$$Y = \frac{W_0}{W} \times 100 \tag{13}$$

where, Y is the oil yield (%), W_0 is the weight of pure oil extracted (g) and W is the weight of the sample of gmelina seed used in the experiment.

Formulation of design matrix

Preliminary data analysis conducted using steepest ascent method shows that curvature effects is possible. In view of curvature, a reduced order quadratic model (ROQM) was fitted over the resulting data as suggested in Eq. (14)

$$Y = \beta_0 + \sum_{i=1}^{k} \beta_i X_i + \sum_{i<}^{k} \sum_{j=2}^{k} \beta_{ij} X_i X_j + \sum_{i=1}^{k} \sum_{j=1}^{k} \delta_{ij} X_i X_j + \varepsilon \tag{14}$$

Equation (14) serves as the global predictive equation from which specific solution may be derived. The determination of the unknown coefficients of β_0, β_i, β_{ij}, and δ_{ij} is accomplished via regression analysis implemented on the statistical analysis software Design-Expert Version 9.1.7.1 trial from the Stat-Ease Inc. using the data recorded from the investigation.

The determination of unknown coefficient of Eq. (14) applies the design matrix of Table 2 formulated by judicious transformation of the actual values of the four control variables at various levels over which the experiments were executed to their coded equivalents using −1 and +1 notations to designate low and high level factor setting and '±α' and '0' for axial and centre points, respectively. The coded values of the independent variables for the design of the experiment for gmelina extraction process are given in Table 1. For statistical analysis, the variables $X_i (i = 1, 2, ..., 4)$ were coded A, B, C and D. The data given in Table 1 were used to formulate a global design matrix of Table 2 below from which further analyses were derived. Y is the response (oil yield) across the various experimental runs. Equation (14) was fitted to the experimental data presented in Table 2 to obtain the final predictive equation for the reaction progress in terms of the coded variables.

Table 1 Experimental range and level of the independent variables for gmelina seed oil extraction

Independent variable	Range and level				
	−α	−1	0	1	+α
Seed particle size (μm) (A)	150	300	600	750	1,000
Temperature (°C) (B)	20	30	40	50	60
Volume of solvent (ml) (C)	50	75	100	125	150
Time (min) (D)	10	20	30	40	60

Table 2 Central composite rotatable design matrix for gmelina seed oil extraction process

Runs	Independent variables				Responses
	A (μm)	B (°C)	C (ml)	D (min)	Y (%)
1	−1.00	1.00	1.00	1.00	31.79
2	1.00	1.00	−1.00	1.00	17.67
3	0.00	0.00	0.00	0.00	31.67
4	0.00	0.00	2.00	0.00	43.56
5	1.00	−1.00	−1.00	1.00	20.14
6	0.00	0.00	0.00	0.00	31.62
7	1.00	1.00	−1.00	−1.00	15.43
8	0.00	−2.00	0.00	0.00	21.54
9	1.00	−1.00	−1.00	−1.00	17.75
10	−1.00	1.00	−1.00	−1.00	32.76
11	0.00	0.00	−2.00	0.00	20.81
12	1.00	−1.00	1.00	−1.00	19.66
13	0.00	0.00	0.00	0.00	31.65
14	−1.00	−1.00	−1.00	1.00	20.55
15	1.00	−1.00	1.00	1.00	22.68
16	1.00	1.00	1.00	1.00	29.61
17	−1.00	−1.00	1.00	−1.00	28.65
18	0.00	0.00	0.00	0.00	31.65
19	−1.00	1.00	1.00	−1.00	52.09
20	1.00	1.00	1.00	−1.00	27.02
21	0.00	2.00	0.00	0.00	47.25
22	0.00	0.00	0.00	2.00	29.25
23	2.00	0.00	0.00	0.00	19.54
24	0.00	0.00	0.00	−2.00	23.87
25	−1.00	1.00	−1.00	1.00	29.11
26	−2.00	0.00	0.00	0.00	50.82
27	−1.00	1.00	1.00	1.00	54.60
28	0.00	0.00	0.00	0.00	31.65
29	−1.00	−1.00	−1.00	−1.00	17.93
30	0.00	0.00	0.00	0.00	31.64

Process analysis

The extraction process was analyzed for the different interaction effects of the process variables on the responses based on solutions derived from Eq. (14) at some assumed experimental conditions so as to reveal possible routes to optimum yield. The popular surface response methodology was suitably utilized for this exercise. The design expert analysis and optimization tool functions provided all the necessary computational assistance. The results were presented as response surface contours.

Results and discussion

Process/statistical analysis

ANOVA for gmelina extraction process

The ANOVA results derived from the predictive model show that the main linear effects due to individual control factors such as particle size (x_1), time (x_2), solvent ratio (x_3), and temperature (x_4) coded as A, B, and C and D, respectively, are all significant process variables, with the observed P values <0.05 in the numerical analysis. This is equally true with the linear interaction effects between particle size and temperature (AD), seed particle size and solvent ratio (AC), and temperature and solvent ratio (CD). The quadratic effect of temperature, denoted by D^2 is significant. The data obtained for specific investigation

were refitted with a modified model obtained by excluding the non-significant variables from the general predictive equation and the results of statistical analysis obtained for the sequence of experimentations are summarized in Table 3. The coefficients of determination R^2 values of 0.9523 obtained for the gmelina seed oil extraction process show that more than 95 % of the overall system variability can be explained by the empirical models of Eq. (14) which is specific case of the general predictive equation derived for the investigation from the multivariate regression analyses implemented on design expert.

$$Y = 31.65 - 6.65A + 5.46B + 5.83C + 1.05D$$
$$- 3.78AB - 2.40AC + 2.65BC - 2.07D^2 \qquad (15)$$

where Y is the predicted value of the dependent variable (oil yield). The coefficients of A, B, C and D are the main linear effects of the independent process variables: seed particle size (x_1), time (x_2), solvent ratio (x_3), and temperature (x_4) respectively, in coded units. AD, DB, and CB represent the linear interaction effects between particle size/temperature, temperature/time, and solvent ratio/time, respectively. A^2, B^2, C^2 and D^2 are the quadratic effects of the respective process variables. The "pred R-squared" of 0.7253 is in reasonable agreement with the "adj R-squared" of 0.9078. "adeq precision" measures the signal

Source	Sum of squares	df	Mean square	F value	P value prob >F
Model	3,187.53	14	227.68	21.40	<0.0001 significant
A-x1	1,062.40	1	1,062.40	99.84	<0.0001
B-x2	714.61	1	714.61	67.15	<0.0001
C-x3	815.03	1	815.03	76.59	<0.0001
D-x4	26.50	1	26.50	2.49	0.1354
AB	229.07	1	229.07	21.53	0.0003
AC	92.16	1	92.16	8.66	0.0101
AD	2.27	1	2.27	0.21	0.6512
BC	112.04	1	112.04	10.53	0.0054
BD	3.13	1	3.13	0.29	0.5954
CD	3.29	1	3.29	0.31	0.5862
A^2	0.22	1	0.22	0.020	0.8881
B^2	0.31	1	0.31	0.030	0.8658
C^2	11.93	1	11.93	1.12	0.3064
D^2	117.06	1	117.06	11.00	0.0047
Residual	159.62	15	10.64		
Lack of fit	159.62	10	15.96	59,857.40	<0.0001 significant
Pure error	1.333E−003	5	2.667E−004		
Cor total	3,347.15	29			
Std. dev.	3.26			R-Squared	0.9523
Mean	29.45			Adj R-squared	0.9078
CV (%)	11.08			Pred R-squared	0.7253
PRESS	919.41			Adeq precision	16.924

Table 3 Analysis of variance table

to noise ratio. A ratio greater than 4 is desirable. The design ratio of 16.924 indicates an adequate signal. This model can be used to navigate the design space and the model F value of 21.40 further indicates that the model is significant. There is only a 0.01 % probability that the "model F value" this large could occur due to noise. P value less than 0.05 indicates model terms are significant.

3-D response surface plots for the optimization process

The 3-D response surface plots are graphical representations of the interactive effects of any two variables. The nature of the response surface curves shows the interaction between the variables. An elliptical shape of the curve indicates good interaction of the two variables and circular shape indicates no interaction between the variables. The 3-D response surface plots shown in Figs. 1, 2, 3, 4, 5 and 6 for the chosen model equation shows the relationship between the independent and the dependent variables. From Fig. 6, the response surface indicates that the

percentage yield of oil increases as temperature and solvent composition increases to optimum condition while further increase leads to decrease of percentage yield of oil. In addition, there was mutual interaction between the temperature and solvent composition. The highest percentage oil yield was obtained when 153 ml of solvent was used. This is in accordance with the result obtained by Meziance and Kadi [27] that studied kinetics and thermodynamics of oil extraction from olive cake. They reported that the positive effect of volume of solvent on oil yield was as a result of increase in the concentration driving force as volume of solvent increases. It was also as a result of increased washing of the oil extracted, away from the particle surface by the solvent as a result of increased volume. The increase in oil yield became less significant at 125 ml because 125 ml hexane was sufficient to bring the oil solute to equilibrium. Similarly, the oil yield increased as the temperature increased from 20 to 60 °C (from the same figure). The highest oil yield was obtained at 60 °C.

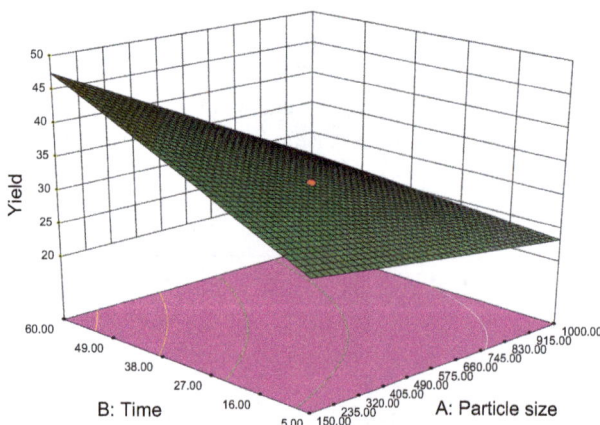

Fig. 1 Response surface plot showing the 3D effect of time and particle size and their interaction effect on the yield of gmelina oil

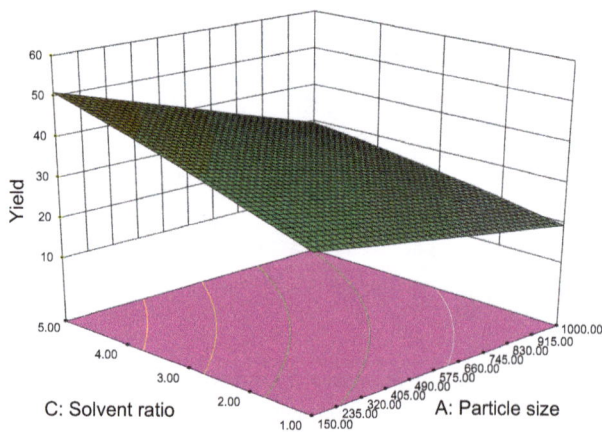

Fig. 3 Response surface plot showing the 3D effect of temperature and seed particle size and their interaction effect on the yield of gmelina oil

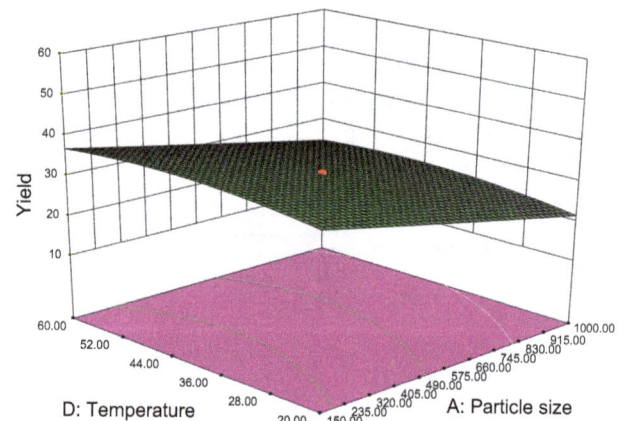

Fig. 2 Response surface plot showing the 3D effect of solvent ratio and particle size and their interaction effect on the yield of gmelina oil

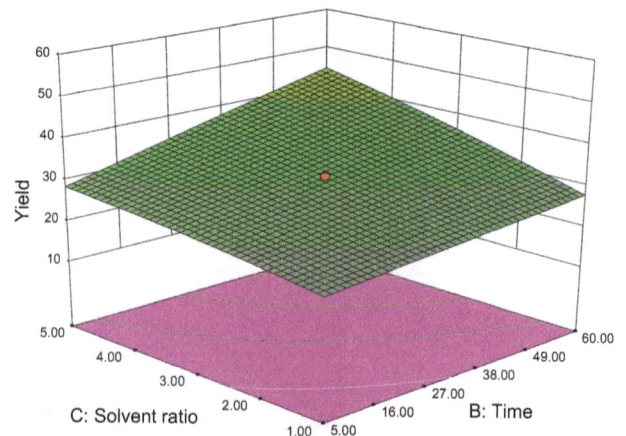

Fig. 4 Response surface plot showing the 3D effect of solvent ratio and time and their interaction effect on the yield of gmelina oil

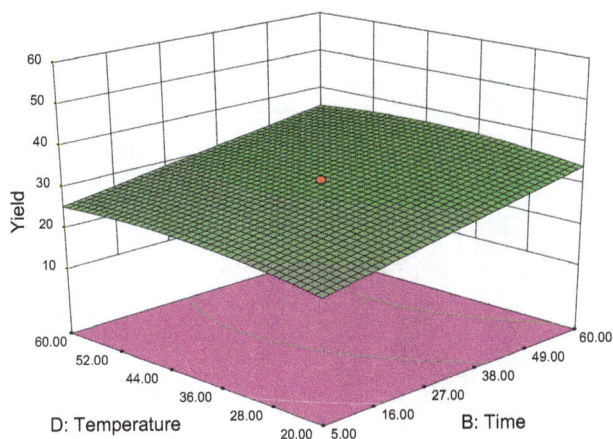

Fig. 5 Response surface plot showing the 3D effect of temperature and time and their interaction effect on the yield of gmelina oil

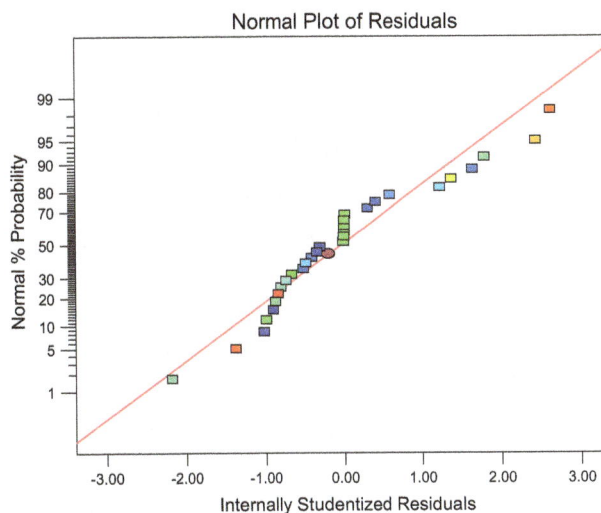

Fig. 7 Normal plots of residuals

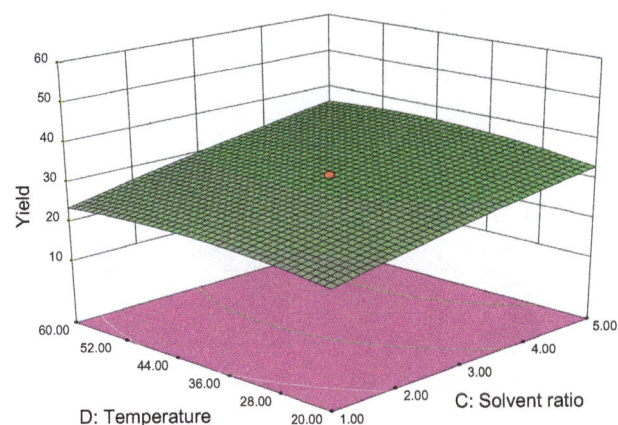

Fig. 6 Response surface plot showing the 3D effect of temperature and solvent ratio and their interaction effect on the yield of gmelina oil

Hickox [28] also reported increase in oil yield with increase in temperature for cotton seed oil. The positive effect of temperature on oil yield is as a result of rupturing of oil cell walls which now creates a void which serves as migratory space for the contents of the oil bearing cells [29]. Temperature influences oil yield and higher extraction is achieved by increasing the temperature, which lowers the viscosity of the oil, releases oil from the intact cells and draws out moisture. Figures 1, 2 and 3 display the responses for the interactive factors; time (x_2) vs seed particle size (x_1), solvent ratio (x_3) vs seed particle size (x_1), and temperature (x_4) vs particle size (x_1), respectively. The 3-D response surface plots shown in Figs. 1, 3 and 4 shows drop in percentage oil yield when seed particle size increases; even at highest setting of temperature (60 °C), time (60 min), solvent ratio (5.0). The negative effect of seed particle size on oil yield could be attributed to the fact that smaller particles have larger amount of surface area coupled with increased number of ruptured cells resulting

in a high oil concentration at the particle surface and low or little diffusion into the particles surface [30]. Sayyar et al. [1], while investigating the extraction of oil from Jatropha seed, suggested also that large particles have smaller amount of surface areas and are more resistant to intrusion of solvent and oil diffusion. Therefore, small amount of oil will be carried from inside the large particles to the surrounding solution. The quadratic effect of temperature (D) is visibly evident from the smooth curve in the response surface plots (see Fig. 3, 6). Dragon et al. [30], reported 46 % oil yield at extraction time of 120 min for gmelina seed using solvent extraction method while the current research recorded 49.50 % oil yield at extraction time of 60 min. Overall, given the long operational time, the earlier report may not be economically advantageous in terms of energy savings.

The normal plot of residuals (Fig. 7) was used to check whether the points will follow a straight line in which we concluded that the residuals follow a normal distribution. Hence, from the figures it is seen that the points were closely distributed to the straight line of the plot, it confirms the good relationship between the experimental values and the predicted values of the response though some small scatter like an "S" shape is always expected. These plots equally confirm that the selected model was adequate in predicting the response variables in the experimental values.

Kinetic study on extraction of gmelina seed oil

Linearization of the Arrhenius law as shown below gives the value of the activation energy (E) and the temperature independent factor (A) from ln(k) against 1/T plot:

$$\ln(k) = \ln(A) + \left(\frac{-E}{R}\right)\frac{1}{T}$$

Fig. 8 A plot of ln(dY/dt) versus ln Y

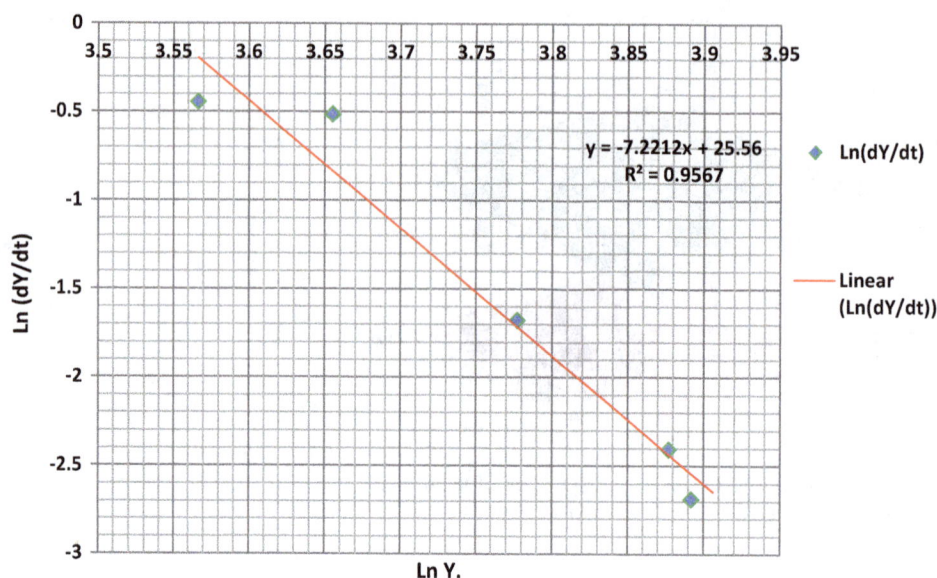

Fig. 9 A plot of ln(K) versus 1/T

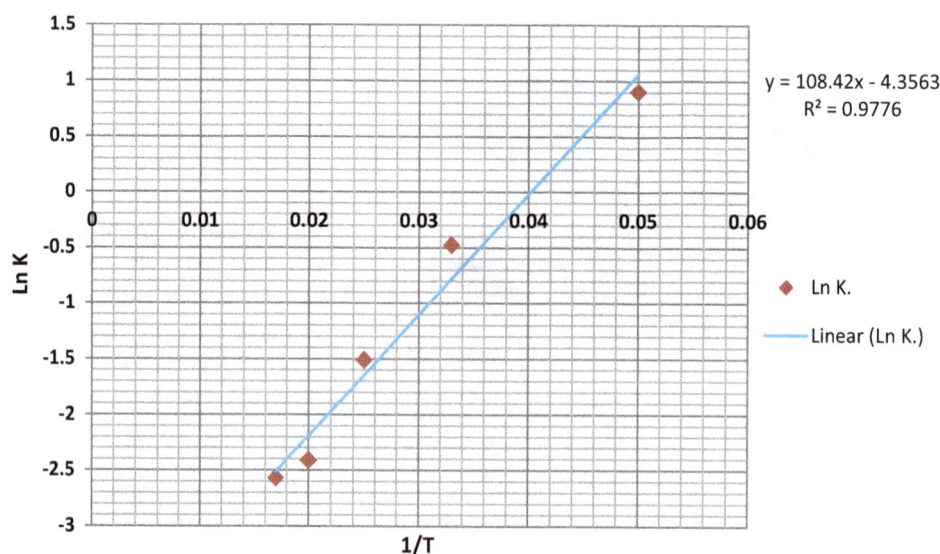

The plot of ln(k) against 1/T gives ln(A) as the slope and $\left(\frac{-E}{R}\right)$ as the intercept. A plot of ln(dY/dt) versus ln Y was found to be linear. A second order kinetics was obtained from the slope of the straight line as shown in Fig. 8. The reaction rate constant was determined from the intercept as 1.26×10^{11} (dm^3 mol^{-1})^3 S^{-1} (Fig. 9).

The positive value of enthalpy change indicates that the process is endothermic and requires energy during process.

Characteristics of gmelina seed oil

The fatty acid composition of gmelina seed oil was analysed by gas chromatography mass-spectrometry (GC–MS). Table 4 and Fig. 10 show the fatty acids present in gmelina seed oil. GC–MS analysis of the oil showed abundance of palmitoleate (31.94 %wt) and arachidic acid

(17.16 %wt). The most abundant unsaturated and saturated fatty acids were palmitoleate (31.94 %wt) and methyl stearate (14.22 %wt), respectively. The oil contains 50.86 % saturated fatty acid and 49.10 % unsaturated fatty acid. The results for the physico-chemical properties of the extracted oil are presented in Table 5. The physico-chemical analysis of the gmelina seed oil indicated physical state of the oil to be liquid and amber yellow at room temperature. The oil content of gmelina arborea seed was found to be 54.6 %wt. The oil content is significant and compares favourably with seed oil of other plants such as *Hevea brasiliensis* (51 %wt), *Hematostaphis berter* (54.5 %wt), *Jatropha curcas* (30–50 %wt), *Sapindus mukorossi* (51 %wt), *Mellia azadirachta* (33–45 %wt), and 55–65 % wt for *Simarouba glauca* [21]. On the basis of the oil content, gmelina arborea seed would be highly suitable economically for industrial applications, as any oil bearing seed that can produce up to 30 % oil are regarded as suitable [7]. The flash, fire, and smoke points were found to be 331.54, 386.12, and 176 °C respectively and were all above the 130 °C minimum ASTM D6751 and 120 °C EN 14214 Biodiesel (B100) Specifications and therefore pose

very negligible or no risk of fire outbreaks in case of accidents (when used as biodiesel) or spills. A value of 1.441 was obtained for the refractive index. The refractive index value obtained falls within the range (1.447–1.490) reported [22] for some other seed oils (1.480 for *Telfairia occidentalis*, 1.468 for *Jatropha curcas*, 1.47 for soybean oil and 1.47 for corn oil) which have myriad industrial applications. The specific gravity of gmelina seed oil was found to be 0.892 at 25 °C and this value is in the range found for other common oils. The SG value is also within the range (0.860–0.900) stipulated by EN 14214 for biodiesel [23]. Iodine value of 31.09 mleq g^{-1} was obtained. Iodine value measures the degree of unsaturation of a particular vegetable oil, oils with iodine value above 135 are classified as drying oil, those with iodine value 110–130 are classified as semi-drying oil and those with iodine value below 90 are non-drying oils. The iodine value of 31.09 mleq g^{-1} for gmelina seed oil is comparable to literature values of castor oils and olive oils which are both non-drying oils. The iodine value suggests the oil contains mainly saturated fatty acids, thus the oil is not

Table 4 Fatty acid composition of gmelina seed oil

Carbon molecules	Name	%wt
C_{10}	Capric acid	1.27
C_{11}	Undecyclic acid	10.49
C_{12}	Lauric acid	6.00
$C_{20:2}$	Arachidic acid	17.16
C_{14}	Myristste	4.59
$C_{16:1}$	Palmitoleate	31.94
C_{17}	Magaric acid	4.84
C_{18}	Methyl stearate	14.22
C_{25}	Pentacosylic acid	3.70
C_{30}	Melissic acid	5.75

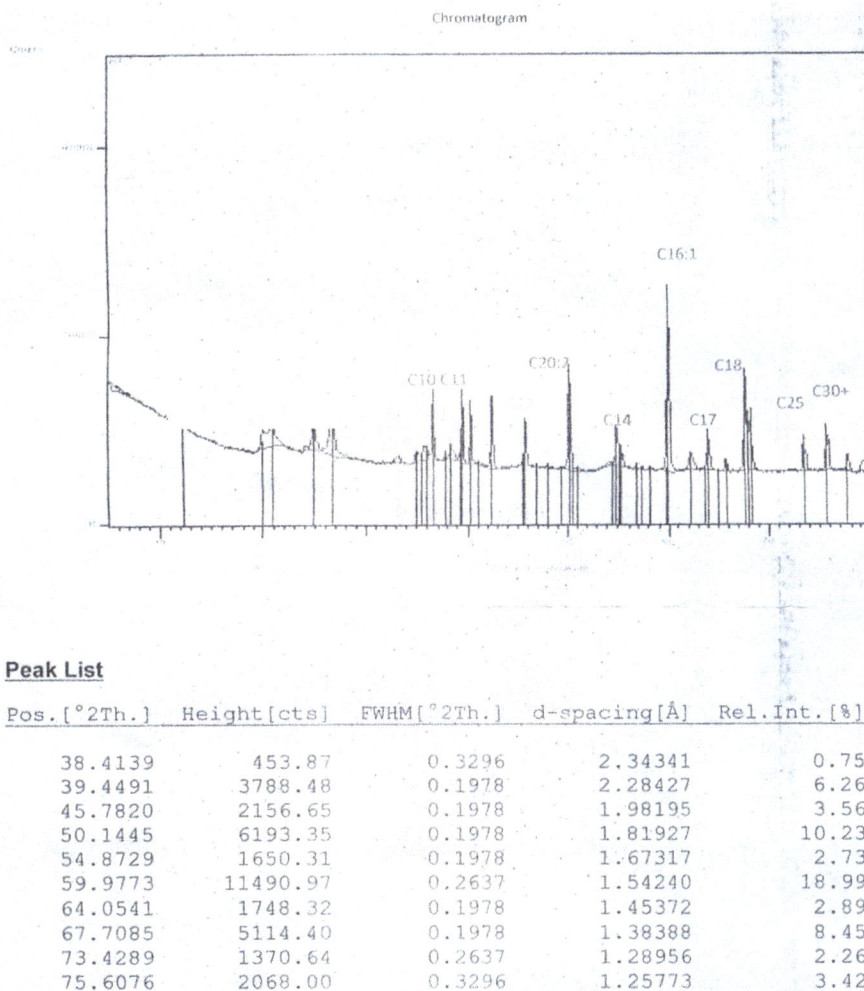

Fig. 10 GC–MS of crude gmelina seed oil

Peak List

Pos. [°2Th.]	Height[cts]	FWHM[°2Th.]	d-spacing[Å]	Rel.Int.[%]
38.4139	453.87	0.3296	2.34341	0.75
39.4491	3788.48	0.1978	2.28427	6.26
45.7820	2156.65	0.1978	1.98195	3.56
50.1445	6193.35	0.1978	1.81927	10.23
54.8729	1650.31	0.1978	1.67317	2.73
59.9773	11490.97	0.2637	1.54240	18.99
64.0541	1748.32	0.1978	1.45372	2.89
67.7085	5114.40	0.1978	1.38388	8.45
73.4289	1370.64	0.2637	1.28956	2.26
75.6076	2068.00	0.3296	1.25773	3.42

suitable as alkyd resin for paint formulation or use as varnishes; it may, however, find uses in conjunction with amino resins as finishes for certain appliances, and in this case the oil may also act as plasticizers. The iodine value is also consistent with the corresponding total saturation of fatty acids (50.86 %) derived from Table 3, thus affirming

the oil is largely consisting of saturated fatty acids. Peroxide value of 9 mleq g^{-1} was obtained for the oil. Peroxide value depends on a number of factors such as oxidation by oxygen, extraction methods, and storage. The low peroxide value suggests that gmelina seed oil is stable to oxidative degradation caused by over exposure to oxygen, heating and improper storage [24]. Saponification value of 32.258 mleq g^{-1} was obtained for the oil. The saponification value is very low when compared with the values for common oils; palm oil (196–205), groundnut oil (188–196), and corn oil (187–196) [25]. The low saponification value is as a result of the abundant long chain fatty acids (found in the oil) which have a relatively fewer number of carboxylic functional groups per unit mass of the oil as compared to short chain fatty acids. Thus the oil is not very suitable to be used for soap production. An acid value of 0.0336 mgKOH g^{-1} gives an indication of the amount of FFA present in the oil at the time of the test. The low acid value is an indication of good non-degraded state of the oil and is within limits for industrial useful oils. The FFA concentration of the oil was low (0.0168) which was consistent with low acid value observed. The low FFA also suggests low levels of hydrolytic and lipolytic activities in the oils. The 5 % total ash content is comparable to 4.56 % for *Jatropha curcas* [26] and it indicates the possible

Table 5 Physico-chemical properties of gmelina seed oil

Physico-chemical parameters	Values
Specific gravity	0.892
Acid value	3.366 %
Viscosity	3.628 pa s
Peroxide value	9 mleq g^{-1}
Saponification value	32.258 mleq g^{-1}
Iodine value	31.09 mleq g^{-1}
Moisture content	7.5 %
Flash point	331.54 °C
Smoke point	176 °C
Refractive index	1.441
Fire point	386.12 °C
Oil content	54.6 %wt
Ash content	5 %
Free fatty acid	1.6833 %

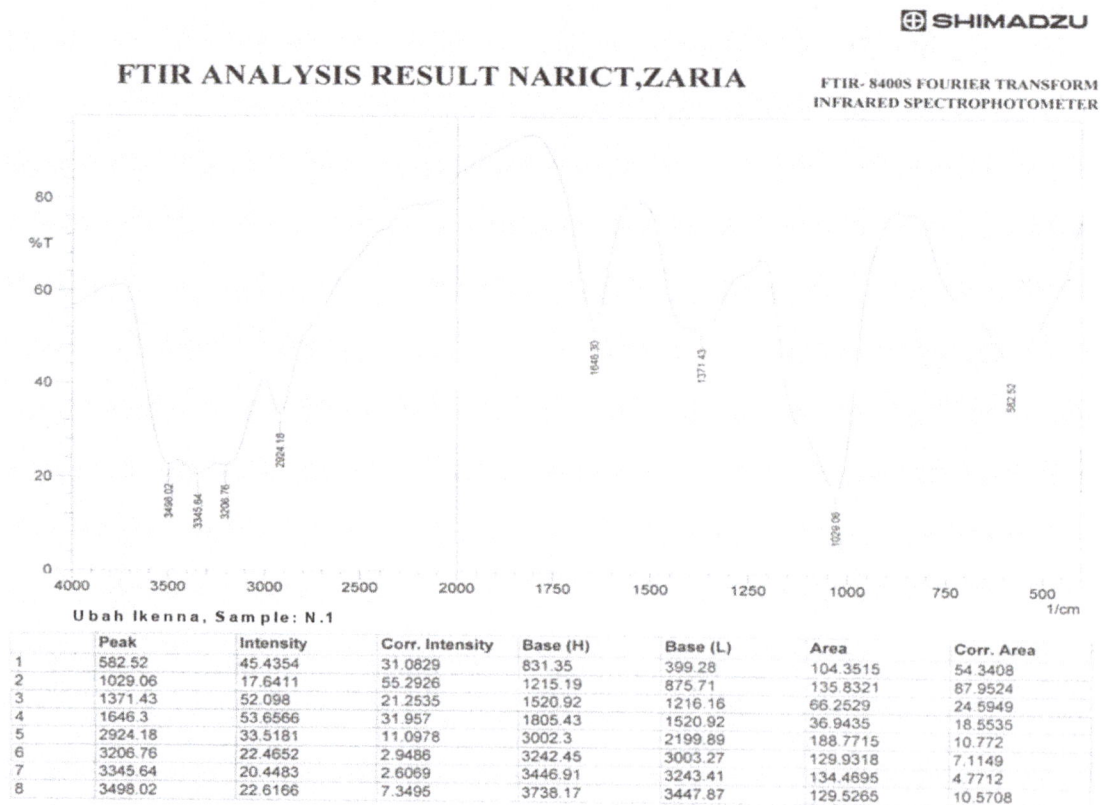

FTIR ANALYSIS RESULT NARICT,ZARIA

Ubah Ikenna, Sample: N.1

	Peak	Intensity	Corr. Intensity	Base (H)	Base (L)	Area	Corr. Area
1	582.52	45.4354	31.0829	831.35	399.28	104.3515	54.3408
2	1029.06	17.6411	55.2926	1215.19	875.71	135.8321	87.9524
3	1371.43	52.098	21.2535	1520.92	1216.16	66.2529	24.5949
4	1646.3	53.6566	31.957	1805.43	1520.92	36.9435	18.5535
5	2924.18	33.5181	11.0978	3002.3	2199.89	188.7715	10.772
6	3206.76	22.4652	2.9486	3242.45	3003.27	129.9318	7.1149
7	3345.64	20.4483	2.6069	3446.91	3243.41	134.4695	4.7712
8	3498.02	22.6166	7.3495	3738.17	3447.87	129.5265	10.5708

Fig. 11 FTIR analysis of gmelina seed oil

presence of abrasive solids, soluble metallic soaps, and silica residue in the seed. The ash content exceeds the ASTM standard of 0.02 maximum for biodiesel production. The moisture content of the oil (7.5 %) was very high, far exceeding the stipulated ASTM standard (0.05 %) for biodiesel production. This high moisture content creates problem in, reduces shelf life and may cause corrosion in internal combustion engine [15]. The structural organization of extracted gmelina nut oil was investigated by FTIR as shown in Fig. 11. In the spectrum of gmelina raw oil, 3,498.02 cm^{-1} correspond to the hydroxyl group (O–H) of the unsaturated fatty acid in the oil. The carboxyl group (C=O) is indicated at 1,646.3 cm^{-1} The straight chain of –CH– stretch in aliphatic compound is found at the band 2,924.18 cm^{-1}. Alkene group (CH=CH) is attributed to the band of 3,206.78 cm.

Conclusion

This study has clearly demonstrated the applicability of RSM in selecting extraction conditions for gmelina oil from its seed. This approach has not only resulted in the maximum oil yield through solvent extraction, but has also guaranteed the fulfillment of the properties requirements of the gmelina oil. The optimum values for yield show that it is an economic source of oil, the low saponification value means that it is not a good ingredient for soap making. The oil is very saturated and can hardly be used for paint making but when combined with other substances it can be used as a finishing agent. Time, seed particle size, temperature, solvent volume, have various effects on the yield of oil. The validation experiments and their accompany quality characteristics were not significantly different from the simulated values at $P < 0.05$. From spectroscopic results, it can be concluded that gmelina seed oil can be used as a source of biodiesel due to its specific gravity and saturated nature. Moreover, as a non-drying oil, it can also find variant uses in the process industry mainly when combined with some resins and compounds.

References

1. Sayyar, S., Abidi, Z.Z., Yunus, R., Mohammed, A.: Extraction of oil from jatropha seeds, optimization and kinetics. Am. J. App. Sci. **6**(7), 1390–1395 (2009)
2. Uzoh, C.F., Onukwuli, O.D., Odera, R.S., Ofochebe, S.: Optimization of polyesterification process for production of palm oil modified alkyd resin using response surface methodology. J. Environ. Chem. Eng. **1**, 777–785 (2013)
3. Uzoh, C.F., Onukwuli, O. D., Odera, R. S., Nwabanne, J. T.: Physico-chemical study of palm kernel oil alkyd resin and optimization of its polyesterification process. Int. J. Chem Eng. (2013).
4. Silva, G.F., Camargo, F.L., Ferreira, A.L.O.: Application of response surface methodology for optimization of biodiesel production by transesterification of soybean oil with ethanol. Fuel Process. Technol. (2010)
5. Hlaing, N.N., Oo, M.M.: Manufacture of alkyd resin from castor oil. World Acad. Sci. Eng. Technol. 48 (2008)
6. Sangay, B., Priyanka, B., Dibakar, O.D.: Gmelina arborea and Tarbernaemontana divaricate seed oils as non-edible feedstock for biodiesel production. Int. J. Chem. Tech. Res. **6**(2), 1440–1445 (2014)
7. Okolie, P.N., Ajekwene, A.E., Uaboi, Egbeni: Extraction and characterization of oil from *Jatropha curcas* seed. World J. Agric. Sci. **8**(4), 359–365 (2012)
8. Tewari, D.N.: A monograph on Gamari (*Gmelina arborea Roxb.*), p 125. International Book Distributors, Dehra Dun. (1995)
9. Adegbehin, J.O., Abayomi, J.O., Nwaigbo, L.B.: Gmelina arborea in Nigeria. Commonw. For. Rev. **67**(2), 159–166 (1988)
10. Topallar, H., Gecgel, U.: Kinetics and thermodynamics of oil extraction from sunflower seeds in the presence of aqeous acidic hexane solution. Turk. J. Chem. **24**, 247–253 (2000)
11. Adeyeye, A.: Composition of seed oils of gmelina arborea and teak tectora-grandis. Pak. J. Sci. Ind. Res. **34**(9), p359 (1971)
12. http://www.dainet.de/fnr/ctvo/paint/2_workshop/ull.doc. Accessed 23 April 2014
13. Sudamalla, P., Saravanan, P., Matheswaran, M.: Optimization of operating parameters using response surface methodology for adsorption of crystal violet by activated carbon prepared from mango kernel. Sustain. Environ. Res. **22**(1), 1–7 (2012)
14. Gunawan, E.R., Suhendra, D.: Four-factor response surface optimization of the enzymatic synthesis of wax ester from palm kernel oil. Indones. J. Chem. **8**(1), 83–90 (2008)
15. Alam, M.Z., Muyibi, S.A., Toramae, J.: Statistical optimization of adsorption processes for removal of 2, 4-dichlorophenol by activated carbon derived from oil palm empty fruit bunches. J. Environ. Sci. China **19**(6), 674–677 (2007)
16. Abali, Y., Colak, S., Yapici, S.: The optimization of the dissolution of phosphate rock with Cl_2–SO_2 gas mixture in aqueous medium. Hydrometallurgy **46**, 27–35 (1997)
17. Narayana, S.K.V., King, P., Gopinadh, R., Sreelakshmi, V.: Response surface optimization of dye removal by using waste prawn shells. Int. J. Chem. Sci. Appl. **2**(3), 186–193 (2011)
18. Umesh, G., Sud, D.: Optimization of process parameters for removal of Cr(VI) from aqueous solutions using modified sugarcane bagasse. E-J. Environ. Agric. Food. Chem. **4**(6), 1150–1160 (2005)
19. Li, D., Park, K., Wu, Z., Guo, X.: Response surface design for nickel recovery from laterite by sulfation-roasting-leaching process. Trans. Nonferrous Met. Soc. China **20**, 92–96 (2010)
20. Basumatary, S., Deka, Dinesh C., Deka, Dibakar C.: Composition of biodiesel from Gmelina arborea seed oil. Adv. Appl. Sci. Res. **3**(5), 2745–2753 (2012)
21. Eckey, E.W.: Vegetable fats and oils. American Chemical Society Monograph Series no. 123. Reinhold Publishing Corporation, New York (1954)
22. Ejikeme, P.M., Anyaogu I.D., Ejikeme, C.L., Nwafor, N.P., Egbuonu, C.A., Ukogu, K., Ihemesie, J.A.: Catalysis in biodiesel by transesterification processes-an insight. J. Chem. **7**(4), 1120–1132 (2010)
23. Ibeto, C.N., Okoye, C.O., Ofoelue, A.U.: Comparative study of the physicochemical characterization of some oil as potential

feedstock for biodiesel production. ISRN Renew. Energy **2012**, 1–5 (2012)

24. Cocks, L.V., Van Rede C.: Laboratory Handbook for Oil and Fat Analysts, pp. 3–37, 107–130. Academic Press, London, New York (1966)

25. Nayak, B.S., Patel, K.N.: Physicochemical characterization of seed and seed Oil of *Jatropha curcas*. Sains Malays. **39**(6), 951–955 (2010)

26. Meziance S., Kadi H., Lamrous O.: Kinetics study of oil extraction from olive foot cake

27. Hickox, G.H.: Some factors affecting the hydraulic extraction of cotton seed oil. J. Am. Oil Chem. Soc. **30**, 481–486 (1953)

28. Adeeko, K.A., Ayibola, O.O.: Processing factors affecting yield and quality of mechanically expressed groundnut oil. J. Agric. Eng. **45**, 31–43 (1990)

29. Ebewele, R.O., Iyayi, A.F., Hymore, F.K.: Considerations of the extraction process and potential technical applications of Nigerian rubber seed oil. Int. J. Phys. Sci. **5**(6), 826–831 (2010)

30. Dragon, R.G., Reyes, A.E.: Proceeding of the International Seminar on Chemistry, pp. 290–293, 30–31 October, Jatinagor (2008)

Permissions

List of Contributors

K. Connelly
Department of Chemistry, University of Aberdeen, Aberdeen AB24 3UE, UK

A.K. Wahab
SABIC T&I and CRI, Riyadh and KAUST, Riyadh, Saudi Arabia

Hicham Idriss
SABIC T&I and CRI, Riyadh and KAUST, Riyadh, Saudi Arabia

Lawrence D'Souza
SABIC Corporate Research and Innovation Center CRI at KAUST, Saudi Basic Industries Corporation, P.O. Box 4545-4700, Thuwal 23955-6900, Saudi Arabia

Satoshi Ujiie
Department of Applied Chemistry, Graduate School of Engineering, Osaka Prefecture University, 1 - 1 Gakuencho, Naka-ku, Sakai, Osaka 599-8531, Japan

Akitoshi Hayashi
Department of Applied Chemistry, Graduate School of Engineering, Osaka Prefecture University, 1 - 1 Gakuencho, Naka-ku, Sakai, Osaka 599-8531, Japan

Masahiro Tatsumisago
Department of Applied Chemistry, Graduate School of Engineering, Osaka Prefecture University, 1 - 1 Gakuencho, Naka-ku, Sakai, Osaka 599-8531, Japan

Daniele Pergolesi
International Research Center for Materials Nanoarchitectonics (WPI-MANA), National Institute for Materials Science (NIMS), 1-1 Namiki, Tsukuba, Ibaraki 305-0044, Japan

Marco Fronzi
International Research Center for Materials Nanoarchitectonics (WPI-MANA), National Institute for Materials Science (NIMS), 1-1 Namiki, Tsukuba, Ibaraki 305-0044, Japan

Emiliana Fabbri
International Research Center for Materials Nanoarchitectonics (WPI-MANA), National Institute for Materials Science (NIMS), 1-1 Namiki, Tsukuba, Ibaraki 305-0044, Japan

Antonello Tebano
CNR-SPIN and Dipartimento di Informatica Sistemi e Produzione, University of Roma Tor Vergata, Rome, Italy

Enrico Traversa
International Research Center for Renewable Energy, State Key Laboratory of Multiphase Flow in Power Engineering, Xi'an Jiaotong University, Xi'an 710049, Shaanxi, China

Satofumi Maruyama
International Center for Materials Nanoarchitectonics (MANA), National Institute for Materials Science (NIMS), 1-1 Namiki, Tsukuba 305-0044, Japan
Department of Applied Physics, Graduate School of Engineering, Tohoku University, 6-6-05 Aoba, Aramaki, Aoba-ku, Sendai 980-8579, Japan

Toshiyuki Nishimura
International Center for Materials Nanoarchitectonics (MANA), National Institute for Materials Science (NIMS), 1-1 Namiki, Tsukuba 305-0044, Japan

Yuzuru Miyazaki
Department of Applied Physics, Graduate School of Engineering, Tohoku University, 6-6-05 Aoba, Aramaki, Aoba-ku, Sendai 980-8579, Japan

Kei Hayashi
Department of Applied Physics, Graduate School of Engineering, Tohoku University, 6-6-05 Aoba, Aramaki, Aoba-ku, Sendai 980-8579, Japan

Tsuyoshi Kajitani
Department of Applied Physics, Graduate School of Engineering, Tohoku University, 6-6-05 Aoba, Aramaki, Aoba-ku, Sendai 980-8579, Japan

Takao Mori
International Center for Materials Nanoarchitectonics (MANA), National Institute for Materials Science (NIMS), 1-1 Namiki, Tsukuba 305-0044, Japan
University of Tsukuba, 1-1-1 Tennodai, Tsukuba, Ibaraki 305-8577, Japan

N. Vinothkumar
Department of Chemical Engineering, Indian Institute of Technology, Guwahati 781 039, Assam, India

Mahuya De
Department of Chemical Engineering, Indian Institute of Technology, Guwahati 781 039, Assam, India

Wilfried Wunderlich
Department of Material Science, Tokai University, Hiratsuka-shi 259-1292, Japan

Takao Mori
Nat. Inst. Mat. Sci. (NIMS), Int. Center Mat. Nanoarchitec. (MANA), Tsukuba 305-0044, Japan

Oksana Sologub
Nat. Inst. Mat. Sci. (NIMS), Int. Center Mat. Nanoarchitec. (MANA), Tsukuba 305-0044, Japan

Kenjiro Fujimoto
Department of Pure and Applied Chemistry, Faculty of Science and Technology, Tokyo University of Science, Yamazaki 2641, Noda, Chiba 278-8510, Japan

Chihiro Yamakawa
Department of Pure and Applied Chemistry, Faculty of Science and Technology, Tokyo University of Science, Yamazaki 2641, Noda, Chiba 278-8510, Japan

Yuki Yamaguchi
Department of Pure and Applied Chemistry, Faculty of Science and Technology, Tokyo University of Science, Yamazaki 2641, Noda, Chiba 278-8510, Japan

Shigeru Ito
Department of Pure and Applied Chemistry, Faculty of Science and Technology, Tokyo University of Science, Yamazaki 2641, Noda, Chiba 278-8510, Japan

N. Gokilamani
Department of Physics, Coimbatore Institute of Technology, Coimbatore, India

N. Muthukumarasamy
Department of Physics, Coimbatore Institute of Technology, Coimbatore, India

M. Thambidurai
Department of Electrical and Computer Engineering, Global Frontier Center for Multiscale Energy Systems, Seoul National University, Seoul 151-744, Republic of Korea

A. Ranjitha
Department of Physics, Coimbatore Institute of Technology, Coimbatore, India

Dhayalan Velauthapillai
Department of Engineering, University College of Bergen, Bergen, Norway

Sushant Kumar
Center for the Study of Matter at Extreme Conditions, College of Engineering and Computing, Florida International University, Miami, FL 33199, USA

Surendra K. Saxena
Center for the Study of Matter at Extreme Conditions, College of Engineering and Computing, Florida International University, Miami, FL 33199, USA

R. B. Rakhi
Materials Science and Engineering, King Abdullah University of Science and Technology (KAUST), Thuwal 23955-6900, Saudi Arabia

Chen
Materials Science and Engineering, King Abdullah University of Science and Technology (KAUST), Thuwal 23955-6900, Saudi Arabia

Dongkyu Cha
Materials Science and Engineering, King Abdullah University of Science and Technology (KAUST), Thuwal 23955-6900, Saudi Arabia

H. N. Alshareef
Materials Science and Engineering, King Abdullah University of Science and Technology (KAUST), Thuwal 23955-6900, Saudi Arabia

Dileep Kumar
VJTI, Matunga (E), Mumbai 400019, MS, India

H. A. Mangalvedekar
VJTI, Matunga (E), Mumbai 400019, MS, India

S. K. Mahajan
Directorate of Technical Education, 3, Mahapalika Marg, Mumbai 400001, MS, India

Misgina Tilahun
Department of Chemical Engineering, Wollo University, SouthWollo, Ethiopia

Omprakash Sahu
Department of Chemical Engineering, Wollo University, SouthWollo, Ethiopia

Manohar Kotha
Department of Electrical and Electronic Engineering, Wollo University, South Wollo, Ethiopia

Hemlata Sahu
Department of Electrical and Electronic Engineering, Wollo University, South Wollo, Ethiopia

Diego C. B. Alves
Materials Science and Engineering, Rutgers University, 607 Taylor Road, Piscataway, NJ 08854, USA
Departamento de Física, Universidade Federal de Minas Gerais, Belo Horizonte, MG 31270-901, Brazil

Rafael Silva
Department of Chemistry and Chemical Biology, Rutgers University, 610 Taylor Road, Piscataway, NJ 08854, USA

Damien Voiry
Materials Science and Engineering, Rutgers University, 607 Taylor Road, Piscataway, NJ 08854, USA

Tewodros Asefa
Department of Chemistry and Chemical Biology, Rutgers University, 610 Taylor Road, Piscataway, NJ 08854, USA
Department of Chemical and Biochemical Engineering, Rutgers University, 98 Brett Road, Piscataway, NJ 08854, USA

Manish Chhowalla
Materials Science and Engineering, Rutgers University, 607 Taylor Road, Piscataway, NJ 08854, USA

Kun Cong
Institute of Organic and Macromolecular Chemistry, Friedrich- Schiller University, Lessingstrasse 12, 07743 Jena, Germany

Mariusz Radtke
Institute of Organic and Macromolecular Chemistry, Friedrich- Schiller University, Lessingstrasse 12, 07743 Jena, Germany

Steffi Stumpf
Jena Center of Soft Mater (JCSM), Friedrich-Schiller University, Philosophenweg 7, 07743 Jena, Germany

Bernd Schröter
Institute of Solid State Physics, Friedrich-Schiller University, Philosophenweg 7, 07743 Jena, Germany

Duncan G. G. McMillan
University Hospital Jena, Friedrich-Schiller University, Bachstrasse 18, 07743 Jena, Germany

Markus Rettenmayr
Institute of Metallic Materials, Friedrich-Schiller University, Löbdergaben 32, 07743 Jena, Germany

Anna Ignaszak
Institute of Organic and Macromolecular Chemistry, Friedrich- Schiller University, Lessingstrasse 12, 07743 Jena, Germany

Asif Rahman
Department of Biological Engineering, Utah State University, 4105 Old Main Hill, Logan, UT 84322, USA
Department of Chemical and Materials Engineering, University of Auckland, Private Bag 92019, Auckland, New Zealand

Michelle E. Dickinson
Department of Chemical and Materials Engineering, University of Auckland, Private Bag 92019, Auckland, New Zealand

Mohammed M. Farid
Department of Chemical and Materials Engineering, University of Auckland, Private Bag 92019, Auckland, New Zealand

M. Marzouki
Laboratoire des Matériaux Utiles, Pôle Technologique de Sidi Thabet, Institut National de Recherche et d'Analyse Physico-chimique, 2020 Sidi Thabet, Tunisia

O. Ghodbane
Laboratoire des Matériaux Utiles, Pôle Technologique de Sidi Thabet, Institut National de Recherche et d'Analyse Physico-chimique, 2020 Sidi Thabet, Tunisia

M. Abdellaoui
Laboratoire des Matériaux Utiles, Pôle Technologique de Sidi Thabet, Institut National de Recherche et d'Analyse Physico-chimique, 2020 Sidi Thabet, Tunisia

E. Goikolea
CIC Energigune, Arabako Parke Teknologikoa, Albert Einstein 48, 01510 Minãno, Spain

E. Goikolea
Université Paul Sabatier, CIRIMAT UMR CNRS 5085, 31062 Toulouse Cedex 4, France

B. Daffos
Université Paul Sabatier, CIRIMAT UMR CNRS 5085, 31062 Toulouse Cedex 4, France

P. L. Taberna
Université Paul Sabatier, CIRIMAT UMR CNRS 5085, 31062 Toulouse Cedex 4, France

P. Simon
Université Paul Sabatier, CIRIMAT UMR CNRS 5085, 31062 Toulouse Cedex 4, France

Q. Yang
School of Materials Science and Engineering, University of Science and Technology Beijing, Beijing 100083, China

H. Zhao
School of Materials Science and Engineering, University of Science and Technology Beijing, Beijing 100083, China

J. Wang
School of Materials Science and Engineering, University of Science and Technology Beijing, Beijing 100083, China

J. Wang
School of Materials Science and Engineering, University of Science and Technology Beijing, Beijing 100083, China

Wang
School of Materials Science and Engineering, University of Science and Technology Beijing, Beijing 100083, China

H. Zhao
Beijing Key Lab of Advanced Energy Materials, Beijing 100083, China

X. Hou
School of Metallurgical and Ecological Engineering, University of Science and Technology Beijing, Beijing 100083, China

A. A. Jazie
Department of Chemical Engineering and Technology, Indian Institute of Technology, (Banaras Hindu University), Varanasi 221 005, UP, India

H. Pramanik
Department of Chemical Engineering and Technology, Indian Institute of Technology, (Banaras Hindu University), Varanasi 221 005, UP, India

A. S. K. Sinha
Department of Chemical Engineering and Technology, Indian Institute of Technology, (Banaras Hindu University), Varanasi 221 005, UP, India

J. J. Lafont
Department of Chemistry, University of Cordoba, Carrera 6, N8 76-103, Montería, Córdoba, Colombia

A. A. Espitia
Department of Chemistry, University of Cordoba, Carrera 6, N8 76-103, Montería, Córdoba, Colombia

J. R. Sodre
Department of Mechanical Engineering, Pontifical Catholic University of Minas Gerais, Av. Dom José Gaspar 500, Belo Horizonte, MG 30535-901, Brazil

C. F. Uzoh
Chemical Engineering Department, Faculty of Engineering, Nnamdi Azikiwe University, Awka, Nigeria

O. D. Onukwuli
Chemical Engineering Department, Faculty of Engineering, Nnamdi Azikiwe University, Awka, Nigeria

J. T. Nwabanne
Chemical Engineering Department, Faculty of Engineering, Nnamdi Azikiwe University, Awka, Nigeria